Springer Handbook of
Auditory Research

Series Editors: Richard R. Fay and Arthur N. Popper

Thomas N. Parks
Edwin W Rubel
Arthur N. Popper
Richard R. Fay
Editors

Plasticity of the Auditory System

With 48 Illustrations

 Springer

Thomas N. Parks
Department of Neurobiology &
    Anatomy
University of Utah School of Medicine
Salt Lake City, UT 84132-3401
USA
Tom.Parks@neuro.utah.edu

Edwin W Rubel
Virginia Merrill Bloedel Hearing Research Center
and
Department of Otolaryngology–Head and Neck
    Surgery
University of Washington
Seattle, WA 98195
rubel@u.washington.edu

Arthur N. Popper
Department of Biology and
    Neuroscience and Cognitive Science
    Program
University of Maryland
College Park, MD 20742-4415, USA
apopper@umd.edu

Richard R. Fay
Department of Psychology and Parmly Hearing
    Institute
Loyola University of Chicago
Chicago, IL 60626, USA
rfay@wpo.it.luc.edu

Series Editors: Richard R. Fay and Arthur N. Popper

Cover illustration: Modified from Figure 2.2, page 27 from Garden GA, Canady KS, Lurie DI, Bothwell M, Rubel EW (1994) A biphasic change in ribosomal conformation during transneuronal degeneration is altered by inhibition of mitochondrial, but not cytoplasmic protein synthesis. J Neurosci 14:1994–2008.

Library of Congress Cataloging-in-Publication Data
Parks, Thomas N.
    Plasticity of the auditory system / Thomas N. Parks, Edwin W. Rubel, Arthur N. Popper.
        p. cm.—(Springer handbook of auditory research ; v. 23)
    Includes bibliographical references and index.
    ISBN 0-387-20986-7 (alk. paper)
    1. Audiometry.   2. Auditory pathways.   3. Plasticity.   4. Hearing.   I. Rubel, Edwin W.   II.
Popper, Arthur N.   III. Title.   IV. Series.
    RF291.P37 2004
    617.8'075—dc22   2004041730

Printed in the United States of America. (MVY)

9 8 7 6 5 4 3 2 1   SPIN 10940556

www.springer-ny.com

Springer is a part of Springer Science+Business Media
springeronline.com

Series Preface

The *Springer Handbook of Auditory Research* presents a series of comprehensive and synthetic reviews of the fundamental topics in modern auditory research. The volumes are aimed at all individuals with interests in hearing research including advanced graduate students, postdoctoral researchers, and clinical investigators. The volumes are intended to introduce new investigators to important aspects of hearing science and to help established investigators to better understand the fundamental theories and data in fields of hearing that they may not normally follow closely.

Each volume presents a particular topic comprehensively, and each serves as a synthetic overview and guide to the literature. As such, the chapters present neither exhaustive data reviews nor original research that has not yet appeared in peer-reviewed journals. The volumes focus on topics that have developed a solid data and conceptual foundation rather than on those for which a literature is only beginning to develop. New research areas will be covered on a timely basis in the series as they begin to mature.

Each volume in the series consists of a few substantial chapters on a particular topic. In some cases, the topics will be ones of traditional interest for which there is a substantial body of data and theory, such as auditory neuroanatomy (Vol. 1) and neurophysiology (Vol. 2). Other volumes in the series deal with topics that have begun to mature more recently, such as development, plasticity, and computational models of neural processing. In many cases, the series editors are joined by a co-editor having special expertise in the topic of the volume.

RICHARD R. FAY, Chicago, Illinois
ARTHUR N. POPPER, College Park, Maryland

Volume Preface

Basic science studies of the auditory system have provided neuroscience with some of its most instructive examples of highly specialized functions arising from unique structures. Working out the developmental mechanisms by which these structures and functions arise continues to provide information of great value to general developmental biology as well as increases our understanding of how the many complex elements coalesce during normal development to form a functional auditory system. Such an understanding of the developing auditory system could potentially impact the optimization of treatment of developmental hearing losses in humans during the first years of life when the auditory system is most plastic, and thus serve in the mitigation of hearing problems in the large numbers of people who experience significant hearing loss.

The chapters in this volume provide a unique understanding of the development and plasticity of the auditory system. In Chapter 1, Parks and Rubel provide an overview of the rest of the volume, as well as discuss the meaning and use of the concept of plasticity. Rubel, Parks, and Zirpel (Chapter 2) focus on development of the cochlear nucleus (CN), among the best understood of the central nervous system (CNS) auditory centers. In reviewing the extensive research literature on normal development of the mammalian superior olivary complex (SOC), Friauf (Chapter 3) notes how many key events occur prior to the onset of hearing: neurogenesis, cell migration, axon and dendrite outgrowth, target selection, and synaptogenesis. Moore and King (Chapter 4) note that adjustments in binaural hearing are probably involved in the natural response to growth-related changes in head size and other natural changes in interaural cues, as well as disease- and injury-induced hearing asymmetries. Although studying plasticity in the auditory cortex is important, it is particularly difficult, for reasons Weinberger considers in Chapter 5. Analysis of auditory neuron function in awake behaving organisms has yielded some of its most impressive results in the study of birdsong. As Brenowitz and Woolley (Chapter 6) discuss, the study of neural plasticity in birdsong has also led neuroscience to recognize the existence of ongoing generation and replacement of projection neurons in the endotherm brain and hormonally induced seasonal changes in the morphology, pharmacology, and physiology of CNS neurons. Finally, Lakes-Harlan

(Chapter 7) approaches plasticity in the insect auditory system by considering four contexts in which it occurs.

As is often the case with chapters in the volumes in the Springer Handbook of Auditory Research, related chapters are often found in other volumes. The structure and function of the CNS discussed in most of the chapters in this volume are considered in great depth for mammals in Vol. 15 (*Integrative Functions in the Mammalian Auditory Pathway*). Related chapters in hearing in birds can be found in Vol. 13 (*Comparative Hearing: Birds and Reptiles*) and on insects in Vol. 10 (*Comparative Hearing: Insects*). Most extensively, issues on development of the auditory system are also found in Vol. 9, which can almost be considered as a companion to this one, *Development of the Auditory System*, and especially in chapters by Sanes and Walsh on development of central auditory processing and Cant on structural development of the central auditory pathways in mammals.

THOMAS N. PARKS, Salt Lake City, Utah
EDWIN W RUBEL, Seattle, Washington
RICHARD R. FAY, Chicago, Illinois
ARTHUR N. POPPER, College Park, Maryland

Contents

Contributors

ELIOT A. BRENOWITZ
Departments of Psychology and Biology and Virginia Merrill Bloedel Hearing
Research Center, University of Washington, Seattle, WA 98195-1525, USA

ECKHARD FRIAUF
Abteilung Tierphysiologie, Fachbereich Biologie, Universität Kaiserslautern,
D-67653 Kaiserslautern, Germany

ANDREW J. KING
University Laboratory of Physiology, Parks Road, Oxford OX1 3PT, UK

REINHARD LAKES-HARLAN
Georg-August Universität Göttingen, Institut für Zoologie und Anthropologie,
37073 Göttingen, Germany

DAVID R. MOORE[1]
MRC Institute of Hearing Research, University Park, Nottingham NG7 2RD,
UK

THOMAS N. PARKS
Department of Neurobiology & Anatomy, University of Utah School of Medicine, Salt Lake City, UT 84132-3401, USA

EDWIN W RUBEL
Virginia Merrill Bloedel Hearing Research Center and Department of
Otolaryngology–Head and Neck Surgery, University of Washington, Seattle, WA
98195, USA

NORMAN M. WEINBERGER
Center for the Neurobiology of Learning and Memory and Department of Neurobiology & Behavior, University of California, Irvine, CA 92697, USA

---

[1]Former address: University Laboratory of Physiology, Parks Road, Oxford OX1 3PT, UK

SARAH M.N. WOOLLEY
Department of Psychology, University of California, Berkeley, CA 94720, USA

LANCE ZIRPEL
Department of Neuroscience, University of Minnesota Medical School, Minneapolis, MN 55455, USA

1

Overview: Development and Plasticity of the Central Auditory System

Thomas N. Parks and Edwin W Rubel

1. Important Reasons for Studying CNS Auditory Development and Plasticity

Basic science studies of the auditory system have provided neuroscience with some of its most instructive examples of highly specialized functions arising from unique structures (e.g., reviews by Carr and Soares 2002; Fettiplace and Ricci 2003; Fuchs et al. 2003; Pollak et al. 2003; Ryugo and Parks 2003). Working out the developmental mechanisms by which these structures and functions arise continues to provide information of great value to general developmental biology (Riley and Phillips 2003) as well as increases our understanding of how the many complex elements coalesce during normal development to form a functional auditory system (Rubel et al. 1998).

Two facts lend additional importance to studies of auditory system development and plasticity. First, very large numbers of people experience significant hearing loss and the often profound disruptions of personal development and social life that follow. Because the human auditory system matures over at least the first decade of life, understanding how early hearing loss affects the development of the central auditory system and auditory processing is of great importance for optimizing the treatment of hearing loss in children (Moore 2002).

Second, because sensorineural hearing loss was the first neural sensory disorder to be treated successfully with a prosthesis, the neurobiological and psychophysical knowledge of human patients' and animal subjects' adaptation to use of cochlear implants forms the intellectual basis for future developments in brain–machine interfaces (Mussa-Ivaldi and Miller 2003). The continuing hope that induced hair-cell regeneration will someday be able to reverse sensorineural hearing loss in people (Bermingham-McDonogh and Rubel 2003; Kawamoto et al. 2003) adds to the importance of understanding how variations in input to the central auditory system can affect its function at all stages of life.

1

1.1 The Many Meanings of "Plasticity"

Although there is no controversy about the meaning of "development," it may indeed be true that, as one prominent scientist was recently quoted as saying, " 'plasticity' is the most abused word in neuroscience" (Holloway 2003, p. 80). The term has been applied to such a broad range of phenomena (recovery of function after injury, adult neurogenesis, synaptic changes associated with learning, experience-dependent reorganization of cortical sensory maps, etc.) that it has sometimes been oversimplified to mean little more than a capacity to change. Because the ability to change or adapt—over millennia, decades, months, and milliseconds—is a well-recognized feature of nervous systems, it is questionable if the term "plasticity" any longer has value as an aid to learning, discovery, or experimental problem solving. Perhaps it has become an example of what Holmes (1921) described as "inadequate catch words . . . phrases which originally were contributions, but which, by their very felicity, delay further analysis for fifty years . . . [and indicate] a slackening in the eternal pursuit of the more exact." Although the editors of this volume maintain some reservations about the excessively plastic meaning of "neural plasticity," they also acknowledge that there is no other commonly used term that describes the same wide range of changes in the nervous system. And, in its broadest sense, the term really does apply aptly to the contents of the present volume, whose expert authors variously define "plasticity" as alterations in structure, connections, or function that occur in response to experimental manipulations such as hearing loss, injury, or overstimulation—but not use-dependent changes during normal development (Friauf, Chapter 3); any change in the structure or function of the binaural auditory system induced by altered inputs, occurring either naturally as a consequence of head growth or as a result of the introduction of abnormal inputs (Moore and King, Chapter 4); or as systematic long-term changes in the response of neurons to sound as a result of experience (Weinberger, Chapter 5). Clearly, with respect to the definition of "plasticity," the editors and authors of this volume have embraced the masterful approach to word meaning presented by Carroll (1872): " 'When I use a word,' Humpty Dumpty said, in rather a scornful tone, 'it means just what I choose it to mean—neither more nor less.' 'The question is,' said Alice, 'whether you *can* make words mean so many different things.' 'The question is,' said Humpty Dumpty, 'which is to be master—that's all.' "

1.2 Accomplishments and Challenges

The ultimate goal of developmental neurobiology with respect to hearing is to provide an understanding, at a predictive causal level, of the biological mechanisms that produce mature function in the auditory system. En route to that destination investigators have labored to establish the normal developmental sequences of behavioral, anatomical, physiological, and molecular events that define the process of maturation (Rubel et al. 1998). When the basic timetable of ontogeny is known, experimental manipulations can be used to determine which of many identified mechanisms are necessary and sufficient to permit normal development of particular features of normal hearing. The seven review chapters in this volume cover a spectrum of issues in the development of audition and also describe experimental manipulations that illuminate ontogenetic mechanisms responsible for particular aspects of central auditory system development.

Rubel, Parks, and Zirpel (Chapter 2) focus on development of the cochlear nucleus (CN), among the best understood of the central nervous system (CNS) auditory centers. These authors first review the molecular mechanisms responsible for specification, selective aggregation, and migration of CN neurons and find that relatively little is known about these critical steps in the development of the central auditory system. Although there is some normal correlative data suggesting that cadherins and ephrins-Eph receptor signaling are important in selective aggregation of the CN, Rubel et al. conclude that further progress will require considerably more experimental study of the molecular mechanisms by which the CN cell groups are specified as auditory neurons in the early rhombencephalon, how these neurons migrate selectively to the lateral brain stem, and how they aggregate in the appropriate pattern to form the striking mosaic pattern of the mature CN. In their review of the innervation of the CN by cochlear nerve axons, Rubel et al. note the substantial amount of normative data on spatiotemporal patterns of innervation and the surprisingly precise nature of these processes at the earliest stages of development. These authors also emphasize the relative paucity of evidence concerning the molecular mechanisms responsible for the bifurcation of ingrowing cochlear nerve axons, their topographic arrangement in the subdivisions of the CN, the formation of highly specialized axon terminals on appropriate target neurons, and the establishment of tonotopic axes that are not aligned with any of the basic embryonic axes. Recent data relating the complex and transient expression patterns of certain ephrins and Eph receptors to the organization of afferent axons in the CN and higher auditory centers are described. Because the deafferented CN was the best early example of the essential role of afferent synaptic input in survival of some CNS neurons, there is a relatively rich literature describing the mechanisms by which cochlear nerve synapses and activity preserve CN neurons. Rubel, Parks, and Zirpel (Chapter 2) review the detailed evidence concerning (1) the natural history of age-dependent deafferentation-induced death and atrophy of CN neurons, (2) the role of presynaptically released glutamate as the trophic agent responsible for mediating afferent-dependent survival of CN neurons, (3) the role of excessive intracellular calcium concentrations produced via calcium-permeable α-amino-3-hydroxy-5-methyl-4-isoxazolepropionic acid (AMPA) receptors in causing CN neuron death, and (4) the ability of calcium-dependent phosphorylation of the cAMP response element binding protein (CREB) transcription factor to prevent death in those CN neurons that survive deafferentation. These authors emphasize the unusual calcium challenges presented to CN neurons, the special calcium-homeostatic mechanisms these neurons employ to survive, and the involvement of anti-apoptotic genes in the sharply age-dependent effects of deafferentation on CN neuron survival.

In reviewing the extensive research literature on normal development of the mammalian superior olivary complex (SOC), Friauf (Chapter 3) notes how many key events occur prior to the onset of hearing: neurogenesis, cell migration, axon and dendrite outgrowth, target selection, and synaptogenesis. The fact that these events do not proceed sequentially from periphery to center but rather in parallel is also noted, as is the relatively precise topography of initial axonal projections, which are then subjected to activity-dependent refinement. One particularly interesting aspect of SOC development is the presence of excitatory and inhibitory synaptic inputs during development of many SOC neurons, and Friauf reviews the evidence for competitive interactions among these developing neurotransmitter systems. The SOC provides dramatic examples of aberrant functional projections induced by deafferentation of immature auditory neurons. Although the capacity of the auditory pathway to make these major structural changes is restricted to the period before the onset of hearing, the size and structural complexity of SOC neurons remain dependent on normal auditory input well past hearing onset.

In their review of plasticity in binaural hearing, Moore and King (Chapter 4) note that adjustments in binaural processing are probably involved in the natural response to growth-related changes in head size and other natural changes in interaural cues, as well as disease- and injury-induced hearing asymmetries. The deleterious effect of early otitis media on development of binaural hearing in children can be persistent, which has important implications for education, pediatrics, and otolaryngology. The persistence of binaural plasticity into later life also suggests that balancing spectra between the two ears should be an important goal in treatment of adult hearing loss. This life-long plasticity prompts Moore and King to conclude that individual variations in CNS structure and behavioral performance, including use-dependent changes within an experiment, should be seen as evidence not of poor experimental control but rather of the malleable response of a living system. Research on binaural plasticity in the computational map of auditory space in the midbrain has created one of the most successful experimental systems for investigating the role of sensory experience in shaping the developing brain. As Moore and King note, some progress has been made in understanding the cellular mechanisms of these phenomena but much remains unclear. These authors also emphasize the importance of extending cellular studies of binaural plasticity to the thalamus and auditory cortex.

Although studying plasticity in the auditory cortex is important, it is particularly difficult, for reasons Weinberger explains in Chapter 5. He focuses on the challenges of designing the most informative experiments, challenges that arise in large part from the differing perspectives of sensory physiologists, who have mainly been interested in isolating sensory processes from experience-dependent effects during experiments, and of learning and memory specialists, who have not directed much attention to the capacity for learning in structures outside the medial temporal lobe. Weinberger notes that there are two disparate views of how learning is involved in the functions of the auditory cortex. One view is that experience-dependent plasticity is only another parameter, along with stimulus parameters such as frequency and sound level, that has to be taken into account during experiments. In this approach, there is often an implicit assumption that auditory function can be understood adequately without regard to behavior, so that learning is regarded merely as a modulator of normal auditory cortical neuron function rather than an embedded influence in all auditory processing. An opposing view holds that an adequate account of the auditory system requires investigation within the context of awake and behaving organisms, a position that makes many interesting experiments impossible or exceptionally difficult. Weinberger argues for a broadly integrative approach to studies of learning in the auditory cortex, using all the available techniques (and inventing new ones) while evaluating the extent to which auditory function studied in the anesthetized state or reduced preparation reflects the reality of the behaving organism.

Analysis of auditory neuron function in awake behaving organisms has yielded some of its most impressive results in the study of birdsong. As Brenowitz and Woolley (Chapter 6) discuss, the study of neural plasticity in birdsong has also led neuroscientists to recognize the existence of ongoing generation and replacement of projection neurons in the endotherm brain and hormonally induced seasonal changes in the morphology, pharmacology, and physiology of CNS neurons. As both birdsong and human speech are learned early in life, young birds and children must hear adult vocalizations to imitate them accurately, and each species shows an innate predisposition to learn conspecific vocalizations. Both song and speech acquisition have an early perceptual phase in which models of the sounds are listened to and memorized. These sensory templates then guide vocal production, with auditory feedback a necessary condition for both the development and maintenance of normal song and speech. After reviewing the considerable body of evidence concerning the neural mechanisms involved in each stage of vocal learning, Brenowitz and Woolley describe the most important open questions for research in this field: the basis of innate auditory preferences for conspecific song; the factors determining the opening and closing of the sensitive period for memorization of song models; the interaction of steroid hormones, neurotrophins, neurotransmitters and their receptors during development and adult plasticity; the functional significance of ongoing neuronal replacement in the song system; and the molecular determinants of song system plasticity at different stages of life.

The realization that many, if not most, *Drosophila* genes have homologs with similar functions in mammals has underscored the value of studying nervous system functions in the most advantageous subjects, which in a number of cases are insects. Lakes-Harlan (Chapter 7) approaches plasticity in the insect auditory system by considering four contexts in which it occurs. Plasticity during development provides several interesting examples. Because in many insects sensory receptor cells are added throughout life, central circuits must be modified continuously to accommodate them. Similarly, there is epigenetically mediated variation in auditory cell number and connections in individual animals of even isogenetic lineages. Activity-dependent modifications of the auditory system during development or in mature animals include habituation, experience-dependent changes in sound-mediated behaviors, and deprivation-sensitive growth of auditory neurons. Lakes-Harlan also discusses several examples of modulatory effects of hormones or behavioral status on auditory system function and reviews the extensive literature on post-injury compensatory plasticity in insects.

1.3 Conclusions

This volume includes representative contributions from most of the areas of current research on plasticity in the developing auditory system. As is perhaps clear from the summaries of the chapters given in the preceding, the field embraces investigations ranging from cellular studies of normal development to experimentally or pathologically induced reorganizations in connectivity and function to the effects of normal sound-related context-dependent learning. Because of the wide range of biological mechanisms encompassed, the field necessarily relies on the full range of neuroscience methods, from molecular genetics to pharmacology to psychophysics and functional brain imaging. The peculiarities of the auditory system differentiate the study of its plasticity from comparable efforts in other neural systems. Although the complexity of subcortical auditory pathways and the incompletely defined organization of the auditory cortex present unique difficulties for experimenters, the richness of behaviorally relevant information conveyed by acoustic stimuli also offers many unique opportunities. Owing to the relatively high prevalence of hearing loss in neonates and children, the clinical relevance of understanding CNS consequences of early deafness will likely continue to engender support for basic science studies of auditory system plasticity. If researchers in this field continue to apply to the auditory CNS the full range of technical and conceptual approaches available for investigating neural development, the study of auditory CNS plasticity should continue to advance general understanding of CNS development as well as lead to better treatment of early hearing loss.

References

Bermingham-McDonogh O, Rubel EW (2003) Hair cell regeneration: winging our way towards a sound decision. Curr Opin Neurobiol 13:119–126.

Carr CE, Soares D (2002) Evolutionary convergences and shared computational principles in the auditory system. Brain Behav Evol 59:294–311.

Carroll L (1872) Through the Looking-Glass. London: Macmillan.

Fettiplace R, Ricci AJ (2003) Adaptation in auditory hair cells. Curr Opin Neurobiol 13:446–451.

Fuchs PA, Glowatzki E, Moser T (2003) The afferent synapse of cochlear hair cells. Curr Opin Neurobiol 13:452–458.

Holloway M (2003) The mutable brain. Sci Am 289:79–85.

Holmes OW (1921) Law in science and science in law. In: Holmes OW (ed), Collected Legal Papers of Oliver Wendell Holmes. New York: Harcourt Brace, pp. 230–231.

Kawamoto K, Ishimoto S, Minoda R, Brough DE, Raphael Y (2003) Math1 gene transfer generates new cochlear hair cells in mature guinea pigs in vivo. J Neurosci 23:4395–4400.

Moore DR (2002) Auditory development and the role of experience. Br Med Bull 63: 171–181.

Mussa-Ivaldi FA, Miller LE (2003) Brain–machine interfaces: computational demands and clinical needs meet basic neuroscience. Trends Neurosci 26:329–334.

Pollak GD, Burger RM, Klug A (2003) Dissecting the circuitry of the auditory system. Trends Neurosci 26:33–39.

Riley BB, Phillips BT (2003) Ringing in the new ear: resolution of cell interactions in otic development. Dev Biol 261:289–312.

Rubel EW, Popper AN, Fay RR (1998) Development of the Auditory System. New York: Springer-Verlag.

Ryugo DK, Parks TN (2003) Primary innervation of the avian and mammalian cochlear nucleus. Brain Res Bull 60:435–456.

2

Assembling, Connecting, and Maintaining the Cochlear Nucleus

EDWIN W RUBEL, THOMAS N. PARKS, AND LANCE ZIRPEL

1. Introduction

The cochlear nucleus (CN) is an essential synaptic intermediary in the ascending auditory pathway and the site of remarkable neuronal specializations that allow this pathway to represent most of the behaviorally relevant information available in sounds (Cant 1992; Rhode and Greenberg 1992; Romand and Avan 1997; Ryugo and Parks 2003). Because of the powerful influence that the developing ear exerts on the developing auditory central nervous system (CNS) (Rubel 1978; Parks 1997; Friauf and Lohmann 1999; Rubel and Fritzsch 2002), considerable research has been directed at understanding the basic events of normal development and the central effects of early deafness. The large literature on normal structural and functional development of the CN has been reviewed in a previous volume of this series (Cant 1998; Sanes and Walsh 1998), and various aspects of abnormal development are discussed in other chapters of the book (Friauf, Chapter 3 and Moore and King, Chapter 4).

This chapter focuses on three areas of research in which results obtained during the past decade have enlarged understanding of CN development. In particular, cellular and molecular aspects of (1) assembly of the CN during development, (2) innervation of the CN by the cochlear nerve, and (3) survival of CN neurons are considered. The significance of cellular plasticity in these key developmental events is emphasized.

2. Assembling the CN

2.1 Hindbrain Patterning and the Development of Neural Circuits

During development, the brain arises from the neural tube. Throughout its length, in response to molecular signals arising mostly from adjacent non-neural structures, the neural tube divides into a series of discrete neuroepithelial domains along both its longitudinal and transverse axes (Sanes et al. 2000). These domains, which include the classic embryologic units of plate (roof, alar, basal, and floor) and vesicle (rhombencephalon, mesencephalon, etc.), have been considered embryonic modules or compartments as they represent histogenetic units within which cells proliferate, migrate, and differentiate into characteristic neurons and glia in relative independence from adjacent domains (Redies and Puelles 2001; Pasini and Wilkinson 2002). Within the rhombencephalon, from which all neurons in the CN are known to arise (Cant 1998), there is a further anteroposterior segmentation into seven rhombomeres (Lumsden and Krumlauf 1996). Many of the genes involved in establishing segmentation in the neural tube have been identified as transcription factors or gene regulatory proteins that also regulate pattern formation in other parts of the embryo (e.g., in the inner ear; see Fritzsch et al. 1998). In the hindbrain, the most important genes guiding rhombomere formation are *Krox20*, *Kriesler*, *Gbx2*, and *Hox* (Pasini and Wilkinson 2002). Longitudinal cell migration becomes restricted just as rhombomere boundaries form and there is quite limited mixing of cells across these boundaries, apparently owing to differential adhesive properties of cells in adjacent rhombomeres (Mathis and Nicolas 2002). Each rhombomere gives rise to a specific set of cranial motor nerves, reflecting the acquisition of a unique identity by each hindbrain segment. In the auditory system, for example, the "motor" neurons of the olivocochlear efferent system all arise from rhombomere (r)4 (Simon and Lumsden 1993).

As the hindbrain develops further, the relatively simple segmental modularity of the rhombomeres must somehow be translated into more complex gene expression patterns that are required for the key events of brain morphogenesis and circuit formation: cell migration and aggregation; axon and dendrite outgrowth; and target recognition and synapse formation. The regulators of these processes include molecules affecting cell–cell and cell–substratum adhesion (e.g., cadherins, integrins, and members of the immunoglobulin superfamily), diffusible molecules that create gradients guiding cell and axon movements (e.g., netrins, semaphorins, slits), and molecules mediating selective attraction and repulsion between neurons and neural processes (e.g., ephrins, Eph receptors, neuropilin) (Sanes et al. 2000; Redies and Puelles 2001).

The influence of the distinct molecular features that define a particular rhombomere early in development on the subsequent maturation of the neurons arising from that rhombomere are poorly understood, although it is clear that some of the key circuit-forming molecules (e.g., Eph receptors, ephrins, and cadherins) are regulated by the transcription factors involved in creating hindbrain segmentation (Pasini and Wilkinson 2002). There is now evidence for a link between expression of developmental patterning genes and specification of individual neuron groups within the vestibular nuclei (Diaz and Glover 2002). This work has also made clear some of the challenges involved in these studies. Glover (2001, p. 691) remarks that because of the highly dynamic patterns of gene expression in the developing hindbrain, "... correlative studies must be detailed and systematic if they are to contribute the information necessary to

understand how genes are linked to the formation of identifiable neuron groups."
Clearly, however, a full molecular understanding of the development (and evolution) of the brain stem auditory system will require such studies. In the following sections we review what has been learned to date about the rhombomeric
origin of the CN and subsequent expression of key molecules involved in selective aggregation and circuit formation.

2.2 Rhombomeric Origins of the CN

Owing to technical limitations, there is no published experimental evidence concerning the rhombomeric origin of the cochlear nuclei in mammals. Two technical approaches, however, have generated extensive data concerning the
rhombomeric origins of the cochlear nuclei in the chick embryo: chick–quail
chimeras and dye labeling. Marin and Puelles (1995) homotopically transplanted individual rhombomeres (r2–r6) from quail embryos into chick embryos
at 2 days of incubation. After survival periods of 9–10 days, the chick embryos
were fixed and alternating sections were stained with an antibody that recognizes
quail cells or with cresyl violet to identify cell groups in the brain. These
authors reported that (1) nucleus angularis (NA), avian homolog of the mammalian dorsal and posteroventral cochlear nuclei, derives from rhombomeres 3–
6; (2) nucleus magnocellularis (NM), homolog of the mammalian anteroventral
cochlear nucleus, derives from r6 and r7, (3) nucleus laminaris (NL), homolog
of the mammalian medial superior olivary nucleus, derives from r5 and r6, and
(4) the superior olivary nucleus (SON) derives from r5.

Cramer et al. (2000a) made very small injections of lipophilic fluorescent
dyes into the hindbrain of chick embryos prior to the birth and migration of the
cells that contribute to the brain stem auditory nuclei. After allowing the embryos to develop until embryonic days (E) 7–13, the investigators examined the
sectioned brains with fluorescence microscopy. Because they were able to doc-
ument precisely the locations of both the original dye injections and the labeled
neurons, Cramer et al. (2000a) produced a detailed fate map for the different
parts of the rhombomeres that contribute to the brain stem auditory nuclei. They
concluded that NA arises from r4 and r5; that NM has contributions from r5,
r6, and r7; that NL arises mostly from r5 with small contributions from r6; and
that the SON arises entirely from r5. Cramer et al. further showed that (1) for
r5, the precursors of NM are located medially and those of NL laterally, and
(2) neurons arising from precursors in a more rostral rhombomere are found
rostrally within each CN.

Taken together, the results of Marin and Puelles (1995) and Cramer et al.
(2000a) show that the avian cochlear nuclei arise from rhombomeres 4–7, with
r5 providing cells to multiple nuclei. Because a single cochlear nucleus, for
example, NM, has precursors in several rhombomeres but no sharp boundaries
within the nucleus corresponding to rhombomeric origin, it is clear there is
considerable cell mixing during migration and nuclear aggregation. Thus, it
appears that the rostrocaudal and mediolateral positions of precursors, but not

rhombomere boundaries per se, affect the positions of their descendent neurons within the CN, suggesting that the precursors are specified in their position at a quite early stage. The fact that precursors of the synaptically connected NM and NL lie, respectively, in the medial and lateral parts of r5 shows that although they are intermixed in the auditory anlage that exists in the rhombic lip region prior to appearance of distinct nuclei (Cramer et al. 2000a), a lineage relationship between NM and NL neurons is highly unlikely. Furthermore, it appears that although some avian brain stem auditory centers may arise from a single rhombomere, others may have origins in several rhombomeres without necessarily exhibiting internal structural differences attributable to the multisegmental origins. Although it may be discovered that rhombomeric origins constrain developmental programs for some aspects of auditory neurons, as they are thought to do for cranial motor neurons (Lumsden and Krumlauf 1996), current evidence suggests that the cochlear nuclei (like some other alar plate derivatives; Marin and Puelles 1995; Glover 2001) develop without major constraints on neuronal differentiation. Thus, after undergoing their final mitotic divisions and while migrating toward their final positions within the rostrocaudal column of neurons that forms the CN, the neurons of the CN must aggregate with neurons of similar type to form the various subdivisions of the CN and prepare to receive specific innervation from the cochlear nerve and other sources.

2.3 Expression of Molecules in the Developing CN that May Affect Aggregation and Target Selection

Cadherins are a large family of cell–cell adhesion molecules with important roles in the morphogenesis of many organs. Several dozen cadherins are expressed in the vertebrate CNS, each with a unique expression pattern, during periods in development when cell groups migrate, aggregate, and form synapses. It has been proposed that cadherins provide a mechanism by which neurons can selectively aggregate and form specific synaptic connections (Redies 2000; Redies and Puelles 2001). Although there is as yet no comprehensive survey of the distribution of various cadherins in the brain, there are several reports of specific patterns of expression in the auditory system. In a study employing in situ hybridization and immunohistochemistry to localize cadherin-6 in the brains of embryonic and postnatal mice, Inoue et al. (1998) found this molecule is strongly expressed throughout most of the auditory pathway: cochlear ganglion, dorsal cochlear nucleus, inferior colliculus, medial geniculate body, and auditory cortex. The ventral CN, in contrast, showed no cadherin-6 expression and the superior olivary nuclei had only weak expression. The expression pattern for another cadherin in the E8–E15 chick brain was described by Arndt and Redies (1996). By means of in situ hybridization and immunohistochemistry, these authors found that R-cadherin was expressed in the nucleus angularis, nucleus magnocellularis, superior olivary nucleus, lateral lemniscal nuclei, and inferior colliculus (Mld) but only in the axons surrounding nucleus laminaris, not in its neurons. In contrast, cadherin-10 is expressed in more rostral portions of the chick auditory pathway but not in the cochlear nuclei (Fushimi et al. 1997).

A number of other molecules thought to be involved in selective adhesion or repulsion between cells at various times during development are expressed with notable strength in the auditory pathway. NB2 is a neural cell recognition molecule of the contactin/F3 group and, by in situ hybridization and immunohistochemistry, has been shown to be strongly expressed throughout the mouse auditory pathway. There is particularly strong immunoreactivity in fibers within the ventral CN and superior olivary nuclei (Ogawa et al. 2001). A genetic inactivation of this gene is reported to result in markedly reduced neuronal activity in the central auditory pathway (Li et al. 2003). Plexins are a family of transmembrane proteins that interact with semaphorins and neuropilins to facilitate repulsive interactions between neurons. Murakami et al. (2001) studied expression of three members of the plexin-A subfamily in embryonic and postnatal mouse brain using immunohistochemistry and in situ hybridization. These authors report strong expression of plexin-A1 in all levels of the auditory pathway, from cochlear ganglion to auditory cortex, including both the dorsal and ventral CN. Plexin-A2 was expressed in other selected brain regions but not in the auditory pathway, and plexin-A3 was expressed in auditory centers as well as most other CNS locations.

Eph receptors are membrane-bound tyrosine kinase receptors that have been implicated in a wide range of developmental processes, including cell migration, axon guidance, and the establishment of topographic maps. The ligands for Eph receptors, the ephrins, are membrane bound and can initiate signal transduction events when bound to Eph receptors (Flanagan and Vanderhaeghen 1998; O'Leary and Wilkinson 1999; Wilkinson 2000). Cramer et al. (2000b, 2002) have studied developmental changes in expression of some Eph receptors and ephrins in the chick brain stem auditory system. Cramer et al. (2000b) used immunohistochemistry to show that EphA4 is expressed in rhombomere 5 (which, as noted in the preceding, contains precursors of both NM and NL) and that, as NM and NL neurons migrate into the auditory anlage around E5–8, EphA4 expression becomes confined to longitudinal strips within the brain stem. At the time in development when synaptic connections are formed between NM and NL (E10–12), EphA4 expression in NL becomes strongly asymmetric, with much higher levels in the dorsal than in the ventral neuropil of this nucleus. At later stages, EphA4 expression in NL becomes symmetric again before disappearing after posthatch day (P) 4. Cramer et al. (2002) went on to study developmental changes in immunoreactivity of the Eph receptors EphB2 and EphB5 and of ephrin-B1 and ephrin-B2 in the chick auditory nuclei. These authors found a complex pattern of expression of these molecules during embryonic life that would allow them to be involved in the maturation of the auditory nuclei and their synaptic connections. Finally, using information derived from the rhombomeric fate map discussed in the preceding, the expression of signaling molecules in the developing chick auditory brain stem is being manipulated experimentally. In the first of these studies, Cramer et al. (2003) showed that misexpression of EphA4 dramatically alters the segregation of ipsilateral and contralateral axons innervating NL.

Although cell adhesion molecules (CAMs) have not yet received much experimental attention from scientists whose main interest is auditory system neurobiology, it is clear from the examples cited above that many CAMs are expressed during key stages in development of the CN. The large number of CAMs expressed in the CNS and their unique and only partially overlapping expression patterns strongly suggest that the adhesive identity of any cell, which must determine the cells with which it aggregates and forms synapses, is likely to be determined by a combinatorial code of CAM expression. Ultimately, it should be determined if a particular pattern of CAM expression is necessary and sufficient to allow a particular neuronal type in the CN to aggregate selectively and to form appropriate synaptic connections with pre- and postsynaptic partners. This goal will require a comprehensive study of CAM expression patterns by developing CN neurons of different types and experimental manipulation of those patterns.

3. Innervating the CN

The development of central projections of eighth nerve ganglion cells has been studied in a variety of ways, ranging from descriptive studies using classical silver staining or the Ramon y Cajal/Golgi methods to more contemporary methods using cell specific markers or axonal tracing. In this section, recent descriptive and mechanistic studies on the development of the eighth nerve projection to the brain stem in avian and mammalian species are summarized. It is useful by way of organization to consider the ontogenetic series of events that take place in the ganglion cells and their surrounding environment.

After the immature neuron has delaminated from the developing otocyst and undergone its final mitotic division, it forms a centrally directed process that traverses the basal lamina surrounding the lateral aspect of the rhombencephalon and enters the brain parenchyma. This protoplasmic process, the eighth nerve axon, bifurcates one or more times to send branches into the presumptive CN subdivisions (Lorente de Nó 1981; Fekete et al. 1984). In mammals, most eighth nerve axons are thought to provide afferents to all three major cochlear nucleus subdivisions: the anteroventral cochlear nucleus (AVCN), the posteroventral cochlear nucleus (PVCN), and the dorsal cochlear nucleus (DCN). In birds, a branch is sent to each of two subnuclei, n. magnocellularis (NM) and n. angularis (NA) (Ryugo and Parks 2003). During development of these projections, the axons must arrange themselves in precisely the order of their peripheral targets in the cochlea. In other words, the frequency/place organization of the sensory epithelium that is mapped onto the population of ganglion cells must be exactly re-created in the organization of projections into each division of the cochlear nucleus. This mapping of the receptor surface onto the cells in each division of the CN establishes the precise tonotopic organization seen physiologically and anatomically in the mature animal.

At the same time or shortly after entering the CN, the eighth nerve axons form different highly stereotyped synaptic specializations that are unique to each target region. The morphology and physiology of the eighth nerve synapse onto postsynaptic cells in AVCN becomes markedly different from those expressed by a collateral of the same axon in the DCN or PVCN. The contacts and synaptic activity transmitted by the cochlear nerve axons can have dramatic influences on the development and maintenance of cells in their targets, the subnuclei of the CN complex.

In the remainder of this chapter, the following topics are considered: (1) eighth nerve growth into the brain parenchyma; (2) development of synaptic contacts between eighth nerve axons and target cells in the CN; (3) emergence of topographic (tonotopic) organization in the CN; and (4) trophic interactions between the eighth nerve and CN cells. In each topic area, the current descriptive information and the level of understanding of cellular and molecular mechanisms is evaluated, and suggestions for future investigation are offered.

3.1 Axon Development

As early cochlear ganglion axons grow through peripheral connective tissue rich in fibronectin and laminin (Hemond and Morest 1991a,b), they are thought to fasciculate with axons of the vestibular nerve, which have already penetrated the developing rhombencephalon. In the chick, the final mitosis of cochleovestibular ganglion cells (CVGs) occurs between E2 and E7, with cochlear cells developing later than vestibular cells (D'Amico-Martel 1982). While cell division is still occurring (i.e., by E3 or H and H stage 19), some axons can be seen entering the medulla (Windle and Austin 1936; Hemond and Morest 1991a). The cochlear processes are probably delayed relative to the vestibular processes by about 1 day. By stage 25–26 (E5), many cochlear axons have penetrated the brain parenchyma (Knowlton 1967; Book and Morest 1990). However, it is not clear from the literature exactly when eighth nerve axons become intercalated among the developing NM and NA neurons (Knowlton 1967; Rubel et al. 1976).

This same close association between the birth date of ganglion cells and central axon formation seen in birds is also present in mammals. For example, in the mouse, most cochlear ganglion cells are born at approximately E13.5 (Ruben 1967), but cochlear axons enter the brain by E13–E14 (Willard 1993, 1995). While the precise timing of ganglion cell birth dates has not been studied in many species, the age at which eighth nerve axons enter the brain has been examined in a variety of mammals, including pig (Shaner 1934), rat (Angulo et al. 1990), human (Moore et al. 1997, Ulatowska-Blaszyk and Bruska 1999), and the Brazilian opossum *Monodelphis domestica* (Willard 1993).

The growth of axons into the brain occurs well before the onset of hearing as defined by physiological responses to acoustic stimuli. In the E16 rat, axons from ganglion cells originating at the basal turn of the cochlea invade both the AVCN and the PVCN (Angulo et al. 1990). Over the next 2–3 days, axons from the middle and apical turns enter the nuclear subdivisions. This is approximately 2 weeks before the rat will hear airborne sounds. Similar data are available for the ferret (*Mustela putorius*, Moore 1991), North and South American opossums (*Didelphis marsupialis*, Willard and Martin 1986; *Monodelphis domestica*, Willard 1993), and hamster (*Mesocricetus auratus*, Schweitzer and Cant 1984). In the human embryo, cochlear nerve fibers invade the VCN by 16 weeks of gestation, whereas physiological and behavioral responses to sound are not apparent until about 26 weeks (Moore et al. 1997). These studies suggest that innervation forms independently of auditory input.

When the axons enter the brain, collaterals form to innervate the subdivisions of the CN. These collaterals must grow to, and then stop growing when they reach, the appropriate target. The cellular and molecular mechanisms underlying the growth and fasciculation, targeting, and branching of these axons are almost completely unknown. In spite of the emerging wealth of information on axonal pathfinding cues in other systems (Flanagan and Vanderhaeghen 1998; Mueller 1999; Brose and Tessier-Lavigne 2000; Raper 2000), the molecules that alter the pathway selection of auditory nerve axons in the brain have not been identified. Similarly, the cellular interactions that induce growing eighth nerve axons to bifurcate once or twice on entering the brain stem and to stop on entering their targets are also unknown. In fact, in most species, it is not agreed on whether auditory nerve axons grow into their final position and provide the attractive signals for postmitotic neuronal precursors to coalesce around them (e.g., see Morest 1969), or if the neuronal precursors of the brain stem auditory nuclei begin their migration and "attract" the growing eighth nerve axonal process. Willard (1990) argues that the auditory nerve grows into the brain stem prior to the migration of auditory neurons. Migrating postmitotic neurons may then be attracted to these axons and cease migrating. Some support for this view is found in both experimental and descriptive studies of the developing chick brain stem. Parks (1979) showed that NA cells migrated into ectopic positions in the brain stem following early otocyst removal at E2.5, which eliminated development of the CVG. In the mouse, neuroblasts forming the main targets of the cochlear nerve leave mitosis on days 10–14 (Taber Pierce 1967; Martin and Rickets 1981). These dates completely coincide with the generation of ganglion cells (Ruben 1967). Therefore, migration of most CN neurons is likely to be occurring at about the same time as most of the eighth nerve axons are arriving. Without experimental manipulations, it is difficult to understand the interactions between migrating neuroblasts that will form the cochlear nucleus and the growing cochlear nerve axons. Some progress has been made toward identifying both intracellular and secreted molecules that may be important for these interactions. For example, Represa and colleagues (San Jose et al. 1997) have begun examining cytoskeletal changes in the growing axons of chicks. In this same species, Morest and colleagues are examining the timing and spatial pattern of fibroblast growth factor-2 (FGF-2) and its receptors in relation to ganglion cell and brain stem development (e.g., Brumwell et al. 2000). Finally, the developmental patterns of expression of neurotrophins and their receptors in the mammalian and chick auditory brain stem are being examined (Hafidi et al. 1996; Hafidi 1999; Cochran et al. 1999).

It is now possible to experimentally address the key issues discussed in the preceding. There is a need for studies combining careful descriptive, developmental methods with experimental manipulations that eliminate either the eighth nerve axons (Ma et al. 1998; 2000) or the hindbrain regions that form the anlagen of the auditory nuclei (e.g., Studer et al. 1998; Cramer et al. 2000a). For example, the target cells within the developing brain stem could be removed to test the role of targets in specifying axonal branching patterns.

3.2 Development of Contacts Between Ganglion Cell Axons and CN Neurons

As axons of the cochlear nerve arrive in the brain stem, they interact with the cell bodies and processes of postmitotic neuroblasts in several important ways. For example, they form synaptic connections to establish the information-processing network of the auditory pathways. One fascinating property of this process is that the different collaterals of the auditory nerve form synaptic connections with very different morphologies. In the AVCN and NM of mammals and birds, respectively, the predominant synaptic morphology is a calyx surrounding much of the cell body, known as the end bulb of Held (Lorente de Nó 1981; Ryugo and Parks, 2003). An example of this type of synapse in the chick NM is shown in Figure 2.1. This presynaptic ending is highly stereotyped and provides a phase-locked, powerful excitatory connection, known to be important for temporal processing. In other regions of the cochlear nuclear complex, more common boutonal synapses are made. Considerable work has been done on the developmental dynamics of end-bulb development in the AVCN and NM.

Jhaveri and Morest (1982a,b) show rather elegantly that postsynaptic NM neurons initially have extensively ramifying dendritic processes among which the ingrowing auditory nerve axons branch and form at least transient synaptic connections. Then, coincident with the early stages of auditory function, the dendritic arbors become resorbed (see also Parks and Jackson 1984; Young and Rubel 1986), and two or three end bulbs form on the cell body of each NM neuron. Formation of the end bulb may be due to coalescence of many terminal arbors or the dramatic expansion of a few of the initial presynaptic structures. Dendritic resorption and end-bulb formation begin at the rostromedial (high-frequency) area of NM and progress caudolaterally along the tonotopic axis of the nucleus (Parks and Jackson 1984; Young and Rubel 1986). The divergence of eighth nerve axons to neighboring cells in NM and the convergence of axons onto single NM neurons have been examined in the chick (Jackson and Parks 1982). The large, complex dendritic arbors of NM neurons at E9–E12 (Young and Rubel 1986) make it difficult to draw firm conclusions about divergence of

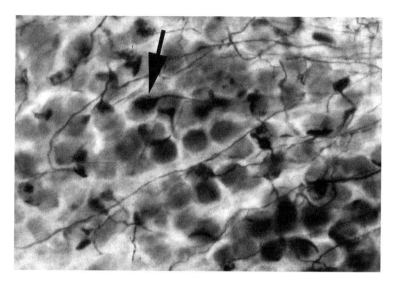

FIGURE 2.1. Calyx of Held synapses on nucleus magnocellularis neurons of the chick. End bulbs were labeled post-fixation with HRP and the tissue counterstained with thionin. The arrowhead indicates a calyx in which the unique morphology is readily apparent. Each NM neuron receives approximately two calyceal synapses that cover two thirds of the cell body. The unique morphology of this glutamatergic synapse helps ensure reliable transmission of auditory information from the eighth nerve to the cochlear nucleus.

presynaptic arbors at the age synaptic connections are forming. But clearly there is a modest decrease in preterminal axonal branching between E14 and E17. In addition, physiological analyses have shown a small decrease (from 4 to 2.4) in the number of unitary EPSPs excited by stimulation of eighth nerve inputs over the same age range (Jackson and Parks 1982). While these results are often cited as supporting the idea of widespread exuberance of axonal connections in the developing nervous system, there is little evidence supporting such an interpretation. The decrease in convergence is quite limited, and there is no evidence that the one or two supernumerary axons come from different cochlear regions. In the next section, evidence is provided that, from the outset, connections appear to form quite precisely in the brain stem auditory pathways.

Developmental studies of the end bulbs of Held have also been performed in the mouse, rat, cat, and barn owl (Mattox et al. 1982; Neises et al. 1982; Ryugo and Fekete 1982; Carr and Boudreau 1996; Limb and Ryugo 2000). While early development has not been studied in detail, the abundance of synaptic profiles in the neuropil and on somatic processes in newborn rats and barn owls suggests that the pattern may be quite similar to that described in the chick. On the other hand, the developmental pattern has been described rather completely in the cat (Ryugo and Fekete 1982) and mouse (Limb and Ryugo 2000). These papers describe a series of changes in end-bulb morphology, from a simple spoon-shaped ending to an elaborate series of filopodia engulfing the somata of AVCN neurons. Interestingly, at all ages described, the ending is elaborated on the somata of the developing AVCN neuron, and the neuron itself is rather adendritic.

The stereotyped structure of the end bulb of Held provides a unique opportunity to consider the relative contributions of the axon collateral vs. the target cell in specifying this presynaptic phenotype. Because other collaterals of the same axons, those terminating in DCN and PVCN, possess boutonal type endings, it would seem logical to speculate that the form is specified by the target. This question was addressed experimentally by Parks et al. (1990), who took advantage of the earlier discovery that NM neurons make ectopic projections to contralateral NM when the contralateral otocyst is removed. NM neurons normally make boutonal synapses onto n. laminaris (NL), the third-order neurons in the avian auditory system. At the light microscopic level, the ectopically projecting NM-to-NM axons formed boutons, suggesting that the cell of origin, not the target cell, specifies synaptic morphology. However, some ultrastructural features resembled the eighth nerve synapse on NM neurons. Thus, it appears from the studies that both axons and target cells determine synaptic morphology.

A second issue is whether eighth nerve action potential generation and synaptic activity influence the development of contacts between the nerve and CN neurons. In the chick, electrophysiological studies have shown that NM neurons are responsive to eighth nerve stimulation at day 10–11 of embryogenesis (Jackson et al. 1982; Pettigrew et al. 1988). Responses to sound are seen in brain stem recordings by E12–E13 (Saunders et al. 1973). Therefore, it is possible that the resorption of dendritic arbors seen in NM neurons and/or the changes in end-bulb morphology are dependent on afferent activity. While end-bulb morphology has not been studied carefully in the chick, neither the time course nor the tonotopic gradient of dendritic changes appears influenced by the presence or activity of eighth nerve axons (Parks and Jackson 1984). In the kitten, however, there is considerable evidence that the presence and activity of the eighth nerve influence the complexity and size of the end bulbs and their ultrastructural characteristics (Saada et al. 1996; Ryugo et al. 1997, 1998; Niparko 1999; Redd et al. 2000).

The presence of spatiotemporal gradients in the relationship between eighth nerve axons and the developing cochlear nucleus has been observed in a variety of studies (Rubel et al. 1976; Jackson et al. 1982; Schweitzer and Cant 1984; Kubke et al. 1999). For example, Schweitzer and Cant found that fibers from the basal portion of the hamster cochlea are the first to enter the DCN, followed by axons from the middle and apical turns, respectively. How such gradients among the axons or the postsynaptic cells in the cochlear nucleus are established remains a mystery awaiting molecular discovery. However, they do appear to be independent of sensory input from the ear (Parks and Jackson 1984).

Finally, it is important to mention that, during the time period when connections are forming between cochlear nerve axons and cochlear nucleus neurons, both elements are likely to be changing in a large variety of cellular and molecular respects, including transmitter and modulator expression and release kinetics, neurotransmitter receptor pharmacology (e.g., Zhou and Parks 1992; Code and McDaniel 1998; Kubke and Carr 1998; Lawrence and Trussell 2000; Parks 2000; Zirpel et al. 2000a), ion channel characteristics (Perney et al. 1992; Garcia-Diaz 1999; Sugden et al. 2002), and other synaptic specializations (e.g., Lurie et al. 1997; Hack et al. 2000). Elucidation of the intracellular and intercellular molecular pathways influencing such changes awaits further research.

3.3 Development of Topographic (Tonotopic) Connections

In the visual, somatosensory, and auditory pathways of most organisms, a highly stereotyped, topographic relationship exists between the receptor surface and the collections of neurons in nuclei or specific brain areas at each level of the ascending sensory pathways. These maps of the receptive surfaces of the organism are defined anatomically by preservation of neighbor relationship projections to each brain region. Physiologically, they are demonstrated by an orderly array of receptive fields seen in postsynaptic responses as one moves an electrode in small increments through a sensory area of the brain. Such maps represent physical space in the visual and somatosensory systems. In the auditory systems of birds and mammals, they provide a representation of a quite different stimulus–response attribute, the "best frequency" or "characteristic frequency" of the neuronal response to acoustic stimulation. This mapping property is a function of the remarkably precise coding of frequency along the cochlea (von Békésy 1960; Rhode 1978; Dallos 1992) and the precise topography of connections between cochlear ganglion cells and hair cells along the sensory epithelium, discussed in the preceding.

When considering the development of topographic (tonotopic) organization of ganglion cell projections to the cochlear nucleus, three issues need to be addressed. First, does the map emerge from relatively indiscriminate connections, or is there a degree of precision as soon as the projection is evident? If some precision is evident from the onset of function, does the "grain" of the map change during further development? Second, a popular belief is that "rough" characteristics initially form and these are "refined" during use. What role, if any, do auditory experience or neuronal activity independent of sound-driven activity have on the development or maintenance of this topography? Finally, and most important, what are the cellular signals responsible for the establishment and maintenance of the tonotopic map?

3.3.1 Precision of Cochlear Nerve Projections

Available evidence suggests that in the developing auditory systems of birds and mammals, the topography of connections between the cochlea, the ganglion cells, and the cochlear nuclei develops quite precisely, well before acoustic information is processed by these cells. Anatomical studies with results relevant to this issue have been provided for the rat (Angulo et al. 1990; Friauf 1992;

Friauf and Kandler 1993), mouse (Fritzsch et al. 1997), opossum (Willard 1993), hamster (Schweitzer and Cant 1984; Schweitzer and Cecil 1992), cat (Snyder and Leake 1997), and chick (Molea and Rubel 2003). No single study has labeled neighboring cells in the spiral ganglion and examined the relative align-ment of terminal fields in the CN or done a similar analysis by retrograde transport (e.g., see Agmon et al. 1995). Indirect evidence for a great deal of initial precision is provided by the demonstrations that terminal arbors in the CN are initially small and precisely oriented; terminal arbors grow as the nucleus expands in volume (Schweitzer and Cecil 1992). Furthermore, well before hear-ing onset in opossum, cat, and chick, small injections of horseradish peroxidase (HRP) into the spiral ganglion label discrete bands of terminals in the CN, and the size of these bands does not change with age (Willard 1993; Snyder and Leake 1997). Although it is impossible to state that the precision or "grain" of the map does *not* change with experience, there is no compelling evidence sup-porting such a view at this time.

3.3.2 Tonotopic Organization of the Early Cochlear Nerve Projection

Physiological studies that have addressed the development of tonotopic organi-zation at the level of the CN or other brain stem nuclei led to similar conclusions. Physiological mapping studies invariably find a precise tonotopic organization early during development (Lippe and Rubel 1985; Sanes et al. 1989; Sterbing et al. 1994; Lippe 1995). Similarly, studies using pure-tone acoustic stimuli to modulate metabolic markers [c-fos, 2-deoxyglucose (2-DG)] have found discrete bands of label in the CN as early as stimuli can elicit a metabolic response (Ryan and Woolf 1988; Friauf 1992; Friauf and Kandler 1993). It appears fashionable to propose that the early topographic organization is somewhat crude or rough (meaning less well ordered, presumably) and that it is "fine-tuned" by auditory experience (e.g., see Friauf and Lohmann 1999). However, little evi-dence exists for any role of auditory experience in shaping the tonotopic orga-nization of connections between the cochlea and the cochlear nuclei. In both birds and mammals, this organization appears before one can readily record responses to acoustic stimuli. There appear to be no gross "mistakes" in the orderly arrangement of connections, and the changes that are seen in the degree of specificity of axonal connections can be easily accounted for by the overall growth of the brain regions. Although the precision of the early eighth nerve to CN projections has not been studied in detail, the pattern has been studied at the next synaptic level. Young and Rubel (1986) examined the topography of the ipsilateral projection between NM and NL, and Sanes and Rubel (1988) studied the development of bilateral connections to the lateral superior olive (LSO) in the gerbil. Young and Rubel used single-cell reconstructions to show that, by E9, which is well before an auditory response can be found, the ipsi-lateral projections from NM to NL are as precise as they will ever be. In fact, subsequent development causes a loss of one dimension of specificity. Sanes and Rubel showed that, at the age responses can first be recorded in the LSO

(P14–P15), the matching of excitatory and inhibitory frequency tuning is virtually perfect. These results again suggest that the tonotopy at the level of the CN must already be mature before the onset of hearing.

3.3.3 Role of Spontaneous Activity

Having established that the tonotopic organization of projections from the cochlear ganglion to the CN emerges prior to auditory function, it becomes important to determine if activity that is independent of acoustic stimulation (spontaneous activity) plays an important role in the establishment and maintenance of appropriate connections. In this case, spontaneous activity is considered as action potential generation in the eighth nerve or CN that is not driven by acoustic stimuli. By this definition, activity of hair cell origin is not precluded. As noted previously, synaptic connections with the CN are formed before the onset of peripheral responses to sound in chicks and mammals (Jackson et al. 1982; Kandler and Friauf 1995). Spontaneous activity can be recorded soon after synaptic connections are seen physiologically or anatomically in chicks (Lippe 1994), wallabies (Gummer and Mark 1994), kittens (Walsh and McGee 1988), and gerbils (WR Lippe, personal communication). At this time, however, there are no convincing data suggesting that the spontaneous activity plays a role in the establishment of topographic connections. Lippe (1994) has described rhythmic activity that is of cochlear origin and shows a gradient in its developmental properties along the tonotopic axis. However, at E14, the age when this gradient is seen, the tonotopically organized projection from the ganglion cells to NM is already well established (Molea and Rubel, 2003).

3.3.4 Evidence for Autonomy of the Tonotopic Axis

Virtually nothing is known about the molecules that determine the tonotopic axis of the cochlear nuclei or guide the establishment of connections in an orderly way along this axis. It is clear, however, that both the presynaptic axons and the postsynaptic target cells must express some sort of signaling molecules that specify the tonotopic axis. Two interesting experiments support this conclusion. First, the resorption of dendrites in the chick NM takes place along a rostromedial to caudolateral spatial "gradient" that matches the tonotopic organization (Rubel and Parks 1975). Remarkably, the dendritic resorption, its time course, and its spatial organization appear independent of presynaptic input from the cochlear nerve (Parks and Jackson 1984). Second, abnormal connections to NM will form a normal orderly array along the tonotopic axis. This was shown by mapping the ectopic connection that forms between the two NMs when a unilateral otocyst removal is performed very early in development (Jackson and Parks 1988). Lippe et al. (1992) recorded from NM neurons while stimulating the contralateral ear in animals in which this projection was induced. Normally, NM axons innervate only NL neurons on the ipsilateral and contralateral sides of the brain (Young and Rubel 1983). When these axons are induced to innervate the contralateral NM, they produce a tonotopic organization indistinguishable from the normal ipsilateral eighth nerve input. This finding again suggests that the tonotopic axis is somehow encoded by the NM neurons and can be communicated to ectopic auditory afferents as well as its normal ipsilateral afferents of the eighth nerve.

3.3.5 Molecular Bases of Tonotopic Axis Formation

Although little is known about the molecules or cellular interactions participating in the establishment of the tonotopic organization of the CN in birds or mammals, developmental gradients in the ingrowth of eighth nerve fibers and of CN properties appear to correspond to the tonotopic axis (Rubel et al. 1976; Rubel 1978; Jackson et al. 1982; Schweitzer and Cant 1984; Willard 1993; Kubke et al. 1999). Timing alone is unlikely to provide the signal (Holt 1984; Holt and Harris 1993, 1998), but these gradients may provide clues to discover candidate molecules. Several growth factors and receptors have been examined in the ganglion cells and CN. Some of those growth factors and receptors appear to be expressed at approximately the time that connections are being established or that auditory function matures (e.g., see Luo et al. 1995; Riedel et al. 1995). However, gradients of expression that match the tonotopic axis at the time topographic connections are forming have not been reported. Understanding gradients of molecules along topographic axes is an important and timely problem in developmental neurobiology, in general, and the auditory pathways may be particularly advantageous for experimentally examining it. Eighth nerve ganglion and cochlear nuclei are derived from entirely separated epithelial compartments that can be separately manipulated. Further, there is a single, functionally defined, axis of orientation.

To adequately address this issue, two areas of research are initially needed. First, detailed analyses of the timing of the development of topographic connections at a single-cell level in a few "model" species are needed. Second, detailed analyses of the spatial and temporal distribution of candidate molecules that have provided important new information in other systems (e.g., Eph receptors and ephrins) are likely to prove important (e.g., see O'Leary and Wilkinson 1999; Wilkinson 2000). For example, recent studies of the developmental distribution of trkB and EphA4 show remarkable and provocative patterns of expression that are likely to be important for determining the laminar specificity of connections between NM and NL (Cochran et al. 1999; Cramer et al. 2000b). Further study of these classes of molecules may be helpful for understanding the development of tonotopy in the cochlear nuclei.

3.4 Influence of Cochlear Nerve on Development of CN

The classic study by Levi-Montalcini (1949) provided one set of fundamental observations underlying our approach to this problem. Levi-Montalcini removed the otocyst, the origin of the sensory cells and ganglion cells of the inner ear, at 2–2½ days of development in chick embryos. This manipulation deprived the embryos of normal input to the developing cochlear and vestibular nuclei of the brain stem. By studying the brain stem in silver-stained sections at various developmental time points, she discovered that the cochlear nuclei (NM and NA) develop normally until approximately E11. After this time, however, the overall volume and the number of neurons in both nuclei decrease dramatically. These observations were later replicated and extended. Parks (1979) carefully followed the progression of events after otocyst removal and found that both NA and NM displayed normal nuclear volume, cell size, and neuron number until E11, after which they rapidly deteriorated. Jackson et al. (1982) then determined that E11 was the first age at which postsynaptic action potentials in NM could be evoked by eighth nerve stimulation. This pair of results has two important implications. The first is that most developmental events take place independent of excitatory afferent activity, even though the eighth nerve fibers are in the vicinity of the cells of the CN earlier in development. Proliferation, early migration, and the establishment of afferent and efferent topographic connections all occur before functional afferent synaptic connections are made. The second implication is that, at the time normal synaptic input occurs, the postsynaptic neurons suddenly become metabolically dependent on the establishment of functional synapses. Without afferent stimulation, cell death, atrophy of the remaining neurons, abnormal migration, and a variety of other abnormalities occur.

The dependence of the postsynaptic neuron on presynaptic input does not seem to be permanent in most species and most sensory systems. For example, if the trophic role of eighth nerve on CN cells is considered, this effect terminates somewhere between 6 weeks and 1 year of age in the chicken (Born and Rubel 1985), at about 14 days after birth (P14) in the mouse (Mostafapour et al. 2000), at about P9 in the gerbil (Hashisaki and Rubel 1989; Tierney et al. 1997), and between P5 and P24 in the ferret (Moore 1990). This differential sensitivity of the postsynaptic neurons to presynaptic manipulations is usually referred to as a critical or sensitive period. In addition to cell death, a large variety of metabolic and structural changes have been examined in neurons and glial cells after cochlear manipulations at different ages in birds and mammals. The reader is referred to earlier reviews by Rubel 1978; Rubel and Parks 1988; Rubel et al. 1990; Moore 1992; Parks 1997; Zirpel et al. 1997; and Friauf and Lohmann 1999 for much of this information. In the remainder of this chapter, such changes are considered only as they relate to the following questions: (1) What is the signal from the presynaptic neuron that maintains the integrity of the postsynaptic neurons? (2) What is the cascade of cellular events in the postsynaptic cell that leads to cell death or cell survival following cochlear removal? (3) What are the biological mechanisms underlying the critical period during which peripheral input is essential for normal development? (4) What is the nature of the variability in cell survival following early deafferentation: why do some cells live and others die?

4. Signals Regulating Neuronal Survival in the CN

4.1 Importance of Integrity of the Cochlear Nerve

What is the nature of the signals transmitted from the cochlear nerve to CN neurons and glia that influence their survival, structure, and metabolism? An extensive literature, beginning with the landmark papers of Wiesel and Hubel (1963, 1965), suggested that patterned sensory information may be of critical importance. A series of papers by Webster and colleagues (Webster and Webster 1977, 1979; Webster 1983a–c, 1988a) and Coleman (Coleman and O'Connor 1979; Coleman et al. 1982) suggested that neonatal acoustic deprivation in mice and rats produced by a conductive hearing impairment (ear plug, closing ear canal, or disarticulation of middle ear bones) causes reduced neuronal size (atrophy) and reduced neuropil volume in the CN. However, in several other species a chronic conductive hearing loss did not cause atrophy of CN neurons, including chick (Tucci and Rubel 1985), ferret (Moore et al. 1989), gerbil (EW Rubel, unpublished observations), or rhesus monkey (Doyle and Webster 1991). Several explanations for this apparent discrepancy have been proposed. The most parsimonious explanation at this time is based on studies comparing both spontaneous eighth nerve activity and cell size changes following purely conductive vs. sensorineural hearing loss. Tucci et al. (1987) showed that a purely conductive hearing loss does not disrupt high levels of spontaneous activity in the auditory nerve and this activity is sufficient to preserve normal neuronal numbers and morphology in the chick NM. However, inner ear manipulations that produce a sensorineural hearing loss always reduce or eliminate spontaneous eighth nerve activity and result in rapid changes in neuronal size in NM. This explanation is supported by studies of experimentally induced sensorineural hearing loss using pharmacological inhibition of eighth nerve spikes with tetrodotoxin (TTX) and aminoglycoside ototoxicity, as well as by studies of animals with congenital hair cell loss (Webster 1985; Born and Rubel 1988; Pasic and Rubel 1989; 1991; Lippe 1991; Sie and Rubel 1992; Dodson et al. 1994; Saada et al. 1996; Saunders et al. 1998). It seems entirely possible, in light of our current knowledge, that the conductive manipulations performed by Webster, Coleman and their colleagues may have resulted in secondary sensorineural damage to the basal part of the rodent cochlea, especially when produced in young animals. Electrophysiological data support this interpretation (Clopton 1980; Evans et al. 1983; Money et al. 1995).

4.2. Identifying the Nature of the Signal

The studies cited above clearly show that the integrity of the auditory nerve is essential for normal development of CN neurons. It is still unclear whether or not patterned activity and/or absolute activity levels are essential signals at this level of the auditory pathways. A long series of studies in chicks and gerbils have attempted to determine the signal or signals that are essential for preserving normal development of CN neurons. The first approach was to ask if eliminating eighth nerve activity without damaging the sensory or neural cells would produce the same postsynaptic changes in the CN as total destruction of the cochlea. This was accomplished by infusion of the sodium channel blocker TTX into the inner ear. Complete blockade of eighth nerve action potentials, in fact, produced rapid changes in NM neurons and AVCN neurons that were indistinguishable from those resulting from complete destruction of the cochlea (Born and Rubel 1988; Pasic and Rubel 1989; Sie and Rubel 1992; Garden et al. 1994). These results strongly suggest that the voltage-dependent release of glutamate or a molecule co-released with glutamate is essential for normal maintenance of CN neurons in young animals. Further support for this conclusion comes from a series of studies on rodents and chicks showing that neuronal atrophy and decreased protein synthesis induced by eighth nerve action potential blockade or sensorineural hearing loss can be reversed by restoration of presynaptic activity (Born and Rubel 1988; Webster 1988; Lippe 1991; Pasic and Rubel 1991; Saunders et al. 1998). In addition, a number of investigators have attempted to use cochlear implants to reverse atrophy of CN cells in cats deafened as neonates or adults. The results are contradictory at this time (Ni et al. 1993; Lustig et al. 1994; Kawano et al. 1997).

Activity in the presynaptic elements during a critical period thus appears to be essential for maintaining cellular integrity and neuronal morphology in NM and AVCN. Is synaptic stimulation necessary? One may recall that the same question was addressed in regard to the neuromuscular system many years ago (Drachman and Witzke 1972; Lømo and Rosenthal 1972). To address this question Hyson and Rubel (1989, 1995) asked if the deprivation-induced changes seen in NM neurons could be prevented by electrical stimulation of the eighth nerve (orthodromic stimulation) and if so, whether they could be equally well prevented by antidromic stimulation of the NM neurons. The results of in vitro orthodromic and antidromic stimulation experiments demonstrated that the early events following deafferentation and activity deprivation, decreased protein synthesis, and ribosomal integrity could be prevented by orthodromic stimulation. However, antidromic stimulation actually exacerbated these degenerative events. More recent experiments have shown that propidium iodide incorporation, a common measure of dying cells, is also prevented by orthodromic stimulation (Zirpel and Rubel 1998). Finally, blocking neurotransmitter release from the eighth nerve fibers by bathing the preparation in low Ca^{2+} concentrations or blocking glutamate receptors reversed the positive effects of orthodromic stimulation (see Rubel et al. 1990; Zirpel et al. 1997). Taken together, these results provide strong evidence that the trophic influences of the eighth nerve on its target neurons in the CN are mediated by voltage-dependent release of glutamate or of a molecule co-released with glutamate and require activation of one or more glutamate receptors on NM neurons. Conversely, deprivation of glutamate release or receptor activation in young animals activates a cascade of events culminating in cell death or atrophy of the postsynaptic neurons. In vitro experiments comparing the effects of antidromic and orthodromic stimulation have not been replicated in the mammalian AVCN. However, the effects of deprivation and of pharmacological blockade of the eighth nerve on deprivation-induced postsynaptic changes in NM and AVCN are strikingly similar. These findings, coupled with the clear homology between NM and AVCN, strongly suggest that similar conclusions can be made for the signal regulating trophic influences on CN neurons in mammals.

4.3 Postsynaptic Events

The immediate and long-term changes in the CN following deafferentation or deprivation have been examined primarily in chicks, rodents, and cats, with differing goals. Most of the studies on cats and guinea pigs have focused on the long-term phenotype of the CN neurons and on whether some or all of the effects of deprivation can be reversed by stimulation through cochlear prostheses. This clinically oriented goal has important implications for interventions in young children suffering serious and profound hearing loss. The second goal is trying to understand the sequelae of events following alterations in afferent activity and determining their causal relationships. This approach can add new information and concepts toward understanding the role of activity in nervous system development and the mechanisms underlying plasticity of the developing nervous system.

4.3.1 Consequences of Activity Deprivation

Before it was appreciated that deprivation of eighth nerve activity produced the same sequence of initial events in the postsynaptic CN neurons as did deafferentation, several investigators removed the cochlea (usually including the ganglion cell bodies) in animals of varying ages and examined the CN weeks or months later (Levi-Montalcini 1949; Powell and Erulkar 1962; Parks 1979; Trune 1982a,b; Nordeen et al. 1983). Large reductions in neuron size, neuropil volume (including dendritic size), nuclear volume, and neuron number, and concomitant increases in neuron packing density, were seen when cochlea were removed in young or embryonic birds and mammals. In general, the changes seen in mature animals were less severe and did not include deafferentation-induced cell death of CN neurons. Changes comparable to those seen in young birds and mammals were also described in frog auditory nuclei after otocyst removal (Fritzsch 1990.)

A series of papers on the chick CN beginning in 1985 led to new ways of thinking about the cascade of cellular events that may lead to these long-term changes. Born and Rubel (1985) carefully examined the time course and age dependence of the morphological changes in NM neurons following cochlea removal. Remarkably, cell death and cell atrophy after cochlear removal occurs extremely rapidly, within 2 days in young chickens. Dramatic cytoplasmic changes in Nissl staining are evident at 12–24 hours. Furthermore, there is no difference in outcome between removing the cochlea alone versus removing the cochlea and the ganglion cells, thereby directly severing the eighth nerve central process. These rapid changes in the responses of the postsynaptic neurons as well as the morphological details described by Born and Rubel suggested that deafferentation evokes an apoptotic-like process in NM neurons. This interpretation has been strengthened by studies showing that protein synthesis, RNA synthesis, ribosome integrity, and ribosomal RNA content all decrease within thirty minutes to a few hours after eliminating eighth nerve activity or removing the cochlea (Steward and Rubel 1985; Rubel et al. 1991; Garden et al. 1994, 1995a,b; Hartlage-Rübsamen and Rubel 1996). These early events are distinctly biphasic (see Fig. 2.2). During the initial 3–4 hours after the onset of deprivation, there appears to be a generalized decrease in synthetic activity with only minor changes in cytoplasmic ultrastructure. This is reflected in quantitative

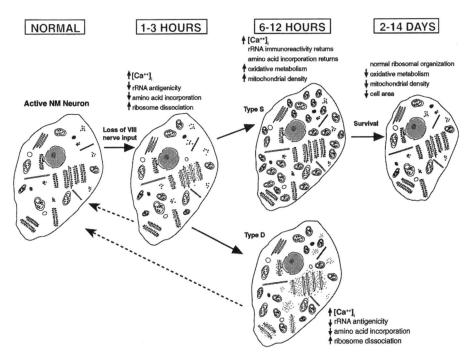

FIGURE 2.2. Schematic illustrating some of the changes undergone by NM neurons following cochlear removal or excitatory input deprivation early in life and showing the alternate pathways followed by the deprived neurons. Type S cells represent those that will survive; type D cells represent those that will degenerate over the next 2 days. Note that all NM neurons, regardless of eventual fate, show similar changes at 3–6 hours following the onset of deprivation, and then split into a bimodal population; one group will remain viable and recover their neuronal phenotype, while the other group will quickly undergo an apoptotic-like cell death. [Modified from Garden et al. (1994).]

measures as an overall, unimodal shift in the distributions of labeling densities. By about 6 hours after deafening, depending on the specific parameter under investigation, a clearly bimodal distribution of NM cells emerges. Approximately 70% of the neurons show partial recovery of protein synthesis and RNA synthesis and no obvious structural alterations in cytoplasmic ribosomes. The remaining 30% of NM neurons show no synthetic activity (by our measures), a complete loss of polyribosomes in their cytoplasm, and loss of staining for ribosomal RNA (see Rubel et al. 1991; Garden et al. 1994, 1995a,b). This latter group represents the neurons that die over the next 2 days, while the approximately 70% of neurons that show less severe changes atrophy, but survive. This series of investigations showed that the effects of deprivation on CN cells are more rapid than expected and that the ultimate fate of the deprived neurons is predictable quite early in the process, by about 6 hours after the beginning of deprivation.

A variety of other rapid and long-term changes in presynaptic and postsynaptic elements of auditory neurons in chicks and mammals have been observed after elimination or reduced eighth nerve activity. These include the expected decrease in glucose uptake in young and adult animals (Lippe et al. 1980; Born et al. 1991; Tucci et al. 1999), dramatic and rapid changes in calcium binding proteins in mature guinea pig and rat (Winsky and Jacobowitz 1995; Caicedo et al. 1997; Forster and Illing 2000; but see Parks et al. 1997), and changes in c-fos protein and mRNA expression (Gleich and Strutz 1997; Luo et al. 1999; see also Zhang et al. 1996). On the other hand, some proteins such as GAP-43 transiently increase expression (Illing et al. 1997), which could be related to some spreading of inhibitory connections after deafferentation (Benson et al. 1997; but see Code et al. 1990). One of the most dramatic and rapid changes in NM neurons that has been observed after activity deprivation in vivo is in the density of staining with antibodies to the cytoskeletal proteins, tubulin, actin, and microtubule associated protein-2 (MAP-2) (Kelley et al. 1997). Within 3 hours after deafening, immunoreactivity of NM neurons to antibodies to these three proteins is dramatically reduced, without a concomitant decrease in mRNA. It was hypothesized that the cytoskeletal proteins change configuration to allow the cells to change shape. Within 4 days, antigenicity in the surviving NM neuron begins to recover and it is entirely normal in appearance when examined about 3 weeks later. Finally, Durham and colleagues have found a biphasic response of Kreb's cycle enzymes and mitochondria density in chick NM neurons (Durham and Rubel 1985; Hyde and Durham 1990, 1994a,b; Durham et al. 1993). During the first 24–36 hours there are increases in enzyme activity and the density of mitochondria, which are followed by a smaller but sustained decrease. These results are discussed further at the conclusion of this section.

The rapid time course and the patterns of structural and ultrastructural changes in CN neurons suggest that the activity-dependent trophic interactions might rely on a rather simple interaction, such as activation of a receptor tyrosine kinase (RTK) and/or maintenance of normal intracellular signaling pathways. The distributions of some RTK receptors during development have recently been described in both birds and mammals (Hafidi et al. 1996; Cochran et al. 1999) and ligands for these receptors are present in the mammal CN (Hafidi 1999). Other families of growth factors are also being examined (e.g., Riedel et al. 1995). However, a role for any of these receptor–ligand pairs has not been tested.

The role of afferent activity on the homeostasis of intracellular calcium concentration ($[Ca^{2+}]_i$) and the importance of $[Ca^{2+}]_i$ for trophic regulation of NM neurons has been studied extensively during the past 5 years (Zirpel et al. 1995, 1997, 1998, 2000a,b; Zirpel and Rubel 1996). Intuitively, it might be expected that deprivation of presynaptic activity would lead to a decrease in $[Ca^{2+}]_i$ in postsynaptic neurons. Surprisingly, just the opposite is true in NM neurons. Elimination of eighth nerve activity leads to a three-fold increase in $[Ca^{2+}]_i$ in NM neurons, which is prevented entirely by electrical stimulation of the auditory nerve or activation of metabotropic glutamate receptors (mGluRs). When the eighth nerve is stimulated in the presence of mGluR antagonists, $[Ca^{2+}]_i$ increases dramatically owing to continued activation of Ca^{2+}-permeable AMPA receptors and activation of voltage-gated Ca^{2+} channels. Furthermore, activation of mGluRs is required for maintenance of ribosomal RNA (Hyson 1998). A direct link between elevated $[Ca^{2+}]_i$ and increased cell death has been established by Zirpel et al. (1998, 2000b). Finally, there is convincing evidence that in the absence of mGluR activation influx of Ca^{2+} through AMPA receptors is involved in creating the hypercalcemic condition (Zirpel et al. 2000a,b; Zirpel and Parks 2001).

In order to understand the relationship between mGluR activation and $[Ca^{2+}]_i$ homeostasis in NM neurons, it is important to remember that most eighth nerve axons, NM neurons, and AVCN neurons have extremely high levels of ongoing "spontaneous" activity, even in silence (Dallos and Harris 1978; Liberman 1978; Tucci et al. 1987; Warchol and Dallos 1990; Born et al. 1991). In addition, Ca^{2+}-permeable AMPA receptors appear to be required for the faithful processing of temporally precise, high-frequency information (Trussell 1998; Parks 2000). This combination seems to place auditory brain stem neurons at high risk for calcium cytotoxicity, a calcium-activated apoptotic-like cascade. Perhaps to adapt to this challenge, auditory neurons are rich in calcium-binding proteins (Takahashi et al. 1987; Braun 1990; Kubke et al. 1999; Hack et al. 2000) and mitochondria, and appear to have specialized (or highly expressed) intracellular pathways by which group 1 and group III mGluRs inhibit cytoplasmic buildup of Ca^{2+}. A series of studies using ratiometric Ca^{2+} imaging suggests that Ca^{2+} permeability and intracellular Ca^{2+} release are regulated by mGluR activation (Lachica et al. 1995, 1996; Kato et al. 1996; Kato and Rubel 1999). These studies suggest a set of mechanisms whereby transmitter release can concomitantly regulate electrical activity and $[Ca^{2+}]_i$ in the postsynaptic neuron. The results to date, therefore, suggest a working model: during normal eighth nerve activity, glutamate release activates Ca^{2+}-permeable AMPA receptors and metabotropic receptors that activate associated downstream signaling pathways including, but not limited to, inhibition of Ca^{2+} permeability of AMPA channels and low voltage activated (L-type) channels as well as Ca^{2+} release from intracellular stores. Each NM neuron then establishes a dynamic balance of these $[Ca^{2+}]_i$ increasing and $[Ca^{2+}]_i$ decreasing mechanisms that result in a relatively stable and healthy level of intracellular calcium. Activity deprivation disrupts this balance, allowing a cascade of events beginning with a rise in $[Ca^{2+}]_i$ that is mediated through the Ca^{2+}-permeable AMPA receptors (Zirpel et al. 2000b). While many more experiments are needed to fill in the details of the intracellular events that lead to cell death or cell phenotype changes, support for this model is emerging (Solum et al. 1997; Caicedo et al. 1998). Further evidence supporting the importance of $[Ca^{2+}]_i$ homeostasis in auditory function comes from homozygous *deafwaddler* mice. These mice walk with a wobbly, hesitant gait and are deaf. They have a mutation in the gene for the plasma membrane calcium ATPase isoform 2 (*PMCA2*) and show structural abnormalities in the spherical cells of the cochlear nucleus (Dodson and Charalabapoulou 2001). These abnormalities are consistent with, and hypothesized to be due to, a deregulation of normal calcium homeostasis resulting from the lack of the *PMCA2* that leads to hypercalcemic pathology. While the *PMCA2* mutation also affects the cochlear hair cells and spiral ganglion neurons, it emphasizes the critical role of calcium homeostasis in the normal, healthy functioning of the auditory system.

4.3.2 Critical Periods

As noted above, some of the transneuronal structural and metabolic interactions between the eighth nerve and CN neurons occur throughout life while others appear limited to a specific period of development. As seen in other developing sensory systems, there appears to be a critical period for trophic regulation of CN neurons by their presynaptic partners. Trune (1982a) showed extensive cell death in mouse CN after neonatal deafferentation but did not test adults. Nordeen et al. (1983), Born and Rubel (1985), Hashisaki and Rubel (1989), and Moore (1990) provided convincing evidence for differential effects of cochlea removal on CN neuronal survival and atrophy in neonatal and adult chicks, gerbils, and ferrets. Young animals were much more susceptible than adults. However, it was not until a recent report by Tierney et al. (1997) that it was appreciated how sharp the "window" of this critical period could be. Tierney et al. found that between P7 and P9 there was an abrupt change in the survival of gerbil CN neurons following deafferentation. Cochlea removals before 7 days of age resulted in 45–88% cell death in the CN; at 9 days of age or later, this same manipulation results in no reliable cell death.

The remarkably rapid changes in susceptibility of CN neurons to deprivation-induced cell death suggest that some rather simple molecular switch may be controlling susceptibility to afferent deprivation. To address this possibility a series of studies examining the critical period for trophic regulation in mice were initiated. The first studies described the temporal boundaries of the critical

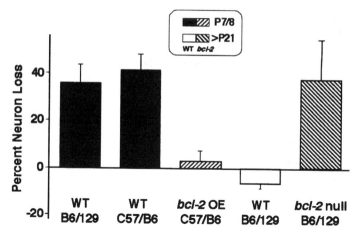

FIGURE 2.3. Summary of data on cell death in ventral cochlear nucleus (VCN) of wild-type (WT) and *bcl-2* manipulated [overexpressing (OE) and knockout (null)] mice after cochlear removal in neonatal (P7/8) and adult (>P21) animals. Note that overexpression of *bcl-2* prevents deafferentation-induced neuronal death in the VCN of susceptible young C57/B6 mice and that a genetic knockout of *bcl-2* permits deafferentation-induced neuronal death in adult B6/129 mice that would otherwise not be susceptible to this treatment. [Data from Mostafapour et al. (2000, 2002).]

period and time course of cell death following deafening (Mostafapour et al. 2000). In addition, experiments with *bcl-2* null and *bcl-2* overexpression mice have shown dramatic modulation of this critical period. Figure 2.3 summarizes the results to date. CN neurons in the mature auditory system of mice lacking the *bcl-2* gene appear to show equivalent susceptibility to deafening as wild-type neonatal mice. Conversely, overexpression of *bcl-2* prevents all transneuronal cell death in neonatal mice (Mostafapour et al. 2002). These results should not be overinterpreted. It is not clear if *bcl-2* modulation is due to a direct role of *bcl-2* (or related gene family members) in determining the critical period or if this protein is playing a role downstream of such a molecule. In any case, these results may provide a beginning toward understanding the biological basis of this critical period.

4.3.3 A Life-or-Death Decision

One of the most intriguing and medically important questions is why some postsynaptic neurons live, while others die, after afferent deprivation in young animals. The proportion of CN neurons that die varies dramatically with species as well as with age. For example, cochlea removals in 3-day-old gerbils result in almost 90% neuron loss within 2 weeks, but the same manipulation at P7 results in only a 50% loss (Tierney et al. 1997). A similar decrease in susceptibility is seen in mice during the first 10 postnatal days (Mostafapour et al. 2000). In chicks, however, only about 30% cell loss is seen at the most vulnerable times (Levi-Montalcini 1949; Parks 1979; Born and Rubel 1985).

What determines which neurons will survive deprivation or deafferentation and which will die? Two major possibilities emerge. The most intuitive hypothesis is that there is a bimodal population of neurons with an intrinsic difference in susceptibility to deafferentation. It is possible that particular differences in receptor phenotypes, for example, will cause two groups of neurons to respond fundamentally differently to deafferentation. While this explanation is particularly attractive for nuclear regions with mixed cell types, such as the mammalian CN, there is, in fact, little supporting evidence (see Tierney et al. 1997). Furthermore, in the avian NM there appears to be only a single neuron type throughout most of the nucleus, and repeated attempts to discover two or more distinct populations on the basis of structure or protein expression have failed (Rubel and Parks 1988; Kubke et al. 1999).

A second hypothesis for explaining the differences in neuronal fate after deafferentation was first explicitly proposed by Garden et al. (1994) and Hyde and Durham (1994a). It was hypothesized that the neuronal populations are not bimodal with respect to susceptibility to afferent deprivation-induced cell death. Instead, it is possible that the deprivation condition elicits two competing intracellular responses. The first response is activation of an apoptotic-like pathway and the second is activation of a survival pathway. This model further suggested that activation of the survival pathways might be delayed by a few hours compared to the apoptotic-like pathway. The resulting amount of cell death would then be a function of the relative effectiveness of these competing pathways, and survival or death of individual cells would be stochastically determined during the period of susceptibility.

There are several lines of evidence supporting this second hypothesis. First, from the time of the initial deafferentation experiments, it was recognized that there is no consistent spatial pattern of cell death in the CN and that there is high variability in the absolute amount of cell death during the period of susceptibility (Born and Rubel 1985; Moore 1990). Second, it has been shown repeatedly that many of the early degradative events following the onset of afferent deprivation are uniform across the population of NM neurons. These events include decreases in protein synthesis, RNA synthesis, ribosomal antigenicity, and cytoskeletal protein antigenicity (Steward and Rubel 1985; Born and Rubel 1988; Garden et al. 1994, 1995a; Hartlage-Rübsamen and Rubel, 1996; Kelley et al. 1997). There are, of course, variations across the population of NM neurons in these responses to deprivation, but there is no hint of a population of neurons that does not respond at all. Third, since 1985 it has been recognized that oxidative enzyme activity actually shows a biphasic response following afferent deprivation. Beginning at about 6 hours and continuing for 24–30 hours, there is a dramatic increase in enzyme activity and this is followed by a long-lasting decrease as has been described in other sensory regions following deprivation (Durham and Rubel 1985; Hyde and Durham

1990; Durham et al. 1993). Coincident with the increase in oxidative enzyme activity is an increase in the density of mitochondria in the cytoplasm of NM neurons (Hyde and Durham 1994b). In addition, NM neuron death 5 days after cochlea removal increases from 30% to 60–80% in chloramphenicol-treated animals (Garden et al. 1994; Hyde and Durham 1994a; Hartlage-Rübsamen and Rubel, 1996). These lines of evidence, in addition to growing bodies of literature showing mitochondrial influences on Ca^{2+} homeostasis and cell survival (Mostafapour et al. 1997; Nicholls and Budd 2000), suggest that the survival mechanisms in deafferented NM neurons may include mitochondrial metabolism.

More recent studies have led to a third hypothesis that is parallel and complementary to the one proposed by Garden et al. (1994) and also incorporates elements of the first hypothesis proposed in this section. All NM neurons show an increase in $[Ca^{2+}]_i$ following activity deprivation, yet 30% die and 70% survive. Preventing this $[Ca^{2+}]_i$ increase reduces neuronal death (Zirpel et al. 1998), suggesting that it is the signal for the "cell death" pathway. However, the surviving neurons also show this same increase in $[Ca^{2+}]_i$, but respond by phosphorylating and activating the transcription factor cAMP response element binding protein (CREB), which is present in all NM neurons (Zirpel et al. 2000b). CREB phosphorylation also occurs in a subpopulation of deafferented AVCN neurons in neonatal mice (Zirpel et al. 2000b) and in rat VCN neurons following decreased electrical activity in the eighth nerve (Illing 2001). In addition, neurons of rat VCN, lateral superior olive, medial nucleus of the trapezoid body, and the inferior colliculus show almost complete dephosphorylation of CREB in response to electrical intracochlear stimulation (Illing 2001). Thus, it would appear that increased $[Ca^{2+}]_i$ also serves as a "survival" signal via CREB-mediated gene transcription. But how can the same signal, increased $[Ca^{2+}]_i$, mediate two such disparate pathways? As proposed by Garden et al. (1994), it is probably a stochastic process, but rather than sequential, competing pathways of survival vs. death, the neurons respond to increased $[Ca^{2+}]_i$ based on their particular state of dynamic calcium buffering at the time of activity deprivation: some neurons implement signal pathways leading to CREB phosphorylation because the calcium dynamics in those neurons *at that time* favor activation of protein kinase A (PKA) and calcium/calmodulin-dependent kinases (Zirpel et al. 2000b), whereas other neurons implement signal pathways leading to cell death because the calcium dynamics in those neurons *at that time* favor up-regulation of mitochondrial cytochrome *c* (Wilkinson et al. 2002), caspase activation (Wilkinson et al. 2003), or other cell death signals. Further support for this hypothesis comes from the observation that rapid buffering of activity deprivation induced calcium increases prevents phosphorylation and activation of CREB, whereas slow buffering of calcium increases does not (Zirpel, unpublished).

What protein expression results from CREB activation that allows the neurons to survive the activity-deprived, hypercalcemic state? While it is unclear at this time, several experiments have provided evidence suggesting that *bcl-2* may be involved. It is well established that the *bcl-2* gene contains a CRE promoter element and is up-regulated by CREB activation (Wilson et al. 1996), and *bcl-2* message is up-regulated in NM neurons within hours following deafferentation (Wilkinson et al. 2002). This is also consistent with the results of the *bcl-2* knockout and overexpressing mice discussed above (Mostafapour et al. 2002).

Figure 2.4 shows a proposed model, consistent with the data, for the response of NM neurons to activity deprivation. Following deafferentation, low concentrations of glutamate remain in the synaptic cleft (Zirpel et al. 2000b). Since

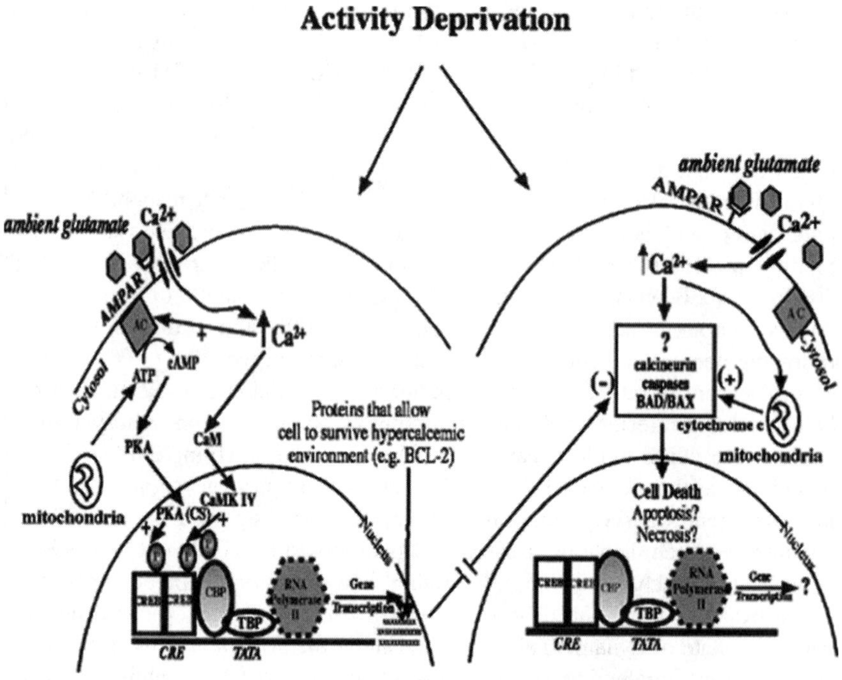

FIGURE 2.4. Schematic diagram of hypothesized model of cochlear nucleus neuron "survival" and "death" pathways. Following deafferentation, ambient glutamate activates Ca^{2+}-permeable AMPA receptors, resulting in an increase in intracellular calcium. In surviving neurons, this increased intracellular calcium activates converging kinase pathways that phosphorylate and activate the transcription factor CREB. CREB then mediates transcription of genes whose protein products enable the neuron to survive the activity-deprived, hypercalcemic state. In neurons that do not survive, the increased intracellular calcium activates pathways that result in caspase cleavage and ultimately cell death. These pathways are necessarily exclusive, but not independent of one another. AC, Adenylate cyclase; PKA, protein kinase A; PKA (CS), protein kinase A catalytic subunit; CaM, calmodulin; CaMK IV, calcium/calmodulin-dependent kinase IV, CREB, cAMP response element binding protein; CBP, CREB binding protein; CRE, cAMP response element TATA, TATA box for transcription initiation; TBP, TATA box binding protein; P, phosphorylation.

the EC_{50} of the AMPA receptors is an order of magnitude lower than for the metabotropic receptors (Raman and Trussell 1992; Raman et al. 1994; Zirpel et al. 1994, 1995a), AMPA receptors are specifically activated by this ambient glutamate. Calcium enters the neurons through the Ca^{2+}-permeable AMPA receptor, and $[Ca^{2+}]_i$ increases. In a subpopulation of neurons, this increased $[Ca^{2+}]_i$ activates adenylate cyclase, thus generating cAMP. The cAMP activates PKA and the catalytic, subunit translocates to the nucleus. In parallel, the increased $[Ca^{2+}]_i$ causes an activation of calmodulin that translocates to the nucleus and activates a calcium/calmodulin-dependent kinase. The activated calcium/calmodulin-dependent kinase and PKA converge and phosphorylate CREB. Through interactions with CREB binding protein, the TATA box binding protein, and RNA polymerase II, CREB initiates transcription of specific genes containing the cAMP response element within their promoter region, such as *bcl-2*. The subsequent protein expression allows the neurons to compensate for and survive in the activity-deprived, hypercalcemic environment. In parallel, the remaining subpopulation of neurons responds to the increased $[Ca^{2+}]_i$ by implementing a cell death pathway that may include mitochondrial degeneration, up-regulation of cytochrome *c*, and activation of various caspases.

While this model is consistent with the current data, some intricacies indicate that rather than being exclusive, an amalgamation of the latter two hypotheses is closer to reality. While it is clear that only the surviving neurons show CREB phosphorylation, they all show increased $[Ca^{2+}]_i$, all express CREB, all show up-regulated *bcl-2*, and all show activated caspase 9. Thus, it would appear that perhaps there is indeed a competition between cell survival and cell death pathways, both mediated by increased calcium, but differentiated not only by the ability to rapidly buffer or spatially restrict calcium signals, but also by the ability of the mitochondria to ramp up ATP production to fuel the active phosphorylation events and subsequent gene transcription mediated by CREB. If the CREB-mediated gene transcripts are not produced, then the caspase/cell death pathway runs its course. Many experiments are needed to fill the gaps within this model.

5. Summary

The vertebrate cochlear nucleus (CN) is composed of several groups of functionally specialized neurons that receive highly specific synaptic input from the cochlear nerve. The physiological response properties of most CN neurons are dominated by cochlear nerve input and many CN neurons require cochlear nerve input during development for their survival and growth. Further understanding of the early development of the CN will require study of the molecular mechanisms by which the CN cell groups are specified as auditory neurons in the early rhombencephalon, how these neurons migrate to the lateral brain stem, and how they aggregate in the appropriate pattern to form the mosaic of the CN. This goal will require discovery of the transcription factors that determine the fate of CN neurons, the factors guiding migration of CN neurons into the auditory anlage, and a comprehensive knowledge of the cell adhesion molecules expressed by the various cell types as they aggregate into the CN.

Innervation of the CN by cochlear nerve axons occurs prior to the onset of activity in the auditory system. The cellular and molecular mechanisms underlying the growth and fasciculation, targeting, and branching of these axons are almost completely unknown. Cochlear nerve axons form a variety of distinct synaptic endings with their CN targets and the available evidence suggests that both pre- and postsynaptic partners contribute to the form of these endings. Contrary to some prevalent generalizations about sensory system development, the topographic precision of cochlear nerve projections to the CN appears to be quite high at early stages and not subject to further improvement by spontaneous activity. There is evidence that the tonotopic axis within the CN develops independently of cochlear nerve input although the molecular mechanisms by which this occurs are unknown.

Once the eighth nerve connections are established in the CN, there is a critical period during which the CN neurons require input from the eighth nerve for maintenance and survival. While the cellular and molecular mechanisms underlying CN neuron responses to deafferentation are becoming elucidated, the main question remains: Why do some neurons survive while others die? To address this issue fully, the genes and gene products being regulated must be identified, their subsequent role in the cellular processes understood, and the ultimate end-effect characterized. To accomplish these goals it will be necessary to characterize and understand not only the specific gene products that are up- or down-regulated, but also the nature of their interactions and how they affect the cellular functioning of the deafferented CN neurons.

Acknowledgments. The authors' laboratories are supported by NIH Grants DC00395, DC03829, and DC04661 (E.W.R.); DC00144 (T.N.P.); and DC04370 and DC05012 (L.Z.).

References

Agmon A, Yang LT, Jones EG, O'Dowd DK (1995) Topological precision in the thalamic projection to neonatal mouse barrel cortex. J Neurosci 15:549–561.

Angulo A, Merchan JA, Merchan MA (1990) Morphology of the rat cochlear primary afferents during prenatal development: a Cajal's reduced silver and rapid Golgi study. J Anat 168:241–255.

Arndt K, Redies C (1996) Restricted expression of R-cadherin by brain nuclei and neural circuits of the developing chicken brain. J Comp Neurol 373:373–399.

Benson CG, Gross JS, Suneja SK, Potashner SJ (1997) Synaptophysin immunoreactivity in the cochlear nucleus after unilateral cochlear or ossicular removal. Synapse 25:243–257.

Book KJ, Morest DK (1990) Migration of neuroblasts by perikaryal translocation: role of cellular elongation and axonal outgrowth in the acoustic nuclei of the chick embryo medulla. J Comp Neurol 297:55–76.

Born DE, Rubel EW (1985) Afferent influences on brain stem auditory nuclei of the chicken: neuron number and size following cochlea removal. J Comp Neurol 231: 435–445.

Born DE, Rubel EW (1988) Afferent influences on brain stem auditory nuclei of the chicken: presynaptic action potentials regulate protein synthesis in nucleus magnocellularis neurons. J Neurosci 8:901–919.

Born DE, Durham D, Rubel EW (1991) Afferent influences on brainstem auditory nuclei of the chick: nucleus magnocellularis neuronal activity following cochlea removal. Brain Res 557:37–47.

Braun K (1990) Calcium-binding proteins in avian and mammalian central nervous system: localization, development and possible functions. Prog Histochem Cytochem 21: 1–64.

Brose K, Tessier-Lavigne M (2000) Slit proteins: key regulators of axon guidance, axonal branching, and cell migration. Curr Opin Neurobiol 10:95–102.

Brumwell CL, Hossain WA, Morest DK, Bernd P (2000) Role for basic fibroblast growth factor (FGF-2) in tyrosine kinase (TrkB) expression in the early development and innervation of the auditory receptor: in vitro and in situ studies. Exp Neurol 162:121–145.

Caicedo A, d'Aldin C, Eybalin M, Puel JL (1997) Temporary sensory deprivation changes calcium-binding proteins levels in the auditory brainstem. J Comp Neurol 378:1–15.

Caicedo A, Kungel M, Pujol R, Friauf E (1998) Glutamate-induced Co^{2+} uptake in rat auditory brainstem neurons reveals developmental changes in Ca^{2+} permeability of glutamate receptors. Eur J Neurosci 10:941–954.

Cant, NB (1992) The cochlear nucleus: Neuronal types and their synaptic organization. In: Webster DB, Popper AN, Fay RR (eds), The Mammalian Auditory Pathway: Neuroanatomy. New York: Springer-Verlag, pp. 66–116.

Cant, NB (1998) Structural development of the mammalian central auditory pathway. In: Rubel EW, Popper AN, Fay RR (eds), Development of the Auditory System. New York: Springer-Verlag, pp. 315–413.

Carr CE, Boudreau RE (1996) Development of the time coding pathways in the auditory brainstem of the barn owl. J Comp Neurol 373:467–483.

Clopton BM (1980) Neurophysiology of auditory deprivation. In: Garlin RJ (ed), Birth Defects: Morphogenesis and Malformations of the Ear. New York: Alan R. Liss, pp. 271–288.

Cochran SL, Stone JS, Bermingham-McDonogh O, Akers SR, Lefcort F, Rubel EW (1999) Ontogenetic expression of trk neurotrophin receptors in the chick auditory system. J Comp Neurol 413:271–288.

Code RA, McDaniel AE (1998) Development of dynorphin-like immunoreactive auditory nerve terminals in the chick. Brain Res Dev Brain Res 106:165–172.

Code RA, Durham D, Rubel EW (1990) Effects of cochlea removal on GABAergic terminals in nucleus magnocellularis of the chicken. J Comp Neurol 301:643–654.

Coleman JR, O'Connor P (1979) Effects of monaural and binaural sound deprivation on cell development in the anteroventral cochlear nucleus of rats. Exp Neurol 64:553–566.

Coleman J, Blatchley BJ, Williams JE (1982) Development of the dorsal and ventral cochlear nuclei in rat and effects of acoustic deprivation. Brain Res 256:119–123.

Cramer KS, Fraser SE, Rubel EW (2000a) Embryonic origins of auditory brain-stem nuclei in the chick hindbrain. Dev Biol 224:138–151.

Cramer KS, Rosenberger MH, Frost DM, Cochran SL, Pasquale EB, Rubel EW (2000b) Developmental regulation of EphA4 expression in the chick auditory brainstem. J Comp Neurol 426:270–278.

Cramer KS, Karam SD, Bothwell M, Cerretti DP, Pasquale EB, Rubel EW (2002) Expression of EphB receptors and EphrinB ligands in the developing chick auditory brainstem. J Comp Neurol 452:51–64.

Cramer KS, Bermingham-McDonogh OM, Krull CE, Rubel EW (2004) EphA4 restricts axonal connections to individual subcellular elements. Devel Biol. In press.

Dallos P (1992) The active cochlea. J Neurosci 12:4575–4585.

Dallos P, Harris D (1978) Properties of auditory nerve responses in absence of outer hair cells. J Neurophysiol 41:365–383.

D'Amico-Martel A (1982) Temporal patterns of neurogenesis in avian cranial sensory and autonomic ganglia. Am J Anat 163:351–372.

Diaz C, Glover JC (2002) Comparative aspects of the hodological organization of the vestibular nuclear complex and related neuron populations. Br Res Bull 57:307–312.

Dodson HC, Charalabapoulou M (2001) PMCA2 mutation causes structural changes in the auditory system in deafwaddler mice. J Neurocytol 30:281–292.

Dodson HC, Bannister LH, Douek EE (1994) Effects of unilateral deafening on the cochlear nucleus of the guinea pig at different ages. Brain Res Dev Brain Res 80: 261–267.

Doyle WJ, Webster DB (1991) Neonatal conductive hearing loss does not compromise brainstem auditory function and structure in rhesus monkeys. Hear Res 54:145–151.

Drachman DB, Witzke F (1972) Trophic regulation of acetylcholine sensitivity of muscle: effect of electrical stimulation. Science 176:514–516.

Durham D, Rubel EW (1985) Afferent influences on brain stem auditory nuclei of the chicken: changes in succinate dehydrogenase activity following cochlea removal. J Comp Neurol 231:446–456.

Durham D, Matschinsky FM, Rubel EW (1993) Altered malate dehydrogenase activity in nucleus magnocellularis of the chicken following cochlea removal. Hear Res 70: 151–159.

Evans WJ, Webster DB, Cullen JK Jr (1983) Auditory brainstem responses in neonatally sound deprived CBA/J mice. Hear Res 10:269–277.

Fekete DM, Rouiller EM, Liberman MC, Ryugo DK (1984) The central projections of intracellularly labeled auditory nerve fibers in cats. J Comp Neurol 229:432–450.

Flanagan JG, Vanderhaeghen P (1998) The ephrins and Eph receptors in neural development. Annu Rev Neurosci 21:309–345.

Forster CR, Illing RB (2000) Plasticity of the auditory brainstem: cochleotomy-induced changes of calbindin-D28k expression in the rat. J Comp Neurol 416:173–187.

Friauf E (1992) Tonotopic order in the adult and developing auditory system of the rat as shown by c-fos immunocytochemistry. Eur J Neurosci 4:798–812.

Friauf E, Kandler K (1993) Cell birth, formation of efferent connections, and establishment of tonotopic order in the rat cochlear nucleus. In: Merchán, MA, Juiz M, Godfrey DA Mugnaini E (eds), The Mammalian Cochlear Nuclei: Organization and Function. New York, Plenum, pp. 19–28.

Friauf E, Lohmann C (1999) Development of auditory brainstem circuitry. Activity-dependent and activity-independent processes. Cell Tissue Res 297:187–195.

Fritzsch B (1990) Experimental reorganization in the alar plate of the clawed toad, Xenopus laevis. I. Quantitative and qualitative effects of embryonic otocyst extirpation. Brain Res Dev Brain Res 51:113–122.

Fritzsch B, Farinas I, Reichardt LF (1997) Lack of neurotrophin 3 causes losses of both classes of spiral ganglion neurons in the cochlea in a region-specific fashion. J Neurosci 17:6213–6225.

Fritzsch B, Barald KF, Lomax MI (1998) Early embryology of the vertebrate ear. In: Rubel EW, Popper AN, Fa, RR (eds), Development of the Auditory System. New York: Springer-Verlag, pp. 80–145.

Fushimi D, Arndt K, Takeichi M, Redies C (1997) Cloning and expression analysis of cadherin-10 in the CNS of the chicken embryo. Dev Dynam 209:269–285.

Garcia-Diaz JF (1999) Development of a fast transient potassium current in chick cochlear ganglion neurons. Hear Res 135:124–134.

Garden GA, Canady KS, Lurie DI, Bothwell M, Rubel EW (1994) A biphasic change in ribosomal conformation during transneuronal degeneration is altered by inhibition of mitochondrial, but not cytoplasmic protein synthesis. J Neurosci 14:1994–2008.

Garden GA, Redeker-DeWulf V, Rubel EW (1995a) Afferent influences on brainstem auditory nuclei of the chicken: regulation of transcriptional activity following cochlea removal. J Comp Neurol 359:412–423.

Garden GA, Hartlage-Rübsamen M, Rubel EW, Bothwell MA (1995b) Protein masking of a ribosomal RNA epitope is an early event in afferent deprivation-induced neuronal death. Mol Cell Neurosci 6:293–310.

Gleich O, Strutz J (1997) Age-dependent effects of the onset of a conductive hearing loss on the volume of the cochlear nucleus subdivisions and the expression of c-fos in the mongolian gerbil (Meriones unguiculatus). Audiol Neurootol 2:113–127.

Glover JC (2001) Correlated patterns of neuron differentiation and Hox gene expression in the hindbrain: A comparative analysis. Br Res Bull 55:683–693.

Gummer AW, Mark RF (1994) Patterned neural activity in brain stem auditory areas of a prehearing mammal, the tammar wallaby (Macropus eugenii). NeuroReport 5:685–688.

Hack NJ, Wride MC, Charters KM, Kater SB, Parks TN (2000) Developmental changes in the subcellular localization of calretinin. J Neurosci 20:RC67.

Hafidi A (1999) Distribution of BDNF, NT-3 and NT-4 in the developing auditory brainstem. Int J Dev Neurosci 17:285–294.

Hafidi A, Moore T, Sanes DH (1996) Regional distribution of neurotrophin receptors in the developing auditory brainstem. J Comp Neurol 367:454–464.

Hartlage-Rübsamen M, Rubel EW (1996) Influence of mitochondrial protein synthesis inhibition on deafferentation-induced ultrastructural changes in nucleus magnocellularis of developing chicks. J Comp Neurol 371:448–460.

Hashisaki GT, Rubel EW (1989) Effects of unilateral cochlea removal on anteroventral cochlear nucleus neurons in developing gerbils. J Comp Neurol 283:5–73.

Hemond SG, Morest DK (1991a) Formation of the cochlea in the chicken embryo: sequence of innervation and localization of basal lamina-associated molecules. Brain Res Dev Brain Res 61:87–96.

Hemond SG, Morest DK (1991b) Ganglion formation from the otic placode and the otic crest in the chick embryo: mitosis, migration, and the basal lamina. Anat Embryol (Berl) 184:1–13.

Holt CE (1984) Does timing of axon outgrowth influence initial retinotectal topography in Xenopus? J Neurosci 4:1130–1152.

Holt CE, Harris WA (1993) Position, guidance, and mapping in the developing visual system. J Neurobiol 24:1400–1422.

Holt CE, Harris WA (1998) Target selection: invasion, mapping and cell choice. Curr Opin Neurobiol 8:98–105.

Hyde GE, Durham D (1990) Cytochrome oxidase response to cochlea removal in chicken auditory brain stem neurons. J Comp Neurol 297:329–339.

Hyde GE, Durham D (1994a) Rapid increase in mitochrondial volume in nucleus magnocellularis neurons following cochlea removal. J Comp Neurol 339:27–48.

Hyde GE, Durham D (1994b) Increased deafferentation-induced cell death in chick brainstem auditory neurons following blockade of mitochondrial protein synthesis with chloramphenicol. J Neurosci 14:291–300.

Hyson RL (1998) Activation of metabotropic glutamate receptors is necessary for transneuronal regulation of ribosomes in chick auditory neurons. Brain Res 809:214–220.

Hyson RL, Rubel EW (1989) Transneuronal regulation of protein synthesis in the brainstem auditory system of the chick requires synaptic activation. J Neurosci 9:2835–2845.

Hyson RL, Rubel EW (1995) Activity-dependent regulation of a ribosomal RNA epitope in the chick cochlear nucleus. Brain Res 672:196–204.

Illing RB (2001) Activity-dependent plasticity in the adult auditory brainstem. Audiol Neurootol 6:319–345.

Illing RB, Horvath M, Laszig R (1997) Plasticity of the auditory brainstem: effects of cochlear ablation on GAP-43 immunoreactivity in the rat. J Comp Neurol 382:116–138.

Inoue T, Tanaka T, Suzuki SC, Takeichi M (1998) Cadherin-6 in the developing mouse brain: expression along restricted connection systems and synaptic localization suggest a potential role in neuronal circuitry. Dev Dynam 211:338–351.

Jackson H, Parks TN (1982) Functional synapse elimination in the developing avian cochlear nucleus with simultaneous reduction in cochlear nerve axon branching. J Neurosci 2:1736–1743.

Jackson H, Parks TN (1988) Induction of aberrant functional afferents to the chick cochlear nucleus. J Comp Neurol 271:106–114.

Jackson H, Hackett JT, Rubel EW (1982) Organization and development of brain stem auditory nuclei in the chick: ontogeny of postsynaptic responses. J Comp Neurol 210:80–86.

Jhaveri SR, Morest DK (1982a) Sequential alterations of neuronal architecture in nucleus magnocellularis of the developing chicken: an electron microscope study. Neurosci 7:855–870.

Jhaveri SR, Morest DK (1982b) Sequential alterations of neuronal architecture in nucleus magnocellularis of the developing chicken: a Golgi study. Neurosci 7:837–853.

Kandler K, Friauf E (1995) Development of glycinergic and glutamatergic synaptic transmission in the auditory brainstem of perinatal rats. J Neurosci 15:6890–6904.

Kato BM, Rubel EW (1999) Glutamate regulates IP3-type and CICR stores in the avian cochlear nucleus. J Neurophysiol 81:1587–1596.

Kato BM, Lachica EA, Rubel EW (1996) Glutamate modulates intracellular Ca^{2+} stores in brainstem auditory neurons. J Neurophysiol 76:646–650.

Kawano A, Seldon HL, Clark GM, Hakuhisa E, Funasaka S (1997) Effects of chronic electrical stimulation on cochlear nuclear neuron size in deaf kittens. Adv Otorhino-laryngol 52:33–35.

Kelley MS, Lurie DI, Rubel EW (1997) Rapid regulation of cytoskeletal proteins and their mRNAs following afferent deprivation in the avian cochlear nucleus. J Comp Neurol 389:469–483.

Knowlton VY (1967) Correlation of the development of membranous and bony laby-rinths, acoustic ganglia, nerves, and brain centers of the chick embryo. J Morph 121: 179–208.

Kubke MF, Carr CE (1998) Development of AMPA-selective glutamate receptors in the auditory brainstem of the barn owl. Microsc Res Techn 41:176–186.

Kubke MF, Gauger B, Basu L, Wagner H, Carr CE (1999) Development of calretinin immunoreactivity in the brainstem auditory nuclei of the barn owl (*Tyto alba*). J Comp Neurol 415:189–203.

Lachica EA, Rubsamen R, Zirpel L, Rubel EW (1995) Glutamatergic inhibition of voltage-operated calcium channels in the avian cochlear nucleus. J Neurosci 15:1724–1734.

Lachica EA, Zirpel L, Rubel EW (1996) Intracellular mechanisms involved in the afferent regulation of neurons in the avian cochlear nucleus. In: Salvi RJ, Henderson D, Coletti V, Fiorino F (eds), Auditory Systems Plasticity and Regeneration. New York: Thieme, pp. 333–353.

Lawrence JJ, Trussell LO (2000) Long-term specification of AMPA receptor properties after synapse formation. J Neurosci 20:4864–4870.

Levi-Montalcini R (1949) The development of the acoustico-vestibular centers in the chick embryo in the absence of the afferent root fibers and of descending fiber tracts. J Comp Neurol 91:209–242.

Li H, Takeda Y, Niki H, Ogawa J, Kobayashi S, Kai N, Akasaka K, Asano M, Sudo K, Iwakura Y, Watanabe K (2003) Aberrant responses to acoustic stimuli in mice deficient for neural recognition molecule NB-2. Eur J Neurosci 17:929–936.

Liberman MC (1978) Auditory-nerve response from cats raised in a low-noise chamber. J Acoust Soc Am 63:442–455.

Limb CJ, Ryugo DK (2000) Development of primary axosomatic endings in the anter-oventral cochlear nucleus of mice. J Assoc Res Otolaryngol 1:103–119.

Lippe WR (1991) Reduction and recovery of neuronal size in the cochlear nucleus of the chicken following aminoglycoside intoxication. Hear Res 51:193–202.

Lippe WR (1994) Rhythmic spontaneous activity in the developing avian auditory sys-tem. J Neurosci 14:1486–1495.

Lippe WR (1995) Relationship between frequency of spontaneous bursting and tonotopic position in the developing avian auditory system. Brain Res 703:205–213.

Lippe WR, Rubel EW (1985) Ontogeny of tonotopic organization of brain stem auditory brain stem auditory nuclei in the chicken: implications for development of the place principle. J Comp Neurol 238:371–381.

Lippe WR, Steward O, Rubel EW (1980) The effect of unilateral basilar papilla removal upon nuclei laminaris and magnocellularis of the chick examined with [^{3}H]2-deoxyglucose autoradiography. Brain Res 196:43–58.

Lippe WR, Fuhrmann DS, Yang W, Rubel EW (1992) Aberrant projection induced by otocyst removal maintains normal tonotopic organization in the chick cochlear nucleus. J Neurosci 12:962–969.

Lømo T, Rosenthal J (1972) Control of Ach sensitivity by muscle activity in the rat. J Physiol 221:493–513.

Lorente de Nó R (1981) The Primary Acoustic Nuclei. New York: Raven Press.

Lumsden A, Krumlauf R (1996) Patterning the vertebrate neuraxis. Science 274:1109–1115.

Luo L, Moore JK, Baird A, Ryan AF (1995) Expression of acidic FGF mRNA in rat auditory brainstem during postnatal maturation. Brain Res Dev Brain Res 86:24–34.

Luo L, Ryan AF, Saint Marie RL (1999) Cochlear ablation alters acoustically induced c-fos mRNA expression in the adult rat auditory brainstem. J Comp Neurol 404:271–283.

Lurie DI, Pasic TR, Hockfield SJ, Rubel EW (1997) Development of Cat-301 immunoreactivity in auditory brainstem nuclei of the gerbil. J Comp Neurol 380:319–334.

Lustig LR, Leake PA, Snyder RL, Rebscher SJ (1994) Changes in the cat cochlear nucleus following neonatal deafening and chronic intracochlear electrical stimulation. Hear Res 74:29–37.

Ma Q, Chen Z, del Barco Barrantes I, de la Pompa JL, Anderson DJ (1998) neurogenin 1 is essential for the determination of neuronal precursors for proximal cranial sensory ganglia. Neuron 20:469–482.

Ma Q, Anderson DJ, Fritzsch B (2000) Neurogenin 1 null mutant ears develop fewer, morphologically normal hair cells in smaller sensory epithelia devoid of innervation. J Assoc Res Otolaryngol 1:129–143.

Marin F, Puelles L (1995) Morpological fate of rhombomeres in quail/chick chimeras: a segmental analysis of hindbrain nuclei. Eur J Neurosci 7:1714–1738.

Martin MR, Rickets C (1981) Histogenesis of the cochlear nucleus of the mouse. J Comp Neurol 197:169–184.

Mathis L, Nicolas JF (2002) Cellular patterning of the vertebrate embryo. Trends Genet 18:627–635.

Mattox DE, Neises GR, Gulley RL (1982) A freeze–fracture study of the maturation of synapses in the anteroventral cochlear nucleus of the developing rat. Anat Rec 204:281–287.

Molea D, Rubel EW (2003) Timing and topography of nucleus magnocellularis innervation by the cochlear ganglion. J Comp Neurol 466:577–591.

Moore DR (1990) Auditory brainstem of the ferret: early cessation of developmental sensitivity of neurons in the cochlear nucleus to removal of the cochlea. J Comp Neurol 302:810–823.

Moore DR (1991) Development and plasticity of the ferret auditory system. In: Altschuler RA, Bobbin RP, Clopton BM, Hoffman DW (eds), Neurobiology of Hearing: The Central Auditory System. New York: Raven Press, pp. 461–475.

Moore DR (1992) Developmental plasticity of the brainstem and midbrain auditory nuclei. In: Romand R (ed), Development of Auditory and Vestibular Systems 2. Amsterdam: Elsevier, pp. 297–320.

Moore DR, Hutchings ME, King AJ, Kowalchuk NE (1989) Auditory brain stem of the ferret: some effects of rearing with a unilateral ear plug on the cochlea, cochlear nucleus, and projections to the inferior colliculus. J Neurosci 9:1213–1222.

Moore JK, Guan YL, Shi SR (1997) Axogenesis in the human fetal auditory system, demonstrated by neurofilament immunohistochemistry. Anat Embryol (Berl) 195:15–30.

Morest DK (1969) The differentiation of cerebral dendrites: A study of the post-migratory neuroblast in the medial nucleus of the trapezoid body. Z Anat Entwicklungs 128:271–289.

Mostafapour SP, Lachica EA, Rubel EW (1997) Mitochondrial regulation of calcium in the avian cochlear nucleus. J Neurophysiol 78:1928–1934.

Mostafapour SP, Cochran SL, Del Puerto NM, Rubel EW (2000) Patterns of cell death in mouse anteroventral cochlear nucleus neurons after unilateral cochlea removal. J Comp Neurol 426:561–571.

Mostafapour SP, Del Puerto NM, Rubel, EW (2002) *bcl-2* overexpression eliminates deprivation-induced cell death of brainstem auditory neurons. J Neurosci 21:4670–4674.

Mueller BK (1999) Growth cone guidance: first steps towards a deeper understanding. Ann Rev Neurosci 22:351–388.

Murakami Y, Suto F, Shimizu M, Shinoda T, Kameyama T, Fujisawa H (2001) Differential expression of plexin-A subfamily members in the mouse nervous system. Dev Dynam 220:246–258.

Neises GR, Mattox DE, Gulley RL (1982) The maturation of the end bulb of Held in the rat anteroventral cochlear nucleus. Anat Rec 204:271–279.

Ni D, Seldon HL, Shepherd RK, Clark GM (1993) Effect of chronic electrical stimulation on cochlear nucleus neuron size in normal hearing kittens. Acta Otolaryngol 113:489–497.

Nicholls DG, Budd SL (2000) Mitochondria and neuronal survival. Physiol Rev 80:315–360.

Niparko JK (1999) Activity influences on neuronal connectivity within the auditory pathway. Laryngoscope 109:1721–1730.

Nordeen KW, Killackey HP, Kitzes LM (1983) Ascending projections to the inferior colliculus following unilateral cochlear ablation in the neonatal gerbil, *Meriones unguiculatus*. J Comp Neurol 214:144–153.

Ogawa J, Lee S, Itoh K, Nagata S, Machida T, Takeda Y, Watanabe K (2001) Neural recognition molecule NB-2 of the contactin/F3 subgroup in rat: specificity in neurite outgrowth-promoting activity and restricted expression in the brain regions. J Neurosci Res 65:100–110.

O'Leary DD, Wilkinson DG (1999) Eph receptors and ephrins in neural development. Curr Opin Neurobiol 9:65–73.

Parks TN (1979) Afferent influences on the development of the brain stem auditory nuclei of the chicken: otocyst ablation. J Comp Neurol 183:665–677.

Parks TN (1997) Effects of early deafness on development of brain stem auditory neurons. Ann Otol Rhinol Laryngol Suppl 168:37–43.

Parks TN (2000) The AMPA receptors of auditory neurons. Hear Res 147:77–91.

Parks TN, Jackson H (1984) A developmental gradient of dendritic loss in the avian cochlear nucleus occurring independently of primary afferents. J Comp Neurol 227:459–466.

Parks TN, Jackson H, Taylor DA (1990) Adaptations of synaptic form in an aberrant projection to the avian cochlear nucleus. J Neurosci 10:975–984.

Parks TN, Code RA, Taylor DA, Solum D, Strauss KI, Jacobowitz D, Winsky L (1997) Calretinin expression in the chick brainstem auditory nuclei develops and is maintained independently of cochlear nerve input. J Comp Neurol 383:112–121.

Pasic TR, Rubel EW (1989) Rapid changes in cochlear nucleus cell size following blockade of auditory nerve electrical activity in gerbils. J Comp Neurol 283:474–480.

Pasic TR, Rubel EW (1991) Cochlear nucleus cell size is regulated by auditory nerve electrical activity. Otolaryngol Head Neck Surg 104:6–13.

Pasini A, Wilkinson DG (2002) Stabilizing the regionalisation of the developing vertebrate central nervous system. BioEssays 24:427–438.

Perney TM, Marshall J, Martin KA, Hockfield S, Kaczmarek LK (1992) Expression of
the mRNAs for the Kv3.1 potassium channel gene in the adult and developing rat
brain. J Neurophysiol 68:756–766.

Pettigrew AG, Ansselin AD, Bramley JR (1988) Development of functional innervation
in the second and third order auditory nuclei of the chick. Development 104:575–
588.

Powell TPS, Erulkar SD (1962) Transneural cell degeneration in the auditory relay nuclei
of the cat. J Anat 96:249–268.

Raman IM, Trussell LO (1992) The kinetics of the response to glutamate and kainate in
neurons of the avian cochlear nucleus. Neuron 9:173–186.

Raman IM, Zhang S, Trussell LO (1994) Pathway-specific variants of AMPA receptors
and their contribution to neuronal signaling. J Neurosci 14:4998–5010.

Raper JA (2000) Semaphorins and their receptors in vertebrates and invertebrates. Curr
Opin Neurobiol 10:88–94.

Redd EE, Pongstaporn T, Ryugo DK (2000) The effects of congenital deafness on auditory nerve synapses and globular bushy cells in cats. Hear Res 147:160–174.

Redies C (2000) Cadherins in the central nervous system. Prog Neurobiol 61:611–
648.

Redies C, Puelles L (2001) Modularity in vertebrate brain development and evolution.
BioEssays 23:1100–1111.

Rhode WS (1978) Some observations on cochlear mechanics. J Acoust Soc Am 64:158–
176.

Rhode WS, Greenberg S (1992) Physiology of the cochlear nuclei. In: Popper AN, Fa,
RR (eds), The Mammalian Auditory Pathway: Neurophysiology. New York: Springer-
Verlag, pp. 94–152.

Riedel B, Friauf E, Grothe C, Unsicker K (1995) Fibroblast growth factor-2-like immunoreactivity in auditory brainstem nuclei of the developing and adult rat: correlation
with onset and loss of hearing. J Comp Neurol 354:353–360.

Romand R, Avan P (1997) Anatomical and functional aspects of the cochlear nucleus.
In: Ehret G, Romand R (eds), The Central Auditory System. New York: Oxford University Press, pp. 97–191.

Rubel EW (1978) Ontogeny of structure and function in the vertebrate auditory system.
In: Jacobson M (ed), Handbook of Sensory Physiology. Vol. IX. Development of
Sensory Systems. New York: Springer-Verlag, pp. 135–137.

Rubel EW, Fritzsch BF (2002) Auditory system development: primary auditory neurons
and their targets. Annu Rev Neurosci 25:51–101.

Rubel EW, Parks TN (1975) Organization and development of brain stem auditory nuclei
of the chicken: tonotopic organization of N. magnocellularis and N. laminaris. J Comp
Neurol 164:411–434.

Rubel EW, Parks TN (1988) Organization and development of the avian brain-stem auditory system. In: Edelman GM, Gall WE, Cowan WM (eds), Auditory Function:
Neurobiological Bases of Hearing, New York: Wiley-Interscience, pp. 3–92.

Rubel EW, Smith DJ, Miller LC (1976) Organization and development of brain stem
auditory nuclei of the chicken: ontogeny of n. magnocellularis and n. laminaris. J
Comp Neurol 166:469–489.

Rubel EW, Hyson RL, Durham D (1990) Afferent regulation of neurons in the brain
stem auditory system. J Neurobiol 21:169–196.

Rubel EW, Falk PM, Canady KS, Steward O (1991) A cellular mechanism underlying the activity-dependent transneuronal degeneration: rapid but reversible destruction of the neuronal ribosomes. Brain Dysfunct 4:55–74.

Ruben RJ (1967) Development of the inner ear of the mouse: a radioautographic study of terminal mitoses. Acta Otolaryngol (Suppl):1–44.

Ryan AF, Woolf NK (1988) Development of tonotopic representation in the Mongolian gerbil: a 2-deoxyglucose study. Brain Res 469:61–70.

Ryugo DK, Fekete DM (1982) Morphology of primary axosomatic endings in the anteroventral cochlear nucleus of the cat: a study of the endbulbs of Held. J Comp Neurol 210:239–257.

Ryugo DK, Parks TN (2003) Primary innervation of the avian and mammalian cochlear nucleus. Brain Res Bull 60:435–456.

Ryugo DK, Pongstaporn T, Huchton DM, Niparko JK (1997) Ultrastructural analysis of primary endings in deaf white cats: morphologic alterations in endbulbs of Held. J Comp Neurol 385:230–244.

Ryugo, DK, Rosenbaum BT, Kim PJ, Niparko JK, Saada AA (1998) Single unit recordings in the auditory nerve of congenitally deaf white cats: morphological correlates in the cochlea and cochlear nucleus. J Comp Neurol 397:532–548.

Saada AA, Niparko JK, Ryugo DK (1996) Morphological changes in the cochlear nucleus of congenitally deaf white cats. Brain Res 736:315–328.

San Jose I, Vasquez E, Garcia-Atares N, Huerta JJ, Vega JA, Represa J (1997) Differential expression of microtubule associated protein MAP-2 in developing cochleovestibular neurons and its modulation by neurotrophin-3. Int J Dev Biol 41:509–519.

Sanes DH, Rubel EW (1988) The ontogeny of inhibition and excitation in the gerbil lateral superior olive. J Neurosci 8:682–700.

Sanes DH, Walsh EJ (1998) The development of central auditory processing. In: Rubel EW, Popper AN, Fay RR (eds), Development of the Auditory System. New York: Springer-Verlag, pp. 271–314.

Sanes DH, Merickel M, Rubel EW (1989) Evidence for an alteration of the tonotopic map in the gerbil cochlea during development. J Comp Neurol 279:436–444.

Sanes DH, Reh TA, Harris WA (2000) Development of the Nervous System. New York: Academic Press.

Saunders JC, Coles RB, Gates GR (1973) The development of auditory evoked responses in the cochlea and cochlear nuclei of the chick. Brain Res 63:59–74.

Saunders JC, Adler JH, Cohen YE, Smullen S, Kazahaya K (1998) Morphometric changes in the chick nucleus magnocellularis following acoustic overstimulation. J Comp Neurol 390:412–426.

Schweitzer L, Cant NB (1984) Development of the cochlear innervation of the dorsal cochlear nucleus of the hamster. J Comp Neurol 225:228–243.

Schweitzer L, Cecil T (1992) Morphology of HRP-labelled cochlear nerve axons in the dorsal cochlear nucleus of the developing hamster. Hear Res 60:34–44.

Shaner RF (1934) The development of the nuclei and tracts related to the acoustic nerve in the pig. J Comp Neurol 60:5–19.

Sie KC, Rubel EW (1992) Rapid changes in protein synthesis and cell size in the cochlear nucleus following eighth nerve activity blockade or cochlea ablation. J Comp Neurol 320:501–508.

Simon H, Lumsden A (1993) Rhombomere-specific origin of the contralateral vestibulo-acoustic efferent neurons and their migration across the embryonic midline. Neuron 11:209–220.

Snyder RL, Leake PA (1997) Topography of spiral ganglion projections to cochlear nucleus during postnatal development in cats. J Comp Neurol 384:293–311.

Solum D, Hughes D, Major MS, Parks TN (1997) Prevention of normally occurring and deafferentation-induced neuronal death in chick brainstem auditory neurons by periodic blockade of AMPA/kainate receptors. J Neurosci 17:4744–4751.

Sterbing SJ, Schmidt U, Rubsamen R (1994) The postnatal development of frequency-place code and tuning characteristics in the auditory midbrain of the phyllostomid bat, Carollia perspicillata. Hear Res 76:133–146.

Steward O, Rubel EW (1985) Afferent influences on brain stem auditory nuclei of the chicken: cessation of amino acid incorporation as an antecedent to age-dependent transneuronal degeneration. J Comp Neurol 231:385–395.

Studer M, Gavalas A, Marshall H, Ariza-McNaughton L, Rijli FM, Chambon P, Krumlauf R (1998) Genetic interactions between Hoxa1 and Hoxb1 reveal new roles in regulation of early hindbrain patterning. Development 125:1025–1036.

Sugden SG, Zirpel L, Dietrich CJ, Parks TN (2002) Development of the specialized AMPA receptors of auditory neurons. J Neurobiol 52:189–202.

Taber Pierce E (1967) Histogenesis of the dorsal ventral cochlear nuclei in the mouse. An autobiographic study. J Comp Neurol 131:27–54.

Takahashi TT, Carr CE, Brecha N, Konishi M (1987) Calcium binding protein-like immunoreactivity labels the terminal field of nucleus laminaris of the barn owl. J Neurosci 7:1843–1856.

Tierney TS, Russell FA, Moore DR (1997) Susceptibility of developing cochlear nucleus neurons to deafferentation-induced death abruptly ends just before the onset of hearing. J Comp Neurol 378:295–306.

Trune DR (1982a) Influence of neonatal cochlear removal on the development of mouse cochlear nucleus: I. Number, size, and density of its neurons. J Comp Neurol 209: 409–424.

Trune DR (1982b) Influence of neonatal cochlear removal on the development of mouse cochlear nucleus: II. Dendritic morphometry of its neurons. J Comp Neurol 209:425–434.

Trussell L (1998) Control of time course of glutamatergic synaptic currents. Prog Brain Res 116:59–69.

Tucci DL, Rubel EW (1985) Afferent influences on brain stem auditory nuclei of the chicken: effects of conductive and sensorineural hearing loss on n. magnocellularis. J Comp Neurol 238:371–381.

Tucci DL, Born DE, Rubel EW (1987) Changes in spontaneous activity and CNS morphology associated with conductive and sensorineural hearing loss in chickens. Ann Otol Rhinol Laryngol 96:343–350.

Tucci DL, Cant NB, Durham D (1999) Conductive hearing loss results in a decrease in central auditory system activity in the young gerbil. Laryngoscope 109:1359–1371.

Ulatowska-Blaszyk K, Bruska M (1999) The cochlear ganglion in human embryos of developmental stages 18 and 19. Folia Morphol (Warsz) 58:29–35.

von Békésy G (1960) Experiments in Hearing. New York: American Institute of Physics.

Walsh EJ, McGee J (1988) Rhythmic discharge properties of caudal cochlear nucleus neurons during postnatal development in cats. Hear Res 36:233–247.

Warchol ME, Dallos P (1990) Neural coding in the chick cochlear nucleus. J Comp Physiol A 166:721–734.

Webster DB (1983a) A critical period during postnatal auditory development of mice. Int J Pediatr Otorhinolaryngol 6:107–118.

Webster DB (1983b) Late onset of auditory deprivation does not affect brainstem auditory neuron soma size. Hear Res 12:145–147.

Webster DB (1983c) Auditory neuronal sizes after a unilateral conductive hearing loss. Exp Neurol 79:130–140.

Webster DB (1985) The spiral ganglion and cochlear nuclei of deafness mice. Hear Res 18:19–27.

Webster DB (1988) Sound amplification negates central effects of a neonatal conductive hearing loss. Hear Res 32:193–195.

Webster DB, Webster M (1977) Neonatal sound deprivation affects brain stem auditory nuclei. Arch Otolaryngol 103:392–396.

Webster DB, Webster M (1979) Effects of neonatal conductive hearing loss on brain stem auditory nuclei. Ann Otol Rhinol Laryngol 88:684–688.

Wiesel TN, Hubel DH (1963) Single-cell responses in striate cortex of kittens deprived of vision in one eye. J Neurophys 26:1003–1017.

Wiesel TN, Hubel DH (1965) Comparison of the effects of unilateral and bilateral eye closure on cortical unit responses in kittens. J Neurophysiol 6:1029–1040.

Wilkinson BL, Sadler KA, Hyson RL (2002) Rapid deafferentation-induced upregulation of bcl-2 mRNA in the chick cochlear nucleus. Brain Res Mol Brain Res 99:67–74.

Wilkinson BL, Elam JS, Fadool DA, Hyson RL (2003) Afferent regulation of cytochrome-c and active caspase-9 in the avian cochlear nucleus. Neurosci 120:1071–1079.

Wilkinson DG (2000) Eph receptors and ephrins: regulators of guidance and assembly. Int Rev Cytol 196:177–244.

Willard F (1990) Analysis of the development of the human auditory system. Sem Hearing 11:107–123.

Willard F (1993) Postnatal development of auditory nerve projections to the cochlear nucleus in *Monodelphis domestica*. In: Merchan MA, Juiz JM, Godfrey DA, Mugnaini E (eds), The Mammalian Cochlear Nuclei: Organization and Function. New York: Plenum Press, pp. 29–42.

Willard FH (1995) Development of the mammalian auditory hindbrain. In: Malhotra S (ed), Advances in Neural Science, Vol. 2. Greenwich, CT: JAI, pp. 205–234.

Willard F, Martin GE (1986) The development and migration of large multipolar neurons into the cochlear nucleus of the North American opossum. J Comp Neurol 248:119–132.

Wilson BE, Mochon E, Boxer LM (1996) Induction of bcl-2 expression by phosphorylated CREB proteins during B-cell activation and rescue from apoptosis. Mol Cell Biol 16:5546–5556.

Windle WF, Austin MF (1936) Neurofibrillar development in the central nervous system of chick embryos up to 5 days' incubation. J Comp Neurol 63:431–463.

Winsky L, Jacobowitz DM (1995) Effects of unilateral cochlea ablation on the distribution of calretinin mRNA and immunoreactivity in the guinea pig ventral cochlear nucleus. J Comp Neurol 354:564–582.

Young SR, Rubel EW (1983) Frequency-specific projections of individual neurons in chick brainstem auditory nuclei. J Neurosci 3:1373–1378.

Young SR, Rubel EW (1986) Embryogenesis of arborization pattern and topography of individual axons in N. laminaris of the chicken brain stem. J Comp Neurol 254:425–459.

Zhang JS, Haenggeli CA, Tempini A, Vischer MW, Moret V, Rouiller EM (1996) Electrically induced fos-like immunoreactivity in the auditory pathway of the rat: effects of survival time, duration, and intensity of stimulation. Brain Res Bull 39:75–82.

Zhou N, Parks TN (1992) Developmental changes in the effects of drugs acting at NMDA or non-NMDA receptors on synaptic transmission in the chick cochlear nucleus (nuc. magnocellularis). Brain Res Dev Brain Res 67:145–152.

Zirpel L, Parks TN (2001) Zinc inhibition of Group I mGluR-mediated calcium homeostasis in auditory neurons. J Assoc Res Otolaryngol 2:180–187.

Zirpel L, Rubel EW (1996) Eighth nerve activity regulates intracellular calcium concentration of avian cochlear nucleus neurons via a metabotropic glutamate receptor. J Neurophysiol 76:4127–4139.

Zirpel L, Nathanson NM, Rubel EW, Hyson RL (1994) Glutamate-stimulated phosphatidylinositol metabolism in the avian cochlear nucleus. Neurosci Lett 168:163–166.

Zirpel L, Lachica EA, Lippe WR (1995) Deafferentation increases the intracellular calcium of cochlear nucleus neurons in the embryonic chick. J Neurophysiol 74:1355–1357.

Zirpel L, Lachica EA, Rubel EW (1997) Afferent regulation of cochlear nucleus neurons: Intracellular mechanisms and signal transduction pathways. In: Berlin C (ed), Neurotransmission and Hearing Loss: Basic Science, Diagnosis, and Management, San Diego, Singular, pp. 47–76.

Zirpel L, Lippe WR, Rubel EW (1998a) Activity-dependent regulation of [Ca2+]i in avian cochlear nucleus neurons: roles of protein kinases A and C and relation to cell death. J Neurophysiol 79:2288–2302.

Zirpel L, Lippe WR, Rubel EW (1998b) Activity-dependent regulation of intracellular calcium in avian cochlear nucleus neurons: roles of protein kinases A and C and correlation with cell death. J Neurophysiol 79:2288–2302.

Zirpel L, Janowiak MA, Taylor DA, Parks TN (2000a) Developmental changes in metabotropic glutamate receptor-mediated calcium homeostasis. J Comp Neurol 421:95–106.

Zirpel L, Janowiak MA, Veltri CA, Parks TN (2000b) AMPA receptor-mediated, calcium-dependent CREB phosphorylation in a subpopulation of auditory neurons surviving activity deprivation. J Neurosci 20:6267–6275.

3

Developmental Changes and Cellular Plasticity in the Superior Olivary Complex

ECKHARD FRIAUF

1. Introduction

The superior olivary complex (SOC) is a conspicuous structure in the mammalian central auditory system. It is located ventrally in the pontine brain stem underneath the cerebellum and consists of several third-order nuclei. The general organization of the SOC is illustrated in Figure 3.1A. Usually three principal SOC nuclei are identified (for abbreviations, see list at end of chapter), the lateral superior olive (LSO), the medial superior olive (MSO), and the medial nucleus of the trapezoid body (MNTB; Irving and Harrison 1967; for review see Schwartz 1992; see also Reuss 2000). These principal nuclei are surrounded by periolivary regions: the superior paraolivary nucleus (SPN), the lateral nucleus of the trapezoid body (LNTB), and the ventral nucleus of the trapezoid body (VNTB). In the rodent brain, the nomenclature occasionally refers to the LNTB and VNTB as the lateroventral (LVPO) and medioventral periolivary region (MVPO), respectively (Osen et al. 1984; Thompson and Thompson 1991). In cats and bats, a dorsomedial periolivary nucleus (DMPO) has been described and appears to be equivalent to the SPN of the rodent brain (Schofield and Cant 1991; Schwartz 1992; Ostapoff et al. 1997; Grothe and Park 2000). Rostrally and caudally to the SOC proper, two additional areas are located that are generally called the rostral and the caudal periolivary region (RPO and CPO, respectively).

Ascending input to the SOC nuclei arises from second-order neurons in the central auditory system, which are the neurons in the cochlear nuclear complex. Several cell types in the ventral cochlear nucleus (VCN) send their axons into the SOC and differentially innervate discrete SOC nuclei (Fig. 3.1B). As is typical for the auditory brain stem, the internuclear connections are highly divergent as well as convergent. In addition, most projections are tonotopically organized. It is in the SOC where the information from the two ears first converges.

As to the function of the SOC, the binaural convergence of input enables the

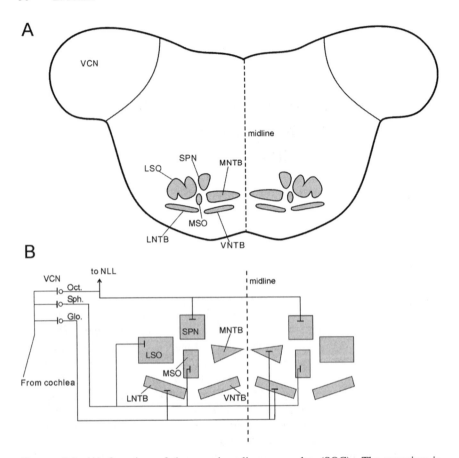

FIGURE 3.1. **(A)** Overview of the superior olivary complex (SOC). The overview is drawn from a transverse section of the *rat* brain, yet it generally illustrates the location of the *mammalian* SOC within the brain stem and the arrangement of the various SOC nuclei. Three principal nuclei form the SOC (Irving and Harrison 1967), namely the lateral superior olive (LSO), the medial superior olive (MSO), and the medial nucleus of the trapezoid body (MNTB). They are surrounded by periolivary regions, among which three regions are particularly prominent and can therefore be designated a nucleus: the superior paraolivary nucleus (SPN), the lateral nucleus of the trapezoid body (LNTB), and the ventral nucleus of the trapezoid body (VNTB). *Note*: Sometimes the LNTB and VNTB are referred to as the lateroventral (LVPO) and medioventral periolivary region (MVPO), respectively (Osen et al. 1984; Thompson and Thompson 1991). In the rodent brain, the SPN is particularly large and has been mistakenly identified as the MSO (which is small) in several older studies. The SPN probably corresponds to the dorsomedial periolivary nucleus (DMPO; Schofield and Cant 1991; Schwartz 1992; Ostapoff et al. 1997), a structure delineated in the cat and bat brain (Osen et al. 1984). Not shown in this cartoon are periolivary regions that are located at the rostral and caudal tip of the SOC; these are generally called rostral and caudal periolivary regions (RPO and CPO). The VNTB and RPO are often considered to be one continuous structure (e.g., Robertson

localization of sound sources in space. In this regard, interaural time differences and interaural intensity differences are the two major cues that are analyzed by the auditory system. The majority of MSO neurons are excited by stimulation of each ear (Fig. 3.2) and are sensitive to interaural time differences in the microsecond range. As maximal excitation is generated when the excitatory signals from the two ears arrive within a narrow time interval of one another, MSO neurons are classically thought to act as coincidence detectors (Jeffress 1948; Moushegian et al. 1967; Joris et al. 1998). Axons of variable length ("delay lines") are believed to contribute to the coincidence detection (Jeffress model). Recently however, the traditional view that coincidence detection is done in a neuronal array of MSO neurons, thus forming a topographic map of horizontal auditory space, was challenged (Brand et al. 2002). The authors also showed that precisely timed glycine-controlled inhibition is a critical part of the mechanism by which the physiologically relevant range of interaural time differences is encoded in the MSO. The topics of binaural localization cues of interaural time differences and the complications to the Jeffress model are covered more thoroughly in reviews by Benedikt Grothe (2000) and Tom Yin (2002).

In contrast to the MSO, the majority of LSO neurons are excited by stimulation of the ipsilateral ear, yet inhibited when the contralateral ear is stimulated (Fig. 3.2; Boudreau and Tsuchitani 1968). MNTB neurons form the relay station in the pathway from the contralateral cochlear nuclear complex to the LSO and convert the excitatory input that they receive into inhibition (Fig. 3.2). Because of the constellation of convergent excitatory and inhibitory synaptic input from the two ears, LSO neurons are sensitive to interaural intensity differences.

Aside from their role in binaural sound localization, SOC neurons, particularly those in the periolivary regions, form a prominent part of the descending auditory system (Warr 1992). They send efferent projections to the cochlear nuclear complex and the cochlea, thereby comprising the olivocochlear system. More elaborate accounts on the function of the SOC in the adult are provided elsewhere (Spangler and Warr 1991; Brugge 1992).

As mentioned in the preceding, one striking feature of the auditory brain

---

1996). (B) The lower panel is a schematic drawing showing the major ascending connections from the ventral cochlear nucleus (VCN) to the SOC nuclei. The input from the VCN arises mainly from spherical bushy cells (Sph), from globular bushy cells (Glo), and from octopus cells (Oct; for review see Helfert et al. 1991). Octopus cells also project to the nuclei of the lateral lemniscus (NLL). Most of these projections are topographically (tonotopically) organized. Axons of most spherical and globular bushy cells travel in the ventral acoustic stria, whereas those from octopus cells generally course in the intermediate acoustic stria. Not shown are multipolar cells, whose predominant projection goes to the contralateral inferior colliculus, but which also send collaterals into principal and periolivary SOC nuclei, for example, the contralateral and ipsilateral VNTB (Adams and Warr 1976; Thompson 1998).

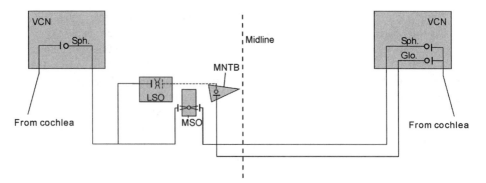

FIGURE 3.2. Schematic drawing depicting the major afferent projections from both ventral cochlear nuclei (VCN) to the three principal SOC nuclei (MSO, LSO, MNTB). For the sake of clarity, the projections to only one side are shown. All projections are excitatory in the adult with the exception of the MNTB-LSO connection, which is inhibitory and drawn with a *dashed line*. MSO neurons are excited from either ear, whereas LSO neurons receive excitatory input from the ipsilateral ear, yet inhibitory input from the contralateral ear. MNTB neurons act as a relay station, converting glutamate-induced excitation into glycine-induced inhibition. Sph., spherical bushy cells; Glo., globular bushy cells.

stem is the divergent and convergent nature of the internuclear connectivity. Another striking feature is the exquisite topography that is present in the great majority of the connections. Several intriguing and developmentally related questions arise from these features: (1) How is the complex, yet highly organized architecture generated during development? (2) How is the neuronal circuitry maintained in the mature brain? (3) Will the circuitry become hard-wired at some time during development or will it be plastic and susceptible to morphological and functional changes throughout life?

   The purpose of the present chapter is to address and discuss these questions. The article is divided into two main parts. First, an overview of the changes that occur during normal development of the SOC is presented. This part covers the period from neurogenesis to the achievement of the mature neuronal circuitry and function (Sections 2–4). Second, the literature on plastic changes becoming evident in the immature and the adult SOC is summarized and evaluated (Section 5). For this purpose, plasticity is defined quite strictly as alterations in structure, connections, or function that occur in response to experimental manipulations, such as artificially produced imbalances in neural activity. These imbalances may be induced by hearing loss, injury, or overstimulation, and the alterations may take place during development as well as in adulthood. The definition excludes alterations that are caused by genetically regulated factors, whose expression is intrinsically determined and independent of experience. It also implies that the use-dependent changes in structure or function occurring during *normal* development will not be considered plasticity related. In the present

chapter, development- and plasticity-related changes are therefore treated in separate sections as far as possible.

Over the past 20 years, several reviews on the development of the auditory
brain stem have been published (Rubel 1978; Brugge 1983; Moore 1983; Rubel
and Parks 1988; Kitzes 1990; Moore 1992a; Willard 1995). Two still timely
reviews that extensively evaluate the literature on the structural and functional
development of the whole central auditory system have been provided in an
earlier volume of this handbook series (Cant 1998; Sanes and Walsh 1998). In
the present chapter, the sections dealing with the developmental aspects may
therefore repeat some of the information available in these reviews, yet they
focus on and *emphasize* the SOC. Finally, related to the present review is a
chapter on plasticity of binaural systems (Chapter 4) by Moore and King in this
book.

Whenever appropriate, the findings on SOC development and plasticity are
discussed in light of the results obtained in other structures, for example, the
retina, the visual and somatosensory cortices, the spinal cord, and the hippocampus. This will be done to demonstrate where general principles of developmentally and plasticity-related changes occur or where the SOC must be set
apart from other structures.

2. Early Development: From Neurogenesis to Axon Outgrowth

2.1 Neurogenesis of SOC Neurons in the Rat

Neuronal development starts with the generation of neurons that are produced
in the neuroepithelial germinal zone. Neurogenesis, defined as the terminal division of precursor cells and associated with the terminal phase of DNA synthesis, begins shortly after the neural tube has closed and the ventricular system
has formed. In the auditory brain stem, neurogenesis has been analyzed mainly
in [³H]thymidine radiographic studies, and the pioneer work in rats was performed by Joseph Altman and Shirley Bayer in the 1980s (Altman and Bayer
1980, 1981). Recently, it was followed up by Motoi Kudo's group, who applied
and immunohistochemically detected the thymidine analog 5-bromodeoxy-
uridine (Kudo et al. 1996, 2000). Concerning the birthdays, the studies achieved
very similar results. SOC neurons belong to the earliest group of auditory neurons generated. In rats, the production period lasts from embryonic day (E) 11
until E15.[1] There appears to be a sequential gradient both in the different nuclei

---

1. There is some confusion in the literature concerning the exact dating of developmental
ages. In the present review, the day of insemination is considered to be equivalent to
embryonic day (E) 0, following the conventional system used by most authors. Likewise,
the day of birth is equivalent to postnatal day (P) 0. Another way of counting which is
still in use is to start prenatal and postnatal development with E1 and P1, respectively.

and in the intranuclear areas (Altman and Bayer 1980; Fig. 3.3). MSO neurons are generated first, namely between E11 and E13, with peak production occurring at E11. SPN neurons are born on E12, the great majority (70–90%) of MNTB neurons at E14 (range: E11–E15), and LSO neurons between E11 and E15.[2] According to Altman and Bayer (1980), neurons within a given nucleus that end up in the high-frequency region are generated earlier than those of the low-frequency region, thus forming a temporal gradient that is related to the tonotopic representation. It is worthwhile to consider here that the rule of thumb which can be deduced from this finding, that high-frequency areas develop earlier than low-frequency areas, is also verified in other aspects of development and appears to be of general validity. This point is returned to in Section 2.4.2.

From a detailed birth dating analysis of rat LSO neurons, which was performed with 5-bromodeoxyuridine in combination with retrograde labeling, it has become apparent that LSO neurons that project into the contralateral inferior colliculus are generated at E11–E12, while those projecting into the ipsilateral inferior colliculus are generated between E13 and E15 (Kudo et al. 1996). In other words, crossed projection neurons in the LSO are generated 1–4 days prior to those giving rise to the uncrossed projection. Interestingly, the authors observed a diffuse distribution of birth-dated neurons in the LSO (Kudo et al. 1996, 2000), therewith contradicting Altman's and Bayer's result of a tonotopic gradient. In another auditory brain stem structure, the inferior colliculus, a tonotopic gradient in neuron birth was also observed (Altman and Bayer 1981). However, others have argued against such a gradient (Faye-Lund and Osen 1985; Coleman 1990) and a cytogenetic gradient perpendicular to the tonotopic organization, rather than congruent with it, was suggested instead (Faye-Lund and Osen 1985). It appears that the issue needs further analysis before the controversy can be solved.

Nothing is known about a correlation of birth date with cell size in the SOC, which would imply that large neurons are produced first, then intermediate-sized neurons, and finally small neurons, which often represent interneurons. This correlation is common to many parts of the CNS. In the auditory system, it was observed in the cochlear nuclear complex (Taber-Pierce 1967; Martin and Rickets 1981; Ivanova and Yuasa 1998) and the inferior colliculus (Altman and Bayer 1981), but it has not been investigated in the SOC.

Whenever necessary, appropriate back-shifting adjustments were made. For example, the data by Altman and Bayer (1980) as well as by Kudo et al. (2000) were thus changed. If relevant information is not provided in a publication, no changes were made, thus leaving the possibility of a 1-day error.

2. Altman and Bayer (Altman and Bayer 1980) made a mistake in the identification of the MSO and SPN. What they called medial superior olivary nucleus (Som) is in fact the SPN, whereas their lateral trapezoid nucleus (TRl) is the MSO. The correct nomenclature is used in the present review (cf. Fig. 18 of their paper and Fig. 3.2).

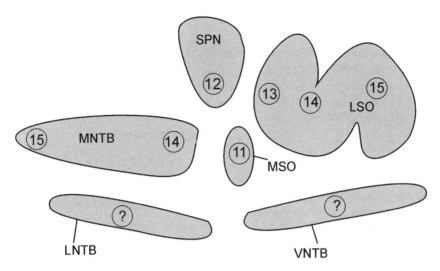

FIGURE 3.3. Neurogenesis in the SOC. *Numbers in circles* refer to embryonic days at which neuron production peaks. The majority of neurons are born between E11 and E15. No data are available for the LNTB and the VNTB. [Redrawn and modified from Figure 18 in Altman and Bayer (1980), reprinted by permission of Alan R. Liss.]

2.2 Neurogenesis of SOC Neurons in Other Species

The general sequence of cell birth that is seen in the nuclei of the rat SOC is also found in other rodents, such as the mouse (Taber-Pierce 1973) and the rabbit (Oblinger and Das 1981). Birth dates in the rabbit appear to be very similar to those in the rat, yet those in mice are about 2 days earlier, namely between E9 and E13.

2.3 Cell Migration

Migration of neurons is another key step in the formation of the central nervous system. After the completion of their terminal mitotic division in the proliferative zone, the still very immature, undifferentiated neurons (i.e., neuroblasts; His 1889) must migrate to their destination. Not much is known about SOC neurons in this regard. Mixing of neurons that are generated in different proliferative zones, yet end up in the same area, occurs in the case of the cochlear nuclear complex (Altman and Bayer 1980; Willard and Martin 1986; Ivanova and Yuasa 1998) as well as other brain stem areas (Harkmark 1954; Ellenberger et al. 1969). Therefore, it is likely that the same holds for the SOC. The reason for the mixing phenomenon, if there is any, is unclear. One can speculate that the different proliferative zones produce different cell types which, after being intermixed, reside in the same nucleus yet give rise to different connectivity.

The process of migration has not been very extensively analyzed in auditory brain stem neurons. The most detailed study was performed in Kent Morest's group (Book and Morest 1990; Hendriks et al. 1999). Their work on neuroblast migration of the chick nucleus magnocellularis (NM), the avian homolog of the mammalian VCN, revealed that these cells first generate leading processes and then translocate their perikarya through these processes from the proliferative zone to the anlage of the magnocellular nucleus. Book and Morest concluded that the cellular elongation and perikaryal translocation constitutes the general mode of neuronal migration in the CNS, despite the fact that the structures in the developing brain stem differ considerably from the laminated organization of other structures, for example, the cerebral cortex. Within the auditory system, neurons destined for the dorsal cochlear nucleus also appear to migrate through perikaryal translocation, as was shown in the North American opossum *Didelphis virginiana* (Willard and Martin 1986) and in the mouse (Ivanova and Yuasa 1998). The arrangement of the leading process is similar to the alignment of radial glial processes, suggesting that, as elsewhere in the brain, the glial scaffolds are involved in the migration and guidance of neuroblasts.

The chick nucleus laminaris (NL, which is probably homologous to the mammalian MSO) and the nucleus magnocellularis (NM, probably homologous to the mammalian VCN) arise from a common anlage, which differentiates as the neuronal precursor cells migrate. NL arises mostly from rhombomere r5, with a few cells arising from r6, whereas NM arises from rhombomeres r5–r7. Thus, the rhombomeres of origin partially overlap. Within r5, however, the progenitors for both nuclei are found in distinct regions, suggesting that they are specified at an early time, namely prior to migration of the precursors to their final position (Cramer et al. 2000a). The issue of early specification and related work are discussed more extensively by Ed Rubel, Tom Parks, and Lance Zirpel in Chapter 2 of this book.

Virtually a "terra incognita" exists concerning the molecular factors that regulate neuroblast migration in the auditory brain stem. Extracellular matrix molecules, such as laminin, tenascin, and chondroitin sulfates (proteoglycan or glycosaminoglycan) are important candidates, but almost nothing is known about their expression or role in the auditory system. The few studies that have analyzed chondroitin sulfates have not investigated the early, prenatal period during which migration takes place (Atoji et al. 1990, 1995; Atoji and Suzuki 1992; Friauf 2000). The data from these studies indicate that chondroitin sulfate proteoglycans are expressed quite late during development, when migration has ceased and synaptic connections are being formed, indicating that these extracellular matrix molecules may be involved in the process of synapse consolidation (Friauf 2000). As expression lasts into adulthood, it is possible that the chondroitin sulfate proteoglycans form a constantly present, repulsive environment for ingrowing axons and thereby restrict the period during which drastic plastic changes, that is, a reorganization of connectivity, can occur in the auditory system.

Cell migration in the auditory brain stem is under the control of electrical activity. This was shown in the embryonic (E5.5) chicken (Hendriks et al.

1999), where migrating neuroblasts in the acoustico-vestibular anlage express functional calcium channels of the N-type and high-threshold potassium channels. Following pharmacological blockade of these channels, migration stopped within minutes and for several hours. Neuroblasts whose potassium channels were tonically blocked showed a surprising plastic response in that they expressed novel, low-threshold potassium channels and resumed their migratory activity within 8 hours. These data provide an interesting example of the rapidity with which auditory neurons react to environmental changes and with which plastic changes can occur. Further examples of such rapidly occurring changes are provided in Section 5.2.

2.4 Axon Outgrowth and Target Finding

Axon outgrowth and target finding are two major steps toward neuronal circuit formation. Three conceptual conclusions can be drawn from the available literature on the SOC. First, axon outgrowth occurs very early and overlaps with the period of cell migration. Neuroblasts that are still in the process of migration, or have not even left their cytogenetic zone, already send their axons out toward the target area. Second, there appears to be no spatial sequence of axon outgrowth in the auditory brain stem in that neurons in caudally located areas (e.g., cochlear nuclear complex) emit their efferent axons earlier than those in more rostrally located areas (e.g., SOC). Consequently, afferent and efferent connections to and from the SOC are laid down simultaneously. This in turn implies that the wiring of the auditory brain stem circuitry is not a "chain reaction" (area A connects to B, then B is induced to connect to C, etc.) but rather a synchronous event. However, it does not imply that there are no interactions between the axonally linked partners and no transneuronal cross-talk (area A may transsynaptically influence the anatomically distant area C). Indeed, the correct circuitry in the auditory brain stem is achieved only when all partners are present and aberrant projections are formed if the cochlear input from one side is ablated by lesions (Jackson and Parks 1988; Kitzes et al. 1995; Russell and Moore 1995). Third, owing to the early emergence of efferent axons, these fibers can arrive at their target areas at a time when the postsynaptic neurons are still migrating and not yet in place. This indicates that it may be some time before these axons can ultimately make contact with their intended target neurons. It is unclear whether the resulting waiting period is associated with transiently formed functional connections to other cells in the SOC, as is the case in the developing input connectivity from the thalamus to the visual cortex, where subplate neurons act as a temporary target and are thought to play an active role in the development of the functional organization of the cortex (Shatz 1992; McAllister 1999).

2.4.1 Projections to the SOC

Because of the prenatal development of axon connectivity in the SOC, intra-uterine tract tracing experiments are technically demanding. Fortunately, however, with the advent of lipophilic carbocyanine dyes (Honig and Hume 1989), tract tracing has become relatively simple because the dyes can be applied to perfusion-fixed tissue, thus bypassing the problems involved with fetal surgery. In both the rat and the gerbil, initial projections from the cochlear nuclear complex to the SOC are generated before birth and laid down with remarkable precision. The conclusion was drawn since no aberrant connections to ectopic nuclei are seen (Kandler and Friauf 1993; Kil et al. 1995). Data obtained in rat fetuses show that axon outgrowth from the cochlear nuclear complex toward the SOC begins as early as E13 (Niblock et al. 1995). At that time, neuroblasts that will eventually migrate into the anlage of the cochlear nucleus are still residing in the ventricular zone, that is, they have not yet begun to migrate. These neuroblasts display efferent axons, equipped with growth cones, that course along the marginal edge of the brain stem, and few axons have already crossed the midline at E13. Moreover, some growth cones are already seen close to their future targets, namely the ipsilateral and contralateral SOC. This occurs at a time when the neurons of the SOC nuclei are still being generated and is in accordance with the assumption that the axons from the cochlear nucleus go through a waiting period. However, results from a detailed time study of the development of the efferent connections from the cochlear nuclear complex question the necessity of a waiting period which is associated with transient postsynaptic targets (Kandler and Friauf 1993). This study showed that axons traverse the developing nuclei in the SOC and the lateral lemniscus between E15 and E17 and grow up into the inferior colliculus, but they do not form collaterals during this period. Such collaterals within the SOC develop only between E18 and E20 and they sprout from the traversing parent axons. The collateral sprouting resembles the mechanism of interstitial budding seen in corticospinal axons and other systems (O'Leary et al. 1990; O'Leary 1992). With a 3-day delay between axonal traversing and collateral sprouting, it is more likely that the SOC target neurons are already in place when the afferent collaterals are formed, and an extended waiting period, if any, may therefore not exist.

An interesting byproduct in the prenatal tracing study by Niblock et al. (1995) was the observation of dorsoventrally oriented glial cells at the midline of the brain stem. At E15, these glial cells appear to be contacted by growth cones of crossing cochlear nucleus axons. It was therefore assumed that the glial cells provide guidance cues for the developing auditory fibers (Niblock et al. 1995), but this interesting aspect of development has not been analyzed experimentally in the auditory system.

As in the case of neuroblast migration, almost nothing is known about the molecular mechanisms that influence axon guidance in the auditory brain stem. Available data from other systems (retino-tectal, spinal cord) suggest that signaling molecules with either attractive or repulsive effects on specific growth cones play a role in guiding axons to their appropriate targets (for reviews see Tessier-Lavigne and Goodman 1996; Mueller 1999; Yu and Bargmann 2001). At least four different mechanisms (contact attraction, chemoattraction, contact repulsion, and chemorepulsion) are involved in axon guidance, and an axon's response to the molecular guidance cues in its immediate environment may depend on the internal state of the growth cone which will dictate whether it detects a cue as repulsive or attractive, thus determining the trajectory it will take. Many guidance cues belong to families of signaling molecules, for example, the semaphorin (fasciclin, collapsin), netrin, ephrin, and slit families. It needs to be seen which members of these families are expressed in the developing auditory brain stem.

Recently, two candidate molecules for establishing the topologic specificity of the projections from the VCN to the SOC have been identified (Cochran et al. 1999; Cramer et al. 2000b). Ephrins are membrane-bound "ligands" that bind to membrane-bound Eph "receptors," on which they mediate bidirectional signaling. Eph receptors belong to the receptor kinase family and have been implicated in a broad range of developmental processes, including axon guidance and establishment of topographic maps (for reviews see Flanagan and van der Haeghen 1998; Cutforth and Harrison 2002; Knöll and Drescher 2002). When the connections from the chick MN to the NL are formed at E10–E11, EphA4 tyrosine kinase expression in the NL is extensive. A very intriguing asymmetric, polarized pattern occurs, with substantially more receptor immunoreactivity on the dorsal side of the nucleus than on the ventral side (Cramer et al. 2000b). This suggests a possible role in guiding growing axons to the appropriate region, namely the axons from the ipsilateral NM to the dorsal dendrites and those from the contralateral NM to the ventral dendrites. Interestingly, EphA4 appears to be distributed complementary to the neurotrophin receptor TrkB, which is confined to the ventral neuropil of NL at that time (Cochran et al. 1999). Thus, EphA4 and TrkB present a possible molecular set by which binaural segregation of axonal terminals can develop in the auditory brain stem (see also Chapter 2 in this book by Rubel, Parks, and Zirpel).

2.4.2 Projections from the SOC

The development of the efferent projections from the SOC has been investigated by means of retrograde transport following tracer injections into the inferior colliculus. In rats, the major projections from ipsilateral (LSO, MSO, SPN, VNTB) as well as contralateral SOC nuclei (LNTB, LSO) are present by birth (Friauf and Kandler 1990). It appears that the dorsal, low-frequency regions in the inferior colliculus are innervated later than the ventral, high-frequency regions, suggesting that the rule "high frequency regions develop earlier than low frequency regions" is true not only for neurogenesis (see Section 2.1), but also for connectivity.

Projections from the LSO to the inferior colliculus are bilateral and topographically organized in the adult mammalian brain. The development of the bilateral projection pattern has been investigated in ferrets, altricial animals whose hearing onset does not occur until the end of the first postnatal month (P27–P32; Moore 1982; Morey and Carlile 1990). In the adult ferret, the contralateral and the ipsilateral LSO contribute about the same proportion of input to the inferior colliculus (57% vs. 43%; Henkel and Brunso-Bechtold 1993). By birth, both the bilaterality and the topography along the frequency axis are present in the projection (Henkel and Brunso-Bechtold 1993). However, in newborns, there is a greater bias toward the contralateral LSO projection than in the adult (at P9, 70% vs. 30%). The data provide further evidence that the bilateral projections from the LSO to the inferior colliculus are formed prenatally and that the general topographic organization is present by birth. Nevertheless, considerable postnatal changes occur in the laterality of the projection, demonstrating that major anatomical modifications take place in the circuitry. At present, it is unknown whether these changes involve the retraction of axons or cell death of projection neurons.

In the North American opossum *Didelphis virgiana*, a marsupial whose postnatal development is quite slow, occurring over a period of 2–3 months (Krause 1998), projections from the auditory hindbrain to the inferior colliculus are present by about P5 (Willard 1995). According to the stage of inner ear development, this age is approximately equivalent to E16 in the mouse (Lim and Anniko 1985). This is consistent with the observation in rats that axons from the primordium of the cochlear nuclear complex arrive in the inferior colliculus at E16/E17 (Kandler and Friauf 1993). Interestingly, the results obtained in *Didelphis virgiana* imply that the projections from the SOC to the inferior colliculus are formed slightly before those from the cochlear nuclear complex. This finding further supports the conclusion that the development of connectivity in the auditory brain stem takes place in parallel and not in a successive, caudorostral fashion.

2.4.3 Projections within the SOC

The ontogeny of internuclear connections within the SOC has not been extensively investigated, most probably owing to technical constraints. Dye injections that are restricted to one of the small SOC nuclei are difficult to make, and spread of tracer and uptake of the tracer by fibers of passage cause further technical problems. A few data from electrophysiological and in vitro labeling studies, however, demonstrate that some of the internuclear SOC projections are present and functional before birth (Kandler and Friauf 1995b; Lohmann et al. 1998) and must, therefore, be laid down during prenatal life.

3. Intermediate Development: Synaptogenesis and Synapse Consolidation

3.1 General Remarks on Excitatory and Inhibitory Synapses

After the outgrowing axons have reached their target areas, synaptogenesis and synapse consolidation subsequently take place. They are the ultimate steps in wiring the nervous system. Our current conceptual understanding of these processes is based predominantly on studies of excitatory synapses. Such studies have been performed in a variety of systems, such as the visual, the somatosensory, and the neuromuscular system (for reviews see Shatz 1990; Dan and Poo 1994; O'Leary et al. 1994; Dan et al. 1995; Katz and Shatz 1996). The problem has always been circling around the question of the relative contribution of genetic and epigenetic factors, that is factors which are intrinsically determined and governed by environmental experience, respectively. In other words, the debate is about the nature–nurture problem. As has become obvious in the visual system, the issue is still controversial despite accumulation of a considerable amount of data and appears far from being resolved (Weliky and Katz 1997; Catalano and Shatz 1998; Crowley and Katz 1999). The situation becomes even more complicated if one considers that synapse formation and formation of connectivity patterns in different brain structures (e.g., brain stem vs. cortex; small vs. large terminals, excitatory vs. inhibitory synapses) may be governed by different rules. Nevertheless, several key elements of synaptogenesis and synapse consolidation have become obvious [see Sanes et al. (2000) for an elaborate treatise on this]. They involve (1) the establishment of an initial set of connections, (2) interactions between synaptically connected neurons and modulation and refinement of the connections by epigenetic factors (e.g., correlated neuronal activity), (3) expression of molecules that enable the recognition between interacting neurons, and (4) an exchange of molecular information that biases the course of subsequent synapse differentiation and ultimately results in the consolidation of appropriate synaptic connections and in correctly functioning connectivity patterns. Genetic loss-of-function experiments, performed mainly in invertebrates, have identified a barrage of target recognition molecules between pre- and postsynaptic partners (for review see Jin 2002). It needs to be seen now which of them act together in the signaling pathways of auditory neurons.

3.2 The SOC, a Favorable System to Analyze Developing Inhibitory (and Excitatory) Connections

Data obtained in the SOC (and elsewhere in the auditory brain stem) have contributed relatively little to the elucidation of the aforementioned key elements of *excitatory* synapse maturation. However, the SOC has become a model system for the investigation of the development of *inhibitory* synapses. This was possible because of the favorable situation that the MNTB provides an anatomically distinct, inhibitory input to the LSO (see Figs. 3.1B, 3.2). This inhibitory input, which is mediated by glycine (Wenthold 1991), is organized highly topographically. It can be selectively analyzed because the efferent axons of MNTB projection neurons project over some distance before terminating on LSO neurons (in the adult rat brain, for instance, the distance between the lateral border of the MNTB and the medial border of the LSO equals approximately 500 μm). The situation is in pleasant contrast to most other brain regions in which the inhibitory input is provided by interneurons which, due to their intrinsic position, are relatively inaccessible for anatomical identification, selective stimulation, and surgical manipulation. As illustrated in Figures 3.1B and 3.2, MNTB neurons receive their major input from globular bushy cells in the VCN on the contralateral side (Held 1893; Warr 1972; Tolbert et al. 1982; Friauf and Ostwald 1988). The axon terminals of these globular bushy cells form large, specialized synapses called calyces of Held (Held 1893) around the somata of MNTB neurons and enable rapid, high-fidelity transmission (Forsythe 1994; Borst and Sakmann 1998). The input mediated via these calyces is glutamatergic. In consequence, MNTB neurons are excited by stimulation of the contralateral ear, which in turn results in inhibition of LSO neurons. Aside from the inhibitory input from the MNTB, LSO neurons receive an excitatory, topographically organized input from spherical bushy cells located in the ipsilateral VCN (Figs. 3.1B, 3.2) and are consequently excited by stimulation of the ipsilateral ear (Cant and Casseday 1986; Cant 1991). The arrangement of excitatory and inhibitory inputs enables LSO neurons to compute interaural intensity differences and to encode information on the azimuthal location of a sound source (for review see Yin 2002). Maximal spike activity in the LSO is produced if a sound source is located in the ipsilateral field directly lateral to the ear. Conversely, a sound source located at the opposite side of the head will produce maximal inhibition in these LSO neurons. By means of the level of excitation, each LSO encodes the location of a sound source on the same side of the body. In Sections 3.3–3.4, the development of synaptic connectivity in the SOC will be discussed in chronological order, starting with changes that occur before hearing onset, and ending with the achievement of mature properties.

3.3 Developmental Changes Before Hearing Onset

Differentiation of the cytoarchitecture of SOC nuclei proceeds prenatally such that by E20, individual nuclei can be distinguished and delineated in the rat (Kandler and Friauf 1993). The period is associated with a tremendous increase in the volume of the nuclei. In gerbils at the day of birth, the SOC has reached only 10% of the adult size and it increases to about 70% until the time when onset of hearing occurs. Thereafter, increase slows down and 80% of the adult size is reached during the fourth postnatal week (Rübsamen et al. 1994). In line with these findings is the observation in rats that the cross-sectional area of the LSO increases about fourfold between P4 and about 1 month postnatal (Rietzel and Friauf 1998). Accompanying the growth of the SOC nuclei are changes in cellular properties that involve structural and functional remodeling. The kind of changes and the question of whether the remodeling is associated with activity-dependent processes and synaptic competition is discussed in the following section.

3.3.1 Changes in Axon Morphology

As mentioned in Section 2.4, the initial set of connections between the cochlear nuclear complex and the SOC nuclei is laid down in prenatal life. This set is formed with remarkable precision, as evidenced by the lack of aberrant projections to wrong, ectopic target nuclei (Kandler and Friauf 1993; Kil et al. 1995). At the cellular level, however, morphological modifications of the connections take place, many of them occurring postnatally. Modifications of the axonal morphology are characterized by an initial growth of terminal processes lasting into postnatal life. Only later are terminal processes lost which leads to more restricted axonal arbors and can be interpreted as a refinement of the connections.

In the neonatal rat, the axon terminals of the globular bushy cells in the contralateral MNTB, the calyces of Held, feature several thin processes that protrude from the immature calyx and extend radially over a distance of about 30 μm (Fig. 3.4; Kandler and Friauf 1993). Both the number and the length of these processes decline with age, and by P14, the earliest age analyzed after hearing onset, the morphology is adultlike. The data are consistent with those obtained in gerbils and cats (Morest 1968) and suggest that morphological maturation of this type of synaptic ending is accomplished without the necessity of acoustic input.

In the neonatal gerbil LSO, axonal arbors of individual MNTB neurons terminate in an ordered projection field, demonstrating that the topography in the principal zone of innervation is already present by birth (Sanes and Siverls 1991). However, the presence of growth cones at the axonal tips indicates that the arbors still sprout and increase in size, and indeed, the arbors do elaborate over the first 10–13 postnatal days and display a diffuse projection pattern along the tonotopic axis of the LSO, presumably forming inappropriate connections distant to the principal zone of innervation. After hearing onset, the specificity of axon arborization is refined and the arbors of individual MNTB neurons end

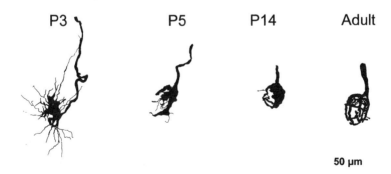

FIGURE 3.4. Morphological changes of calyces of Held in the developing MNTB of rats. The changes take place postnatally before hearing onset and are characterized by a loss of fine processes. In neonates, the processes are numerous and emanate from the immature calyx. The loss of the processes results in the focal synaptic ending which terminates on a single MNTB neuron in adults. [Modified from Figure 13 in Kandler and Friauf (1993) and reprinted by permission of Wiley-Liss.]

up occupying a smaller fraction of the LSO (see Section 3.3.2). These data are in some contrast to those obtained in the calyx synapse in the MNTB, because they indicate that the maturation of presynaptic terminals may well be under the influence of acoustic experience.

3.3.2 Changes in Dendritic Morphology

Like the axonal arbors, the dendritic trees of SOC neurons display a growth period during the time before hearing onset (the first 2 postnatal weeks in rats). In newborn rats, LSO neurons have complex dendritic trees with many dendritic branches and endpoints (Rietzel and Friauf 1998). The dendritic trees become broader along the tonotopic axis between P4 and P12 and the territory occupied by the dendrites of a single neuron becomes slightly larger (Fig. 3.5). The situation changes abruptly and drastically with hearing onset after which the number of dendritic endpoints is reduced and the relative dendritic width (i.e., the ratio of the dendritic width and the length of the S-shaped transverse axis of the LSO) becomes significantly smaller (Fig. 3.5).

In gerbils, changes in the dendritic morphology before hearing onset have not been analyzed as to the present. Data are available in the ferret SOC, where

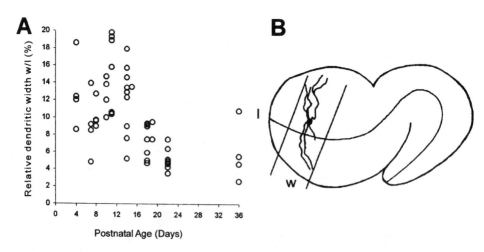

FIGURE 3.5. Summary diagram illustrating the quantitative changes that take place in the extent of the dendritic arbors of developing rat LSO principal neurons. The figure provides information about the relative amount of space occupied by the dendritic tree of individual cells within the coronal plane. The size of the LSO increases with age, the relative dendritic width (in %) is plotted vs. age in **A**; it was calculated as the ratio (w/l) of the absolute dendritic width (w) and the length of the curved transverse axis of the LSO (l). The procedure is illustrated in (**B**). The data demonstrate an age-dependent refinement of dendritic morphology and a reduction of the extent of dendritic width between P12 and P22, probably resulting in sharper topography. [Taken and adapted, with permission, from Rietzel and Friauf (1998).]

both MSO and LSO principal neurons feature the typical dendritic bipolar morphology seen in the adult already at birth. Prior to hearing onset, there is a transient appearance of dendritic appendages which disappear until the end of the first postnatal month (Henkel and Brunso-Bechtold 1990, 1991). These data show that the basic morphological phenotype and the dendritic territories of SOC neurons are generated very early during development, yet that morphological modifications take place thereafter. In contrast to the results in rats and gerbils, the disappearance of exuberant dendritic branches in ferrets does not coincide with the onset of acoustically evoked activity.

3.3.3 Changes in Functional Characteristics

The synapses in the SOC become functional shortly after the afferent axons have invaded the target zone. In rats, glutamatergic and glycinergic synaptic responses can be evoked via electrical stimulation in fetal life, that is at a time well before acoustic responses can be elicited (Kandler and Friauf 1995b). The fact that these responses can be evoked at E18, when axon collaterals have just emerged from the parent axons originating in the cochlear nuclear complex, is indicative for a short delay between axon ingrowth and formation of functional synapses. The data also imply that if neurotransmission at these synapses occurs in vivo, it must be activated through spontaneous activity present in the system. As yet, spontaneous activity in the auditory brain stem before hearing onset has not been thoroughly analyzed. The only report on SOC neurons comes from the chick NL, in which spontaneous firing occurs rhythmically in synchronous bursts at periodic intervals between E14 and E19 (Lippe 1994).[3] Between E14–E15 and E18, the mean interburst interval decreases from 4.9 s to 2.1 s. Shortly before hatching, synchronous bursting becomes replaced by unpatterned, maintained firing activity which is also present in 1–2-week-old chickens (Fig. 3.6). The spontaneous spike activity is abolished if the sodium channel blocker tetrodotoxin is applied to the oval window, implying that its origin is in the inner ear. Studies in other auditory brain stem areas of prehearing animals have also identified patterned spontaneous action potential activity. Recordings have been made in the cochlear nucleus of gerbils (Woolf and Ryan 1985), cats (Walsh and McGee 1988), and tammar wallabies (*Macropus eugenii*; Gummer and Mark 1994) as well as in the gerbil inferior colliculus (Kotak and Sanes 1995). It needs to be seen whether the temporal pattern of the spontaneous activity provides some cues that contribute to the dynamic changes seen in other parameters at that time.

Very fundamental functional modifications occur in the synaptic response properties of SOC neurons in prehearing animals. Electrophysiological results

---

3. In a strict sense, Lippe (1994) performed his study in animals shortly *after* the onset of hearing which takes place between E12 and E14 in chicks (Saunders et al. 1973). However, as auditory thresholds are still very high and normally occurring levels of airborne sound are unlikely to elicit spike activity in ovo, the reference is cited here and not below.

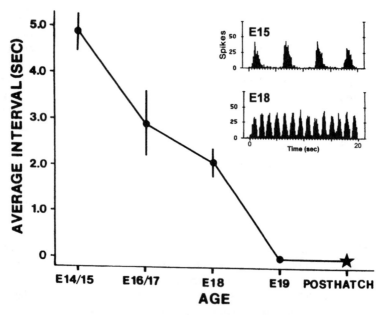

FIGURE 3.6. Developmental changes in the spontaneous activity pattern in third-order and second-order nuclei of the chick auditory brain stem (NL and NM, respectively). Spontaneous action potential activity is present prenatally, and the average interval between bursts of spiking activity decreases with age, demonstrating that the rate of rhythmic bursting increases during development. *Insets* show two examples of spontaneous activity recorded at E15 and E18. [Reproduced with permission and modified from Figures 4 and 5 in Lippe (1994).]

from in vitro experiments in brain stem slices demonstrated that the nature of synaptic transmission in neonates (rats and gerbils) is quite different from that in the adult. This is particularly true for the inhibitory input. As a consequence, neuronal activity in the immature SOC is likely to have strikingly different effects than in the adult.

An unexpected result was found when the characteristics of the glycinergic projection from the MNTB to the LSO were analyzed in acute slice preparations of the rat SOC. During fetal and early postnatal life, activation of glycine receptors (via electrical stimulation of the MNTB or ligand application) does not evoke inhibitory, hyperpolarizing responses, but rather depolarizing responses in the majority of LSO neurons (Kandler and Friauf 1995b; Ehrlich et al. 1999). At perinatal ages, the amplitude of these depolarizing responses is sufficient to surpass spike threshold and to elicit action potentials. Therefore, the glycine-induced responses can be considered excitatory rather than resulting in shunting inhibition (Ehrlich et al. 1999). Throughout development, glycine receptors on LSO neurons appear to be Cl^- channels, because a transiently high permeability to other ions, such as HCO_3^-, is not observed. An active Cl^-

regulation mechanism, which generates and maintains a high intracellular Cl^- concentration and, in consequence, a relatively positive Cl^- equilibrium potential (Lu et al. 1999; Rivera et al. 1999), is responsible for the depolarizing responses evoked by glycine (Ehrlich et al. 1999; Kakazu et al. 1999). The results imply that there is a perinatal stage during which both components of the convergent input system from the two ears excite their target neurons in the LSO. It is conceivable that the action of the two components is synergistic: they may achieve an effect that each component alone cannot achieve. Moreover, based on the observation that the intracellular Ca^{2+} concentration in LSO neurons of neonatal rats and mice is increased following glycine or γ-aminobutyric acid (GABA) application (Kullmann et al. 2002), like in other systems (e.g., hippocampus: Cherubini et al. 1990; Leinekugel et al. 1997; hypothalamus: Obrietan and van den Pol 1995; spinal cord: Reichling et al. 1994; inferior colliculus: Frech et al. 1999), one can hypothesize that the "inhibitory" inputs to the LSO affect similar, Ca^{2+}-dependent signal transduction cascades during the depolarizing phase as those known to be activated by glutamatergic synapses (Johnston et al. 1992; Malenka and Nicoll 1993). It is possible that glycinergic transmission, by means of the intracellular Ca^{2+} increase, activates a biochemical cascade that is involved in the maturation of the LSO circuit at the cellular level. The growth of dendritic arbors before hearing onset (see Section 3.3.2) or a strengthening of synaptic connections could be regulated by this cascade.

The dynamic properties of both the excitatory and the inhibitory input system to LSO neurons change considerably in early postnatal development prior to hearing onset (Sanes 1993). For example, the duration of postsynaptic potentials is much longer (up to 100-fold) in newborn gerbils compared to P15 animals (Fig. 3.7). Likewise, the rising slope of the potentials is shallow by birth and the steepness increases significantly with age. These properties enable the summation of individual postsynaptic potentials in response to repetitive stimuli in newborns and may result in long-lasting effects that last far beyond the stimulus period.[4]

An interesting aspect about the nature of the factors that can lead to functional changes during SOC development is provided by the finding that the amount of excitatory and inhibitory activity in the SOC can be influenced by neuromod-

---

4. The paper by Sanes (1993) also reports that the reversal potential of MNTB-evoked inhibitory postsynaptic potentials at P2–P6 (-69 mV to -73 mV) is nearly identical to that seen at P17–P23. This result is in contrast to the findings in the rat LSO (Kandler and Friauf 1995b; Ehrlich et al. 1999; Balakrishnan et al. 2003), where relatively positive glycine reversal potentials are seen throughout the nucleus in neonates (~ -47 mV at P1–P4), becoming more negative with age (~ -82 mV at P9–P11). Recently, however, Kotak et al. (1998) noted that a progressive negative shift of the reversal potential of MNTB-evoked inhibitory postsynaptic responses occurs in the *medial* limb of the gerbil LSO (~ -35 mV at P3–P5; ~ -60 mV at P12–P16). They performed their study with whole-cell recordings, thereby unavoidably dialyzing the neurons and changing their ionic composition, including the internal Cl^- concentration. Further patch clamp recordings in the perforated patch configuration, and not in the whole-cell modus, should help to shed more light on this issue.

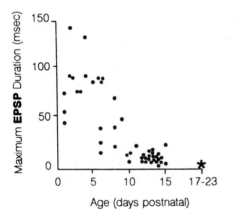
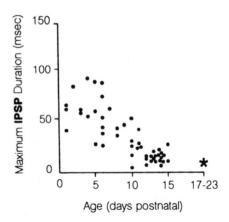

FIGURE 3.7. Early postnatal changes in the duration of excitatory and inhibitory post-synaptic potentials (EPSP and IPSP, respectively) in the gerbil LSO. There is a dramatic decrease between P1 and P12 and a moderate decrease thereafter. Asterisk illustrates mean value from animals at P17–P23. [Reproduced with permission and modified from Figure 2B in Sanes (1993).]

ulators prior to hearing. Such influences were shown for serotonin which can act via multiple receptors, thus having some actions that are similar to conventional neurotransmitters and some actions that are more like those of neuro-modulators (Tecott and Julius 1999). Serotonin induces prolonged inhibitory activity in LSO neurons of P6–P8 gerbils and depresses excitatory postsynaptic responses, thus boosting the inhibitory strength (Fitzgerald and Sanes 1999). Because the effects were rarely observed between P8 and P13, it appears that neuromodulator activity is limited to restricted time periods. This conclusion is corroborated by the fact that another neuromodulator, namely the neuropeptide somatostatin, shows a transient expression in the developing SOC, as do its receptors (Kungel and Friauf 1995; Thoss et al. 1996). During that time, so-matostatin influences the dendritic development in the LSO (Kungel et al. 1997).

3.3.4 Changes in Molecular Composition

Because of the great number of morphological and physiological changes oc-curring in the SOC during the period of synaptogenesis, it is a matter of course that these changes go along with prominent changes in the molecular repertoire. Indeed, such changes occur in most if not all SOC neurons during the prehearing phase, and the previous section (3.3.3) has already elucidated the dynamics in the distribution and function of neuropeptides. In general, the changes comprise the appearance as well as the disappearance of proteins and thus provide evi-dence that progressive and regressive events participate in SOC development. For example, the calcium-binding proteins calbindin, calretinin, and parvalbumin are up- and down-regulated differentially in specific SOC nuclei of rats (Friauf

1993, 1994; Lohmann and Friauf 1996), and their transient presence in some nuclei is likely to be of age-related functional significance, such as to buffer intracellular Ca^{2+} at the time when glutamatergic and glycinergic inputs to the LSO are excitatory. The neurotrophin receptor TrkB, known to mediate trophic actions between neurons and to support neuron survival via binding of the brain-derived neurotrophic factor BDNF, is virtually absent in the gerbil SOC at the day of birth but becomes strongly expressed by P7, suggesting a role in the postnatal maintenance of function (Hafidi et al. 1996b). Glia-associated molecules become present during the first postnatal life in the gerbil LSO; their developmental time course is in line with the idea that glia cells influence the developmental refinement of dendritic and axonal arbors in the LSO (Hafidi et al. 1996a).

Modifications in the set of molecules that participate in glycinergic neurotransmission take place during the first two postnatal weeks in the rat SOC: the glycine receptor molecules change from an immature isoform present in neonates to an adult isoform, which contains α_1 subunits (Sato et al. 1992, 1995; Friauf et al. 1997). The functional consequence of this change is a longer open time of the chloride channels and, thus, a longer duration of postsynaptic potentials in the newborns, as seen in the developing spinal cord (Takahashi et al. 1992) and the LSO (Sanes 1993). Not only do the open kinetics of glycine receptor molecules change, but the machinery for the reuptake of glycine from the synaptic cleft, which is mediated by the glycine transporter GLYT2, is installed during perinatal life (Friauf et al. 1999). This is also likely to contribute to a more efficient termination of glycine activity and to shorter postsynaptic responses. Another striking change in a neurochemical phenotype related to inhibitory neurotransmission is a fundamental transition from $GABA_A$ receptor-associated to glycine receptor-associated molecules between P4 and P14 in the medial limb of the gerbil LSO (Kotak et al. 1998; Korada and Schwartz 1999). As the synaptic responses also change from predominantly GABAergic to glycinergic with age, the observations suggest that GABAergic transmission may be important in early maturation of inhibitory contacts in the LSO. As no transition from GABAergic to glycinergic was seen in the lateral LSO, differential effects are to be expected within this nucleus.

3.4 Developmental Changes After Hearing Onset

The onset of hearing is associated with profound changes in the neuronal activity in the auditory system. Whereas spontaneous activity prior to hearing onset is quite low and results in low activity levels (Section 3.3.3), acoustic input will drive the system to much higher discharge levels. Therefore, hearing onset, although it may not occur suddenly, must be considered a rather drastic event with far-reaching consequences. In rats, the time of hearing onset has been determined in several ways, and P12–P14 was identified as the time when reliable responses to naturally occurring sound levels can be evoked (Tokimoto et al. 1977; Uziel et al. 1981; Blatchley et al. 1987; Puel and Uziel 1987; Kelly

1992; Geal-Dor et al. 1993). Before P12, an epidermal flap still covers the external meatus (Silverman and Clopton 1977), the middle ear is still filled with mesenchyme and fluid (Saunders et al. 1983), and the structures in the inner ear are still immature (Lenoir et al. 1980, 1987; Roth and Bruns 1992a,b).

3.4.1 Changes in Axon Morphology

Until P10–P13, which is before hearing onset, there is an increase in the extent of terminal axon arbors of MNTB neurons in the gerbil LSO. Thereafter, however, the territory occupied by a single axon arbor becomes restricted and by the end of the third postnatal week, the diffuse projections to inappropriate zones in the LSO have been eliminated and exuberant boutons (presumably representing synaptic contacts) have disappeared (Sanes and Siverls 1991). Although direct evidence is missing, the number of synapses of an individual axon that innervates its target cells is likely to be reduced by the pruning of axonal branches. The data show that remodeling of inhibitory arborizations occurs during normal development like it does in excitatory projections in other systems (e.g., auditory nerve: Jackson and Parks 1982; neuromuscular junction: Balice-Gordon et al. 1993; retino-thalamic projection: Shatz 1996).

3.4.2 Changes in Dendritic Morphology

Whereas growth of dendritic trees occurs in the prehearing period, the opposite happens in hearing animals before maturity is achieved. In the rat and gerbil SOC, the complexity of dendritic trees decreases during the third postnatal week and the territory occupied by the dendritic arbor of a single neuron becomes smaller and spatially constrained. The changes are observed in the LSO (Sanes et al. 1992b; Rietzel and Friauf 1998) and the MSO (Rogowski and Feng 1981). In the rat LSO, the changes appear to be completed at about P22, when the adult morphological properties are obtained (Rietzel and Friauf 1998). The result of the smaller territory is a topographic refinement of the morphology along the tonotopic axis which is in line with the changes seen in the presynaptic axon arbors of MNTB neurons and in accordance with a functional refinement, namely the sharpening of frequency tuning curves. In ferrets, terminal tufts of dendritic tendrils form after the onset of hearing, that is after the first postnatal month (Henkel and Brunso-Bechtold 1991). The function of these tufts is not known, yet they are likely to receive synaptic input.

3.4.3 Changes in Functional Characteristics

Only a few in vitro studies have been performed from hearing animals and therefore changes of functional characteristics at the cellular level are not well documented. Intrinsic membrane properties and synaptic integration of SOC neurons continue to develop after the onset of hearing (Sanes 1993; Kandler and Friauf 1995a), but whether acoustic experience influences these processes has not been investigated. Fos immunoreactivity indicates that the tonotopic organization within the rat SOC is adultlike shortly after hearing onset, indicating that auditory experience, as expected, does not contribute to the maturation of tonotopy/cochleotopy (Friauf 1992).

3.4.4 Changes in Molecular Composition

Several types of molecules change after hearing onset in the SOC. An example is provided by the extracellular matrix molecule Cat-301. Although it can first be immunohistochemically detected at P7–P11 in the gerbil SOC, age-related changes occur predominantly after hearing onset (Lurie et al. 1997). As seen with other molecules of the chondroitin sulfate type (Friauf 2000), prominent staining develops around the somata of SOC neurons. By P21, Cat-301 immunoreactivity has reached an adultlike pattern and intensity in the LSO, the MSO, and the VNTB. Cat-301 development appears to parallel many aspects of physiological and morphological maturation. Immunohistochemical staining for the neurotrophin receptor TrkC reaches a maximum level during the third postnatal week and declines until P30 (Hafidi et al. 1996b). Calbindin-D_{28k} immunoreactivity in the somata of rat LSO neurons declines rapidly during the third postnatal week, whereas it increases in MNTB neurons (Friauf 1993).

4. Late Development: Achievement of Mature Properties

Maturity of the auditory brain stem stations is usually determined in terms of functional properties, and auditory brain stem responses are often analyzed to determine when adult values of auditory sensitivity are reached. In gerbils, there is a "refinement period" from P15–P30, during which thresholds of auditory brain stem responses (waves III and IV are thought to be generated in the SOC; Melcher and Kiang 1996) improve steadily until they reach adultlike values by P30 (McFadden et al. 1996). The improvement is accompanied by the emergence of V-shaped frequency-threshold curves. Other functional properties, such as the sensitivity of LSO neurons to interaural intensity differences, also mature toward the fourth and fifth postnatal week (Smith and Kraus 1987; Sanes and Rubel 1988). Auditory brain stem responses in rats reach adultlike amplitudes by the end of the third postnatal week, that is about 10 days after hearing onset (Tokimoto et al. 1977). In ferrets, the thresholds and the latencies of auditory brain stem responses become adultlike by P40, which is also about 10 days after hearing onset (Moore 1991). Nothing is known about the achievement of adult properties in individual SOC nuclei of humans. Human hearing begins approximately at the 25th week of gestation (thus, infants at birth have 12–14 weeks of fetal auditory experience) and achieves its full expression postnatally in the first 5–10 years of life, including the latency of brain stem auditory evoked potentials (Eggermont 1992; Ruben 1992; Jiang and Tierney 1995). In the altricial species, whose hearing onset is postnatal, it is most likely that the morphological changes underlying the achievement of mature functional properties in the SOC and the other brain stem nuclei take place in the peripheral auditory system (middle and inner ear) as well as the brain stem nuclei themselves, such as through increases in myelination. There is, however, also a centrifugal influence from the central structures to the periphery because the development of normal cochlear sensitivity is impaired after lesioning the olivocochlear system in neonates: in cats, this results in a decreased sharpness of tuning of auditory nerve fibers in the adult (Walsh et al. 1998). The impaired development of the active mechanical processes in the cochlea, which involve the proper function of outer hair cells, is presumably the cause of the abnormal tuning curves.

5. Plasticity in the Superior Olivary Complex

As defined in the beginning of this review, plasticity implies alterations in structure, connections, or function that occur in response to induced imbalances in neural activity. These imbalances comprise a loss or an excess of activity in response to deprivation or overstimulation, and the resulting alterations may occur both during development and in adulthood. Because of their relative easiness, the majority of experiments performed in this area involved artificially induced changes of input activity, such as manipulations of the cochlea to affect the rate of peripheral input activity or acoustical or electrical stimulation paradigms.

Without much doubt, the classical experiment analyzing the role of peripheral input signals on the development of the auditory brain stem was performed more than 50 years ago by Rita Levi-Montalcini (1949). She unilaterally ablated the otocyst (the primordium of the inner ear) in chick embryos and found a decrease in the number of second-order auditory neurons, thus providing first evidence that afferent activity is a relevant signal in the regulation of cell survival. Her pioneer experiments initiated a series of further investigations that contributed significantly to our current understanding of the role that input from presynaptic sources exerts on the normal development of the auditory brain stem.

5.1 Cellular Plasticity Before Hearing Onset

One of the most dramatic effects that cochlear ablation has on the development of the auditory brain stem becomes evident in a drastic anatomical reorganization of the ascending input to the SOC of prehearing animals (Kitzes et al. 1995; Russell and Moore 1995). Following unilateral cochlea removal, the LSO on the ipsilateral side, the dendrites of both MSOs on the homolateral side, as well as the MNTB on the contralateral side are deafferented and deprived of excitatory input. When the ablation is performed in neonatal gerbils (P2–P5), a number of novel, aberrant, and permanent projections originate from the VCN of the untreated side and terminate in these deafferented areas (Fig. 3.8). Interestingly, these novel projections terminate in topographically appropriate

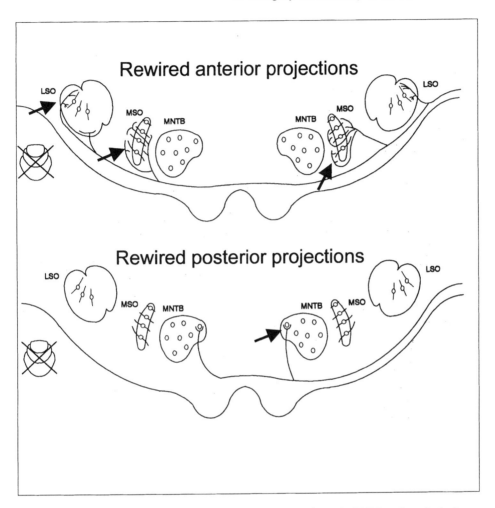

FIGURE 3.8. Schematic representation of the projections from the VCN to the principal SOC nuclei that occur after unilateral removal of the cochlea in neonatal gerbils. Aberrant projections terminate in the nuclei that were functionally deafferented. From the viewpoint of the unmanipulated VCN, these are the *contralateral* LSO, the *heterolateral* sides of both MSOs, and the *ipsilateral* MNTB (marked by *arrows*). [Modified from Figure 11 in Russell and Moore (1995) and reprinted by permission of Wiley-Liss. Basically identical results are described in Kitzes et al. (1995).]

regions in the deafferented targets and appear to establish the same type of axon terminals that they do in the normal counterparts (e.g., calyx terminals are induced in the ipsilateral MNTB). Interestingly and of so far unknown implication for sound localization, the abnormal projection to the ipsilateral MNTB becomes functional (Kitzes et al. 1995). As cochlea ablation at P10 (hearing onset is at about P12 in gerbils; Finck et al. 1972; McFadden et al. 1996) or later does not produce gross changes in connectivity (Russell and Moore 1995), the predisposition for such drastic plastic changes appears to be lost before hearing onset.

Aside from the effects on axonal connectivity, cochlear ablation before hearing onset leads to changes in the number of SOC neurons and in their morphology. After unilateral cochlea removal in ferrets at P5, the number of LSO neurons on the lesioned side (which normally receive excitatory input from there; Fig. 3.2) decreases by more than 30% (Moore 1992b). In contrast, the number of LSO neurons on the contralateral side (which normally receive inhibitory input from the ablated side) is even greater than in normal controls and the somata become slightly larger. These findings confirm the trophic effects of excitatory input on SOC integrity and also demonstrate the inability of intact inhibitory neurotransmission to support cell survival. The data even suggest that a lack of inhibition leads to hypertrophic effects, possibly due to an increased activation level, and that a balance between excitation and inhibition is required to achieve normal cell morphology. As Moore did his morphometric analysis in adult animals (>3 months), it cannot be determined whether the trophic excitatory effects are important before and/or after hearing onset. It is also unclear whether the effects in the contralateral LSO would be different if the analysis were done at a time when the inhibitory input is most likely still depolarizing because of the relatively positive Cl^- equilibrium potential seen in early development (see Section 3.3.3).

The influence of inhibition on dendrite development in the LSO becomes apparent from work done in Dan Sanes's laboratory. Pharmacological blockade of glycine receptors (Sanes and Chokshi 1992) or unilateral cochlea removal (Sanes et al. 1992a) in newborn gerbils both result in a hypertrophic response as evidenced by the lack of dendritic pruning which normally occurs during the third postnatal week. The functionally deafferented LSO neurons display dendrites with a significantly greater number of branches and their arbors are more widespread along the frequency axis. Again, although deprivation started before hearing onset, the morphometric analysis was done at a time when the animals were able to hear (at P21), thus making it impossible to distinguish between prehearing and posthearing effects or between the depolarizing and hyperpolarizing periods of inhibitory transmission. However, Sanes and coworkers also analyzed the influence of inhibitory afferents on dendrite development in the gerbil LSO prior to hearing onset (Aponte et al. 1996); cochlear ablation was done at P7 and dendrite morphology was analyzed 1–6 days later). LSO neurons located contralateral to the ablation side (which were deprived from their inhibitory input) had hypertrophied dendritic trees, with the effects occurring with a short delay of 2–3 days. Ipsilaterally located LSO neurons (which were deprived from their excitatory input) developed smaller somata, yet their dendrites neither shrank nor expanded. The data indicate that spontaneous inhibitory activity modulates the growth of developing SOC neurons and contributes to the pruning of postsynaptic dendritic branches. The same conclusion can be drawn from results obtained in organotypic slice cultures of the gerbil SOC, where blockade of (spontaneous) glycine neurotransmission with strychnine leads to an increase in dendritic branching and length in LSO neurons (Sanes and Hafidi 1996).

Neurotransmission does not only influence the development of *dendritic* morphology, it also exerts some action on *axon* development, as evidenced by the finding that axon arbors of MNTB neurons remain expanded in the LSO when deafferented by contralateral cochlear ablation, thus mirroring the effects seen on their postsynaptic targets, the dendrites of LSO neurons (Sanes and Takács 1993). In these experiments, however, the MNTB neurons were deprived of their excitatory input and it is unclear whether the decreased activity in this input or a presumably decreased release of glycine from the MNTB axon terminals caused the trophic effects in these arbors. If the first was the case, then this would be a unique example where a lack of excitation has a trophic effect in the developing SOC.

Concerning functional effects of cochlear ablation on SOC neurons prior to hearing, the synaptic physiology changes considerably in response to functional denervation of the glycinergic afferents to the LSO. Glycine deprivation of LSO neurons by contralateral removal of the cochlea or by strychnine rearing results in the up-regulation of functional NMDA receptors and in enhanced excitatory postsynaptic potentials from the unmanipulated, ipsilateral afferent fibers (Kotak and Sanes 1996). Thus, decreased inhibitory activity appears to be associated with a long-term strengthening of the excitatory input to the LSO. This is another example of the consequences of imbalanced activity on the development of SOC neurons. When the deprived excitatory input (the input to the *ipsilateral* LSO in which the spontaneous glutamatergic transmission was presumably reduced) was stimulated electrically in such experiments (Kotak and Sanes 1997), fewer LSO neurons were seen to respond with EPSPs (60% vs. 91% in controls) and the excitatory synaptic strength was reduced throughout the 6-day period analyzed (<7 mV vs. >11 mV in controls). In contrast, inhibitory responses became more prominent in the ipsilateral, functionally denervated input (60% vs. 6% in controls).

The conclusion from all the results evaluated in this section is that both inhibitory activity and excitatory activity are involved in generating the functional connectivity in the developing SOC. As the reported effects were all found in prehearing animals, it is the spontaneous activity, even though it may be of relatively low rate, that is necessary and sufficient to produce the normal morphological and functional cellular properties seen before hearing onset. The effects of inhibitory and excitatory activity are converse and probably competitive, and a balance of both types is a prerequisite for normal development.

Further support that activity is causally involved in the normal development of SOC structure before hearing onset comes from in vitro experiments done in our lab on organotypic slice cultures (Ehrlich et al. 1998; Lohmann et al. 1998; Löhrke et al. 1998). The slice culture preparation is a coronal section through the brain stem of neonatal rats (P3–P5) and contains the SOC nuclei plus part of the afferent projections from the cochlear nuclear complex. It is deprived from peripheral input originating in the inner ear and represents a system comparable to binaural deafferentation in vivo. When the slice cultures are incubated in standard culture medium for up to 2 weeks, the LSO neurons develop an abnormal appearance. Organotypic features develop only when voltage-operated Ca^{2+} channels of the L-type are activated, for example, by K^+-induced depolarization (Lohmann et al. 1998). The K^+-induced effect is optimal when 25 mM KCl is applied, whereas concentrations of 10 mM or 50 mM do not produce organotypic features. These findings suggest that an optimal amount of input activity is required for the viability of the LSO neurons and point to an increased intracellular Ca^{2+} concentration $[Ca^{2+}]_i$ as a downstream factor that follows excitation. Apparently, $[Ca^{2+}]_i$ must stay within a critical range to support neuron viability (cf. Hegarty et al. 1997). An adequately balanced $[Ca^{2+}]_i$ within this range seems to be of general importance for neuron survival throughout the central nervous system (Collins et al. 1991; Koike and Tanaka 1991; Ghosh et al. 1994; Franklin et al. 1995). As the regulation involves activity-dependent processes, action potential activity acts as the initial signal which triggers a trophic, survival-promoting cascade. All the available evidence implies that during early development, the trophic effect is attributed to spontaneous activity.

An intriguing question is whether activity-dependent, plasticity-related modification of synaptic physiology occurs in the developing SOC and whether this results in long-lasting functional changes. The question has been tackled in in vitro experiments employing electrical and pharmacological stimulation to affect the neuronal activity level. Low-frequency stimulation of the excitatory afferents to the gerbil LSO, or short tetanic activation with a stimulus pattern that resembled the spontaneous bursting activity seen in vivo, led to prolonged depolarizations which lasted for minutes (Kotak and Sanes 1995). Metabotropic glutamate receptors seem to mediate the effect as agonists of this receptor type elicited comparable long-lasting responses, and a prolonged Ca^{2+} influx seems to be involved as well. The data demonstrate a novel form of functional plasticity that differs from long-term potentiation phenomena seen elsewhere (Constantine-Paton and Cline 1998; Bortolotto et al. 1999). Another form of plasticity is seen in the inhibitory input to the LSO neurons, where low-frequency stimulation results in long-lasting depression in slices of P7–P12 gerbils (Kotak and Sanes 2000). This type of long-lasting synaptic plasticity involves an increase of the intracellular free Ca^{2+} concentration and is age dependent, as much smaller effects are observed at P17–P19. The period during which synaptic depression of inhibitory input to the LSO can be elicited coincides with the period of axonal refinement of the inhibitory input to the LSO, yet a causal relationship between the functional and morphological changes seen in the inhibitory connection needs to be demonstrated. Moreover, it will be interesting to see if there is an interdependence between the prolonged depolarization and the long-lasting depression seen in the excitatory input and inhibitory input, respectively.

5.2 Cellular Plasticity Shortly After Hearing Onset

Smaller somata of MNTB neurons are observed in mice that are deprived of airborne sound stimulation during development (Webster and Webster 1977). Short-term responses to deafferentation occur not only in soma size, but also at the level of dendritic morphology. The changes of dendritic morphology do not become manifested globally; rather they are highly input specific and occur at defined dendrites on a given neuron leaving other dendrites unaffected. Convincing evidence for this was quite elegantly provided in the NL (= MSO) of the chicken. As in the mammalian MSO, NL neurons have dendrites that are spatially segregated into two domains. Dorsally oriented dendrites receive input from the *ipsilateral* NM, whereas ventrally oriented dendrites are innervated by axons from the *contralateral* NM (Parks et al. 1983). When the input to the ventral dendrites was selectively removed in 9–12-day-old chickens (hearing onset in chickens is at E12–E14; Saunders et al. 1973), these dendrites atrophied, became shorter, and had lost about 60% of their length 16 days after surgery (Deitch and Rubel 1984). No such changes were observed in the dorsal, non-deafferented dendrites, showing that the effects are specific, subcellular, and restricted to the input-deprived postsynaptic structures. Remarkably, the atrophy began very rapidly after deafferentation, as evidenced by the finding that the dendrites were 10% shorter by 1 hour and 16% shorter by 2 hours following input removal.

Taken together, the above data demonstrate highly localized and input-specific alterations of cell morphology in response to changing presynaptic activity and indicate that plastic changes, in this case the restructuring of dendritic arbors in association with the resorption of postsynaptic surface area, can occur at a time scale of minutes to hours. Activity-dependent restructuring of dendritic processes was very recently analyzed in real time within the visual system. Using time-lapse confocal imaging of GFP-labeled retinal ganglion cells during the period of synaptogenesis, it was shown that structural changes are prominent and even occur in seconds (Wong et al. 2000). Moreover, dendritic motility is regulated by neurotransmission, actin-dependent, and controlled by Rho family GTPases. It will be interesting to see whether a similar mechanism participates in the dendritic plasticity observed in the various auditory brain stem nuclei.

There is another interesting observation on plastic changes in the dendritic morphology of SOC neurons that should be mentioned. In the adult chicken, the dendrites of NL neurons display a systematic length change along a spatial axis in that neurons in the caudolateral, low frequency region have much longer dendrites than neurons in the rostromedial, high-frequency region (Smith and Rubel 1979). The spatial gradient begins to emerge after E15 and continues to develop for some time after hatching (Smith 1981). To test whether its development is controlled by acoustic input, Parks (1981) removed an otocyst at E3 and found a drastic decrease of the dendritic length by E17. However, the spatial gradient was unaffected. After bilateral otocyst removal (Parks et al. 1987), the dendritic length decreased also, although to a lesser extent. Again, the spatial gradient was unaffected. The data are consistent with the idea that the spatial gradient of dendritic length is genetically determined in the NL and independent of input signals from the avian cochlear, yet that dendritic length is influenced by input activity.

Like cells in the chicken NL, the neurons of the rat MSO display a substantial, yet input-specific amount of dendritic retraction in response to auditory deprivation. Monaural and binaural occlusion experiments in rats in which the ear canal was closed with silicone plugs between P12 and P60 to dampen the acoustic input by at least 40 dB resulted in increased retraction of the deafferented dendrites of MSO bipolar neurons (Feng and Rogowski 1980). Monaural occlusion led to asymmetric dendritic trees, with the nondeprived side becoming dominant. Basically the same result was found in gerbils following unilateral destruction of the cochlea (Russell and Moore 1999). When the cochlea was removed at P18, a significant, rapid (<3 days), and sustained dendritic atrophy and a reduction in the number of distal dendritic branches was observed. The changes occurred in both MSOs but they were confined to the dendritic branches that received excitatory input from the ablated side, again demonstrating the input-specific effect in a given neuron.

5.3 Cellular Plasticity in the Adult SOC

The results obtained in the early lesioning studies before and shortly after the onset of hearing point to the crucial influence that ongoing synaptic activity exerts on the integrity of cell structure and function in the auditory brain stem. To address the question of whether this is true throughout life, the method of choice is again to deprive adult animals of input activity by either surgically or pharmacologically silencing the auditory nerve. Several studies have been performed with this approach and the conclusion that can be drawn is that plastic changes in response to deprivation do indeed occur throughout life in the auditory brain stem, yet the severity of these changes becomes less with age.

5.3.1 Plastic Changes in Cell Morphology

In general, there have been few studies on transsynaptic degeneration following mechanical ablation of the cochlea in the adult mammalian auditory system and even fewer following acoustic overstimulation (noise damage). The majority of these studies looked for effects in the cochlear nuclear complex, yet some evidence is also available from the SOC. Following cochlear damage in adult cats, Jean-Baptiste and Morest (1975) analyzed the terminals of globular bushy cells in the MNTB, that is, the calyces of Held. The subcellular changes they observed included reductions in the size and number of vesicles in the calyces. The postsynaptic elements, the MNTB principal cells, were also affected in that their soma size was reduced. The changes were apparent several weeks after cochlear ablation, confirming the results originally obtained by Powell and Erulkar (1962) in adult cats. Recently, the pattern of transsynaptic axonal degeneration was further investigated in adult chinchillas following mechanical lesions of the cochlea (Morest et al. 1997). After a survival time of 6 days, the terminal parts of the axons in the LSO, MSO, and MNTB were fragmented, consistent with a rapid transneuronal atrophy associated with decreased neural activity in the projection from the VCN. However, the relevant fibers in the trapezoid body did not degenerate even if the survival time was prolonged to 14 days. These data suggest that the atrophic effects are restricted to the synaptic ending and do not involve the whole axonal length. The paper also described degenerative effects in the dendrites of the MSO (no information on MNTB and LSO), providing further evidence that the integrity of the cellular morphology in the adult SOC is dependent on activity and on trophic support from the periphery. The same conclusion can be drawn from the results of a similar study in adult chinchillas involving acoustic trauma by a brief period of overstimulation (105 min, 108 dB SPL; Kim et al. 1997), but excitotoxic mechanisms due to prolonged hyperactivity could also have contributed to the axonal degeneration in this case. Concerning the time course of degeneration, the latter study found effects after 4 days, yet no effects if the survival time was 2 days or shorter, demonstrating that the latency period is in the range of several days.

As mentioned in Section 5.2, the soma size of developing SOC neurons is regulated by afferent input activity. That the same holds for adult SOC neurons was demonstrated by Pasic et al. (1994), who ablated the cochlea or pharmacologically blocked action potential activity unilaterally in 4–6-week-old gerbils to permanently or reversibly eliminate afferent electrical activity. To describe the dynamics in more detail, the authors followed the time course of changes at a shorter time scale (1–7 days) than the previous reports. Forty-eight hours after cochlear ablation, the soma size in MNTB principal cells contralateral to the ablation side had decreased by 13% (Fig. 3.9). When action potential activity was suppressed unilaterally in the auditory nerve with the sodium channel blocker tetrodotoxin for 48 hours, the change in cell size was about the same (17%). No statistically significant effect was seen 24 hours after ablation. The effects caused by tetrodotoxin treatment were reversible, as a significant difference was no longer evident 7 days after a 48-hour blockade of auditory nerve activity. The results imply that trophic substances of presently unknown nature are continually released by electrically active afferent axons and contribute to the maintenance of cell structure, even in the adult brain stem. What was surprising about the results obtained in the MNTB was that the decrease in cell size did not occur earlier than 48 hours following ablation and was thus 24 hours delayed compared to the effects seen in the VCN. This may be explained by an activity-independent release of trophic molecules which eventually ceases because VCN neurons decrease the synthesis and/or transport of these molecules.

5.3.2 Plastic Changes in Functional Properties

One interesting question that relates to functional changes induced by silencing the peripheral input in the adult is whether the strength of synaptic transmission is altered in auditory brain stem nuclei. This question was initially addressed

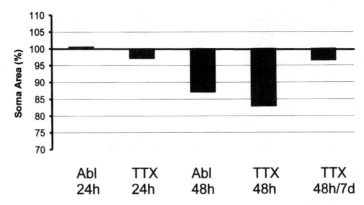

FIGURE 3.9. Decreases in cell size of MNTB principal cells in young adult gerbils (4–6 weeks) following cochlear ablation (Abl) or blockade of spike activity in the auditory nerve with tetrodotoxin (TTX). Soma area decreases significantly to values around 85% 2 days (48 hours) after manipulation. The effect is not obvious after a survival time of 1 day (24 hours). The change in cell body size is reversible as 7 days after a 48-hour treatment with tetrodotoxin (48 hours/7 days), normal values are obtained again. [Modified from Figure 3C in Pasic et al. (1994) and reprinted by permission of Wiley-Liss.]

in cats, and a hyperexcitability was seen in the LSO and the MSO following bilateral destruction of the cochleae, suggesting that the efficacy of synaptic transmission may indeed be changed (Gerken 1979). The question was recently followed up in guinea pigs in which unilateral removal of the middle ear ossicles (malleus and incus) or unilateral cochlear ablation was performed; subsequently, neurotransmitter release and uptake activity were determined in a quantitative radioactivity assay for glutamate, glycine, and GABA. Both types of lesion induced long-term regulatory changes in transmitter release from presynaptic terminals in the LSO, MSO, and MNTB, but the resulting spatial and temporal profiles were complex and dependent on the kind of lesion (Potashner et al. 1997; Suneja et al. 1998). The authors concluded that the induced changes are consistent with abnormally strengthened glutamatergic transmission in the LSO and MSO, providing evidence for adjustments of synaptic efficacy in the SOC after perturbation of acoustic input in the adult. In contrast, the glycinergic transmission in the SOC appeared to be barely affected, indicating that the enhanced excitatory transmission was not associated with deficient inhibitory transmission and disinhibition. Further studies are necessary to clarify the contribution of presynaptic and postsynaptic components to long-term changes in synaptic transmission.

5.3.3 Plastic Changes in Protein Composition

Plasticity-related changes in protein (or mRNA) composition in the adult auditory brain stem are only now beginning to be explored (cf. Illing et al. 2000;

Illing 2001). Ben Illing's group looked for evidence of synaptogenesis and synaptic remodeling in the SOC following cochlear ablation (1997) and used the growth-associated protein GAP-43 as an indicator for such remodeling (Benowitz and Routtenberg 1997). High levels of GAP-43 mRNA and protein are found during the first postnatal week in the SOC of normal rats (Horvath et al. 1997; Illing et al. 1999), whereas only low levels of protein persist into adulthood. When a unilateral cochlear ablation is performed in adult rats, a reexpression of GAP-43 occurs. Somata in the ipsilateral LSO display a maximum of GAP-43 expression 7 days after surgery ($\sim$ 370 strongly immunoreactive cells were counted) and a decline thereafter (at 8 weeks after surgery, $\sim$ 12 cells remained, being equivalent to preoperative levels). In the other SOC nuclei, such changes are not observed. The authors interpreted their findings that the axons of those LSO neurons which form the olivocochlear bundle were transected by the surgery, leading to the reexpression of GAP-43 in their cell bodies. An alternative explanation is that the effect was transsynaptic and mediated by afferent neurons. Whatever may have caused the increase in immunoreactivity, it is conceivable that LSO neurons in adults are temporarily capable of reacting to cochleotomy by sprouting and forming new synapses. Whether this is associated with a transient increase in regenerative potential is not known but it stands as an intriguing possibility.

6. Summary

This review has analyzed two aspects of dynamic processes taking place in the SOC. One aspect is developmentally related changes that lead to the maturation of the highly precise structure and function of the SOC circuitry; the other aspect is plastic changes caused by perturbations of input from the peripheral sensory organ, the cochlea.

6.1 Developmental Changes

Most developmentally related steps in the SOC occur during prenatal and early postnatal life when the auditory system is still devoid of acoustic input. Neurogenesis, cell migration, axon outgrowth from the neurons that will provide ascending input, and target finding are completed before hearing onset. Likewise, many aspects of synaptogenesis are accomplished before acoustic stimuli can drive the system. The aforementioned major events do not follow a strictly sequential pattern, yet they overlap considerably in time; for example, a substantial amount of axon outgrowth already takes place during the period of cell migration. A corollary to be inferred from this is that the developing system takes advantage of the minuteness of the growing brain stem and establishes internuclear connections when the distances are still very short. Furthermore, there is no spatial sequence (e.g., afferent projections from the cochlear nuclear complex to the SOC develop at the same time when efferent projections from the SOC to the inferior colliculus develop).

A topographic, cochleotopic order in connectivity becomes present as early as the connections form in the SOC. Nevertheless, at the cellular level the connectivity is refined during the period of synapse consolidation, as evidenced by morphological and functional restructuring both before and after hearing onset. The restructuring is activity dependent, and proper synaptic neurotransmission is required to shape the pre- and postsynaptic elements (i.e., the axonal and dendritic arbors) and to contribute to functional maturation. Early on, the synaptic neurotransmission is mediated through spontaneous activity. Both the excitatory and the inhibitory input systems in the SOC are refined by activity-dependent processes. Each input system can affect the other in a way that appears to involve some competition, and a balance between excitation and inhibition is required for normal development, but the mechanisms are not known in any detail. Excitatory input is trophic whereas inhibitory activity leads to a retraction of cellular processes. Paradoxically, there is an early period in synapse consolidation during which the inhibitory input is excitatory; an active regulation of the internal Cl^- concentration is the cause for this phenomenon. Although the implications of this transient excitatory effect in the inhibitory projection are unknown, it has been posited that it participates in a trophic process and leads to expansion of cell morphology and stabilization of the synapses.

A point of knowledge has been reached where construction of models of activity-dependent inhibitory changes in the developing SOC has come within reach (Elliott and Shadbolt 1998). However, with more surprising results to be expected from laboratory experiments, it is very likely that our present understanding as well as the currently available models will need substantial modification.

6.2 Cellular Plasticity

Various forms of plasticity-related changes are obvious in the SOC in response to perturbed input activity, both during development and adulthood. During intermediate development, when synapse consolidation takes place, the changes are most dramatic. In prehearing animals, perturbation of input during the period of synapse formation results in profound pathway reorganization, evident in the rewiring of input into the SOC nuclei. Those nuclei that are functionally denervated become innervated by aberrant projections from second-order nuclei. During the same time, a substantial amount of neuron loss occurs when the activity in the excitatory afferents is perturbed. The remaining input activity in the inhibitory projections is unable to support cell survival. It even appears that removal of inhibitory input results in higher cell number, further corroborating the suggestion that the amount of excitatory and inhibitory inputs must be balanced to develop normal SOC circuitry.

As seen in other sensory systems, the susceptibility for gross anatomical modifications is lost before hearing onset in the SOC, indicating that the bulk of connections become hard-wired before the auditory brain stem reaches maturity. Likewise, the influence of auditory nerve activity on cell number appears to disappear with age. In contrast, the cellular integrity remains dependent on input activity throughout life. This is evidenced by plastic changes in cellular morphology, such as shrinkage of somata and dendritic and axonal atrophy, which occur after cochlea ablation, acoustic deprivation, or overstimulation. The plastic changes occur on a time scale of hours to days and are restricted to the specific input region, meaning that in a given neuron, only those dendritic branches that are deprived of input become atrophic, while the others develop normally. Spatially limited metabolic changes within the cells are likely to induce the discrete plastic responses.

Some clinical implications become obvious from the results of animal research summarized in this review. The consequences of conductive hearing loss in childhood on the development of auditory brain stem structures are intriguing. Likewise, humans suffering from age-related hearing loss will have to contend with the problem that the maintenance of cell integrity in the adult auditory brain stem depends on acoustic input.

Since the seminal treatise on plasticity of the developing auditory system released more than 20 years ago (Ruben and Rapin 1980), considerable progress has been made in our understanding of the kind of cellular changes that take place following perturbations of input, their time course, and their manifestations. Nevertheless, the mechanisms of cellular plasticity are still poorly understood. Of particular importance (and of general interest) is the elucidation of the nature of interactions between excitatory and inhibitory afferents innervating the same population of target cells and of the signal transduction cascades that couple synaptic activation with molecular and structural changes. Undoubtedly, a golden age for cellular neurobiology and molecular neurobiology is ahead of us.

Acknowledgments. The author was supported by several grants from the Deutsche Forschungsgemeinschaft (Fr 772/1, Fr 772/5, SFB 269). During the preparation of this book chapter, support was provided by the Sonderforschungsbereich SFB 530 and the University of Kaiserslautern.

Abbreviations

CPO	caudal periolivary region
DMPO	dorsomedial periolivary nucleus
E	embryonic day
EPSP	excitatory postsynaptic potential
GAP-43	growth associated protein 43
Glo	globular bushy cells

IPSP	inhibitory postsynaptic potential
LSO	lateral superior olive
LNTB	lateral nucleus of the trapezoid body
LVPO	lateroventral paraolivary region
MNTB	medial nucleus of the trapezoid body
MSO	medial superior olive
MVPO	medioventral paraolivary region
NL	nucleus laminaris (avian nucleus homologous to mammalian MSO)
NLL	nuclei of the lateral lemniscus
NM	nucleus magnocellularis (avian nucleus homologous to mammalian VCN)
Oct	octopus cells
P	postnatal day
Sph	spherical bushy cells
TTX	tetrodotoxin
VCN	ventral cochlear nucleus
RPO	rostral periolivary region
SOC	superior olivary complex
SPL	sound pressure level
SPN	superior paraolivary nucleus
VNTB	ventral nucleus of the trapezoid body

References

Adams JC, Warr WB (1976) Origins of axons in the cat's acoustic striae determined by injection of horseradish peroxidase into severed tracts. J Comp Neurol 170:107–122.

Altman J, Bayer SA (1980) Development of the brain stem in the rat. III. Thymidine-radiographic study of the time of origin of neurons of the vestibular and auditory nuclei of the upper medulla. J Comp Neurol 194:877–904.

Altman J, Bayer SA (1981) Time of origin of neurons of the rat inferior colliculus and the relations between cytogenesis and tonotopic order in the auditory pathway. Exp Brain Res 42:411–423.

Aponte JE, Kotak VC, Sanes DH (1996) Decreased synaptic inhibition leads to dendritic hypertrophy prior to the onset of hearing. Audit Neurosci 2:235–240.

Atoji Y, Suzuki Y (1992) Chondroitin sulfate in the extracellular matrix of the medial and lateral superior olivary nuclei in the dog. Brain Res 585:287–290.

Atoji Y, Kitamura Y, Suzuki Y (1990) Chondroitin sulfate proteoglycan in the extracellular matrix of the canine superior olivary nuclei. Acta Anat 139:151–153.

Atoji Y, Yamamoto Y, Suzuki Y (1995) The presence of chondroitin sulfate A and C within axon terminals in the superior olivary nuclei of the adult dog. Neurosci Lett 189:39–42.

Balakrishisan V, Becker M, Löhrke S, Nothwang HG, Güresir E, Friauf E (2003) Expression and function of chloride transporters in the developing auditory brainstem. J Neurosci 23:4134–4145.

Balice-Gordon RJ, Chua CK, Nelson CC, Lichtman JW (1993) Gradual loss of synaptic cartels precedes axon withdrawal at developing neuromuscular junctions. Neuron 11: 801–815.

Benowitz LI, Routtenberg A (1997) GAP-43: an intrinsic determinant of neuronal development and plasticity. Trends Neurosci 20:84–91.

Blatchley BJ, Cooper WA, Coleman JR (1987) Development of auditory brainstem response to tone pip stimuli in the rat. Dev Brain Res 32:75–84.

Book KJ, Morest DK (1990) Migration of neuroblasts by perikaryal translocation: role of cellular elongation and axonal outgrowth in the acoustic nuclei of the chick embryo medulla. J Comp Neurol 297:55–76.

Borst JG, Sakmann B (1998) Calcium current during a single action potential in a large presynaptic terminal of the rat brainstem. J Physiol (Lond) 506:143–157.

Bortolotto ZA, Fitzjohn SM, Collingridge GL (1999) Roles of metabotropic glutamate receptors in LTP and LTD in the hippocampus. Curr Opin Neurobiol 9:299–304.

Boudreau JC, Tsuchitani C (1968) Binaural interaction in the cat superior olive S segment. J Neurophysiol 31:442–454.

Brand A, Behrend O, Marquardt T, Mcalpine D, Grothe B (2002) Precise inhibition is essential for microsecond interaural time difference coding. Nature 417:543–547.

Brugge JF (1983) Development of the lower brainstem auditory nuclei. In: Romand R (ed), Development of Auditory and Vestibular Systems. New York: Academic Press, pp. 89–120.

Brugge JF (1992) An overview of central auditory processing. In: Popper AN, Fay RR (eds), The Mammalian Auditory Pathway: Neurophysiology. New York: Springer-Verlag, pp. 1–33.

Cant NB (1991) Projections to the lateral and medial superior olivary nuclei from the spherical and globular bushy cells of the anteroventral cochlear nucleus. In: Altschuler RA, Bobbin RP, Clopton BM, Hoffman DW (eds), Neurobiology of Hearing: The Central Auditory System. New York: Raven Press, pp. 99–119.

Cant NB (1998) Structural development of the mammalian auditory pathways. In: Rubel EW, Popper AN, Fay RR (eds), Development of the Auditory System. New York: Springer-Verlag, pp. 315–413.

Cant NB, Casseday JH (1986) Projections from the anteroventral cochlear nucleus to the lateral and medial superior olivary nuclei. J Comp Neurol 247:457–476.

Catalano SM, Shatz CJ (1998) Activity-dependent cortical target selection by thalamic axons. Science 281:559–562.

Cherubini E, Rovira C, Gaiarsa JL, Corradetti R, Ben-Ari Y (1990) GABA mediated excitation in immature rat CA3 hippocampal neurons. Int J Dev Neurosci 8:481–490.

Cochran SL, Stone JS, Bermingham-McDonogh O, Akers SR, Lefcort F, Rubel EW (1999) Ontogenetic expression of trk neurotrophin receptors in the chick auditory system. J Comp Neurol 413:271–288.

Coleman JR (1990) Development of auditory system structures. In: Coleman JR (ed), Development of Sensory Systems in Mammals. New York: Wiley, pp. 205–247.

Collins F, Schmidt MF, Guthrie PB, Kater SB (1991) Sustained increase in intracellular calcium promotes neuronal survival. J Neurosci 11:2582–2587.

Constantine-Paton M, Cline HT (1998) LTP and activity-dependent synaptogenesis: the more alike they are, the more different they become. Curr Opin Neurobiol 8:139–148.

Cramer KS, Fraser SE, Rubel EW (2000a) Embryonic origins of auditory brain-stem nuclei in the chick hindbrain. Dev Biol 224:138–151.

Cramer KS, Rosenberger MH, Frost DM, Cochran SL, Pasquale EB, Rubel EW (2000b) Developmental regulation of EphA4 expression in the chick auditory brainstem. J Comp Neurol 426:270–278.

Crowley JC, Katz LC (1999) Development of ocular dominance columns in the absence of retinal input. Nat Neurosci 2:1125–1130.

Cutforth T, Harrison CJ (2002) Ephs and ephrins close ranks. Trends Neurosci 25:332–334.

Dan Y, Poo M (1994) Retrograde interactions during formation and elimination of neuromuscular synapses. Curr Opin Neurobiol 4:95–100.

Dan Y, Lo Y, Poo M (1995) Plasticity of developing neuromuscular synapses. Prog Brain Res 105:211–215.

Deitch JS, Rubel EW (1984) Afferent influences on brain stem auditory nuclei of the chicken: time course and specificity of dendritic atrophy following deafferentation. J Comp Neurol 229:66–79.

Eggermont JJ (1992) Development of auditory evoked potentials. Acta Otolaryngol (Stockh) 112:197–200.

Ehrlich I, Lohmann C, Ilic V, Friauf E (1998) Development of glycinergic transmission in organotypic cultures from auditory brain stem. NeuroReport 9:2785–2790.

Ehrlich I, Löhrke S, Friauf E (1999) Shift from depolarizing to hyperpolarizing glycine action in rat auditory neurons is due to age-dependent CI$^-$ regulation. J Physiol (Lond) 520:121–137.

Ellenberger C, Hanaway J, Netsky MG (1969) Embryogenesis of the inferior olivary nucleus in the rat: a radioautographic study and a re-evaluation of the rhombic lip. J Comp Neurol 137:71–88.

Elliott T, Shadbolt NR (1998) A model of activity-dependent anatomical inhibitory plasticity applied to the mammalian auditory system. Biol Cybern 78:455–464.

Faye-Lund H, Osen KK (1985) Anatomy of the inferior colliculus in rat. Anat Embryol 171:1–20.

Feng AS, Rogowski BA (1980) Effects of monaural and binaural occlusion on the morphology of neurons in the medial superior olivary nucleus of the rat. Brain Res 189:530–534.

Finck ACD, Schneck CD, Hartman AF (1972) Development of cochlear function in the neonate Mongolian gerbil (*Meriones unguiculatus*). J Comp Physiol Psychol 78:375–380.

Fitzgerald KK, Sanes DH (1999) Serotonergic modulation of synapses in the developing gerbil lateral superior olive. J Neurophysiol 81:2743–2752.

Flanagan JG, van der Haeghen P (1998) The ephrins and Eph receptors in neural development. Annu Rev Neurosci 21:309–345.

Forsythe ID (1994) Direct patch recording from identified presynaptic terminals mediating glutamatergic EPSCs in the rat CNS, in vitro. J Physiol (Lond) 479:381–387.

Franklin JL, Sanz-Rodriguez C, Juhasz A, Deckwerth TL, Johnson EM Jr (1995) Chronic depolarization prevents programmed death of sympathetic neurons *in vitro* but does not support growth: requirement for Ca^{2+} influx but not Trk activation. J Neurosci 15:643–664.

Frech MJ, Deitmer JW, Backus KH (1999) Intracellular chloride and calcium transients evoked by γ-aminobutyric acid and glycine in neurons of the rat inferior colliculus. J Neurobiol 40:386–396.

Friauf E (1992) Tonotopic order in the adult and developing auditory system of the rat as shown by c-fos immunocytochemistry. Eur J Neurosci 4:798–812.

Friauf E (1993) Transient appearance of calbindin-D_{28k}-positive neurons in the superior olivary complex of developing rats. J Comp Neurol 334:59–74.

Friauf E (1994) Distribution of calcium-binding protein Calbindin-D_{28k} in the auditory system of adult and developing rats. J Comp Neurol 349:193–211.

Friauf E (2000) Development of chondroitin sulfate proteoglycans in the central auditory system of rats correlates with acquisition of mature properties. Audiol Neurootol 5: 251–262.

Friauf E, Kandler K (1990) Auditory projections to the inferior colliculus of the rat are present by birth. Neurosci Lett 120:58–61.

Friauf E, Ostwald J (1988) Divergent projections of physiologically characterized rat ventral cochlear nucleus neurons as shown by intraaxonal injection of horseradish peroxidase. Exp Brain Res 73:263–284.

Friauf E, Hammerschmidt B, Kirsch J (1997) Development of adult-type inhibitory glycine receptors in the central auditory system of rats. J Comp Neurol 385:117–134.

Friauf E, Aragón C, Löhrke S, Westenfelder B, Zafra F (1999) Developmental expression of the glycine transporter GLYT2 in the auditory system of rats suggests involvement in synapse maturation. J Comp Neurol 412:17–37.

Geal-Dor M, Freeman S, Li G, Sohmer H (1993) Development of hearing in neonatal rats: air and bone conducted ABR thresholds. Hearing Res 69:236–242.

Gerken GM (1979) Central denervation hypersensitivity in the auditory system of the cat. J Acoust Soc Am 66:721–727.

Ghosh A, Carnahan J, Greenberg ME (1994) Requirement for BDNF in activity-dependent survival of cortical neurons. Science 263:1618–1623.

Grothe B (2000) The evolution of temporal processing in the medial superior olive, an auditory brainstem structure. Prog Neurobiol 61:581–610.

Grothe B, Park TJ (2000) Structure and function of the bat superior olivary complex. Microsc Res Techn 51:382–402.

Gummer AW, Mark RF (1994) Patterned neural activity in brain stem auditory areas of a prehearing mammal, the tammar wallaby (*Macropus eugenii*). NeuroReport 5:685–688.

Hafidi A, Katz JA, Sanes DH (1996a) Differential expression of MAG, MBP and L1 in the developing lateral superior olive. Brain Res 736:35–43.

Hafidi A, Moore T, Sanes DH (1996b) Regional distribution of neurotrophin receptors in the developing auditory brainstem. J Comp Neurol 367:454–464.

Harkmark W (1954) Cell migrations from the rhombic lip to the inferior olive, the nucleus raphe and the pons. A morphological and experimental investigation on chick embryos. J Comp Neurol 100:115–209.

Hegarty JL, Kay AR, Green SH (1997) Trophic support of cultured spiral ganglion neurons by depolarization exceeds and is additive with that by neurotrophins or cAMP and requires elevation of $[Ca^{2+}]_i$ within a set range. J Neurosci 17:1959–1970.

Held H (1893) Die zentrale Gehörleitung. Arch Anat Physiol, Anat Abt 17:201–248.

Helfert RH, Snead CR, Altschuler RA (1991) The Ascending Auditory Pathways. In: Altschuler RA, Bobbin RP, Clopton BM, Hoffman DW (eds), Neurobiology of Hearing: The Central Auditory System. New York: Raven Press, pp. 1–25.

Hendriks R, Morest DK, Kaczmarek LK (1999) Role in neuronal cell migration for high-threshold potassium currents in the chicken hindbrain. J Neurosci Res 15:805–814.

Henkel CK, Brunso-Bechtold JK (1990) Dendritic morphology and development in the ferret medial superior olivary nucleus. J Comp Neurol 294:377–388.

Henkel CK, Brunso-Bechtold JK (1991) Dendritic morphology and development in the ferret lateral superior olivary nucleus. J Comp Neurol 313:259–272.

Henkel CK, Brunso-Bechtold JK (1993) Laterality of superior olive projections to the inferior colliculus in adult and developing ferret. J Comp Neurol 331:458–468.

His W (1889) Die Neuroblasten und deren Entstehung im embryonalen Mark. Abh Ges Wissensch Math Phys Kl 15:311–372.

Honig MG, Hume RI (1989) DiI and DiO: versatile fluorescent dyes for neuronal labeling and pathway tracing. Trends Neurosci 12:333–341.

Horvath M, Forster CR, Illing RB (1997) Postnatal development of GAP-43 immunoreactivity in the auditory brainstem of the rat. J Comp Neurol 382:104–115.

Illing RB (2001) Activity-dependent plasticity in the adult auditory brainstem. Audiol Neuro-Otol 6:319–345.

Illing RB, Horvath M, Laszig R (1997) Plasticity of the auditory brainstem: effects of cochlear ablation on GAP-43 immunoreactivity in the rat. J Comp Neurol 382:116–138.

Illing RB, Cao QL, Forster CR, Laszig R (1999) Auditory brainstem: development and plasticity of GAP-43 mRNA expression in the rat. J Comp Neurol 412:353–372.

Illing RB, Kraus KS, Michler SA (2000) Plasticity of the superior olivary complex. Microsc Res Techn 51:364–381.

Irving R, Harrison JM (1967) The superior olivary complex and audition: a comparative study. J Comp Neurol 130:77–86.

Ivanova A, Yuasa S (1998) Neuronal migration and differentiation in the development of the mouse dorsal cochlear nucleus. Dev Neurosci 20:495–511.

Jackson H, Parks TN (1982) Functional synapse elimination in the developing avian cochlear nucleus with simultaneous reduction in cochlear nerve branching. J Neurosci 2:1736–1743.

Jackson H, Parks TN (1988) Induction of aberrant functional afferents to the chick cochlear nucleus. J Comp Neurol 271:106–114.

Jean-Baptiste M, Morest DK (1975) Transneuronal changes of synaptic endings and nuclear chromatin in the trapezoid body following cochlear ablations in cats. J Comp Neurol 162:111–134.

Jeffress LA (1948) A place theory of sound localization. J Comp Physiol Psychol 41:35–39.

Jiang ZD, Tierney TS (1995) Development of human peripheral hearing revealed by brainstem auditory evoked potentials. Acta Paediat 84:1216–1220.

Jin Y (2002) Synaptogenesis: insights from worm and fly. Curr Opin Neurobiol 12:71–79.

Johnston D, Williams S, Jaffe D, Gray R (1992) NMDA-receptor independent long-term potentiation. Annu Rev Physiol 54:489–505.

Joris PX, Smith PH, Yin TCT (1998) Coincidence detection in the auditory system: 50 years after Jeffress. Neuron 21:1235–1238.

Kakazu Y, Akaike N, Komiyama S, Nabekura J (1999) Regulation of intracellular chloride by cotransporters in developing lateral superior olive neurons. J Neurosci 19:2843–2851.

Kandler K, Friauf E (1993) Pre- and postnatal development of efferent connections of the cochlear nucleus in the rat. J Comp Neurol 328:161–184.

Kandler K, Friauf E (1995a) Development of electrical membrane properties and discharge characteristics of superior olivary complex neurons in fetal and postnatal rats. Eur J Neurosci 7:1773–1790.

Kandler K, Friauf E (1995b) Development of glycinergic and glutamatergic synaptic transmission in the auditory brainstem of perinatal rats. J Neurosci 15:6890–6904.

Katz LC, Shatz CJ (1996) Synaptic activity and the construction of cortical circuits. Science 274:1133–1138.

Kelly JB (1992) Behavioral development of the auditory orientation response. In: Romand R (ed), Development of Auditory and Vestibular Systems 2. Amsterdam, London, New York, Tokyo: Elsevier, pp. 391–418.

Kil J, Kageyama GH, Semple MN, Kitzes LM (1995) Development of ventral cochlear nucleus projections to the superior olivary complex in gerbil. J Comp Neurol 353: 317–340.

Kim J, Morest DK, Bohne BA (1997) Degeneration of axons in the brainstem of chinchilla after auditory overstimulation. Hear Res 103:169–191.

Kitzes LM (1990) Development of auditory system physiology. In: Coleman JR (ed), Development of Sensory Systems in Mammals. New York: Wiley, pp. 249–288.

Kitzes LM, Kageyama GH, Semple MN, Kil J (1995) Development of ectopic projections from the ventral cochlear nucleus to the superior olivary complex induced by neonatal ablation of the contralateral cochlea. J Comp Neurol 353:341–363.

Knöll B, Drescher U (2002) Ephrin-As as receptors in topographic projections. Trends Neurosci 25:145–149.

Koike T, Tanaka S (1991) Evidence that nerve growth factor dependence of sympathetic neurons for survival *in vitro* may be determined by levels of cytoplasmic free Ca^{2+}. Proc Natl Acad Sci USA 88:3892–3896.

Korada S, Schwartz IR (1999) Development of GABA, glycine, and their receptors in the auditory brainstem of gerbil: a light and electron microscopic study. J Comp Neurol 409:664–681.

Kotak VC, Sanes DH (1995) Synaptically evoked prolonged depolarizations in the developing auditory system. J Neurophysiol 74:1611–1620.

Kotak VC, Sanes DH (1996) Developmental influence of glycinergic transmission: regulation of NMDA receptor-mediated EPSPs. J Neurosci 16:1836–1843.

Kotak VC, Sanes DH (1997) Deafferentation weakens excitatory synapses in the developing central auditory system. Eur J Neurosci 9:2340–2347.

Kotak VC, Sanes DH (2000) Long-lasting inhibitory synaptic depression is age- and calcium-dependent. J Neurosci 20:5820–5826.

Kotak VC, Korada S, Schwartz IR, Sanes DH (1998) A developmental shift from GABAergic to glycinergic transmission in the central auditory system. J Neurosci 18:4646–4655.

Krause WJ (1998) A Review of Histogenesis/Organogenesis in the Developing North American Opossum (*Didelphis virginiana*). Berlin, Heidelberg: Springer-Verlag.

Kudo M, Kitao Y, Okoyama S, Moriya M, Kawano J (1996) Crossed projection neurons are generated prior to uncrossed projection neurons in the lateral superior olive of the rat. Dev Brain Res 95:72–78.

Kudo M, Sakurai H, Kurokawa K, Yamada H (2000) Neurogenesis in the superior olivary complex in the rat. Hear Res 139:144–152.

Kullmann PHM, Ene FA, Kandler K (2002) Glycinergic and GABAergic calcium responses in the developing lateral superior olive. Eur J Neurosci 15:1093–1104.

Kungel M, Friauf E (1995) Somatostatin and leu-enkephalin in the rat auditory brainstem during fetal and postnatal development. Anat Embryol 191:425–443.

Kungel M, Piechotta K, Rietzel H-J, Friauf E (1997) Influence of the neuropeptide somatostatin on the development of dendritic morphology: a cysteamine-depletion study in the rat auditory brainstem. Dev Brain Res 101:107–114.

Leinekugel X, Medina I, Khalilov I, Ben-Ari Y, Khazipov R (1997) Ca^{2+}-oscillations mediated by the synergistic excitatory actions of GABA and NMDA receptors in the neonatal hippocampus. Neuron 18:243–255.

Lenoir M, Shnerson A, Pujol R (1980) Cochlear receptor development in the rat with emphasis on synaptogenesis. Anat Embryol 160:253–262.

Lenoir M, Puel J-L, Pujol R (1987) Stereocilia and tectorial membrane development in the rat cochlea. A SEM study. Anat Embryol 175:477–487.

Levi-Montalcini R (1949) The development of the acoustico-vestibular centers in the chick embryo in the absence of the afferent root fibers and of descending fiber tracts. J Comp Neurol 91:209–242.

Lim DJ, Anniko M (1985) Developmental morphology of the mouse inner ear. A scanning electron microscopic observation. Acta Otolaryngol Suppl 422:1–69.

Lippe WR (1994) Rhythmic spontaneous activity in the developing avian auditory system. J Neurosci 14:1486–1495.

Lohmann C, Friauf E (1996) Distribution of the calcium-binding proteins parvalbumin and calretinin in the auditory brainstem of adult and developing rats. J Comp Neurol 367:90–109.

Lohmann C, Ilic V, Friauf E (1998) Development of a topographically organized auditory network in slice culture is calcium dependent. J Neurobiol 34:97–112.

Löhrke S, Kungel M, Friauf E (1998) Electrical membrane properties of trapezoid body neurons in the rat auditory brainstem are preserved in organotypic slice cultures. J Neurobiol 36:395–409.

Lu J, Karadsheh M, Delpire E (1999) Developmental regulation of the neuronal-specific isoform of K-Cl cotransporter KCC2 in postnatal rat brains. J Neurobiol 39:558–568.

Lurie DI, Pasic TR, Hockfield SJ, Rubel EW (1997) Development of Cat-301 immunoreactivity in auditory brainstem nuclei of the gerbil. J Comp Neurol 380:319–334.

Malenka RC, Nicoll RA (1993) NMDA-receptor-dependent synaptic plasticity: multiple forms and mechanisms. Trends Neurosci 16:521–527.

Martin MR, Rickets C (1981) Histogenesis of the cochlear nucleus of the mouse. J Comp Neurol 197:169–184.

McAllister AK (1999) Subplate neurons: a missing link among neurotrophins, activity, and ocular dominance plasticity? Proc Natl Acad Sci USA 96:13600–13602.

McFadden SL, Walsh EJ, McGee J (1996) Onset and development of auditory brainstem responses in the Mongolian gerbil (*Meriones unguiculatus*). Hear Res 100:68–79.

Melcher JR, Kiang NYS (1996) Generators of the brainstem auditory evoked potential in cat. III. Identified cell populations. Hear Res 93:52–71.

Moore DR (1982) Late onset of hearing in the ferret. Brain Res 253:309–311.

Moore DR (1983) Development of inferior colliculus and binaural audition. In: Romand R (ed), Development of Auditory and Vestibular Systems. New York: Academic Press, pp. 121–166.

Moore DR (1991) Development and plasticity of the ferret auditory system. In: Altschuler RA, et al. (eds), Neurobiology of Hearing: The Central Auditory System. New York: Raven Press, pp. 461–475.

Moore DR (1992a) Developmental plasticity of the brainstem and midbrain auditory nuclei. In: Romand R (ed), Development of Auditory and Vestibular Systems. Amsterdam: Elsevier, pp. 298–320.

Moore DR (1992b) Trophic influences of excitatory and inhibitory synapses on neurones in the auditory brain stem. NeuroReport 3:269–272.

Morest DK (1968) The growth of synaptic endings in the mammalian brain: a study of the calyces of the trapezoid body. Z Anat Entwickl-Gesch 127:201–220.

Morest DK, Kim JN, Bohne BA (1997) Neuronal and transneuronal degeneration of auditory axons in the brainstem after cochlear lesions in the chinchilla: cochleotopic and non-cochleotopic patterns. Hear Res 103:151–168.

Morey AL, Carlile S (1990) Auditory brainstem of the ferret: maturation of the brainstem auditory evoked response. Dev Brain Res 52:279–288.

Moushegian G, Rupert A, Whitcomb MA (1967) Stimulus coding by medial superior olivary neurons. J Neurophysiol 30:1239–1261.

Mueller BK (1999) Growth cone guidance: first steps towards a deeper understanding. Annu Rev Neurosci 22:351–388.

Niblock MM, Brunso-Bechtold JK, Henkel CK (1995) Fiber outgrowth and pathfinding in the developing auditory brainstem. Dev Brain Res 85:288–292.

Oblinger MM, Das GD (1981) Neurogenesis in the brain stem of the rabbit: an autoradiographic study. J Comp Neurol 197:45–62.

Obrietan K, van den Pol AN (1995) GABA neurotransmission in the hypothalamus: developmental reversal from Ca^{2+} elevating to depressing. J Neurosci 15:5065–5077.

O'Leary DDM (1992) Development of connectional diversity and specificity in the mammalian brain by the pruning of collateral projections. Curr Opin Neurobiol 2:70–77.

O'Leary DDM, Bicknese AR, De Carlos JA, Heffner CD, Koester SE, Kutka LJ, Terashima T (1990) Target selection by cortical axons: alternative mechanisms to establish axonal connections in the developing brain. In: Cold Spring Harb Symp Quant Biol Cold Spring Harbor, NY: Cold Spring Harbor Laboratory Press, pp. 453–468.

O'Leary DDM, Ruff NL, Dyck RH (1994) Development, critical period plasticity, and adult reorganizations of mammalian somatosensory systems. Curr Opin Neurobiol 4: 535–544.

Osen KK, Mugnaini E, Dahl A-L, Christiansen AH (1984) Histochemical localization of acetylcholinesterase in the cochlear and superior olivary nuclei. A reappraisal with emphasis on the cochlear granule cell system. Arch Ital Biol 122:169–212.

Ostapoff EM, Benson CG, Saint Marie RL (1997) GABA- and glycine-immunoreactive projections from the superior olivary complex to the cochlear nucleus in guinea pig. J Comp Neurol 381:500–512.

Parks TN (1981) Changes in the length and organization in nucleus laminaris dendrites after unilateral otocyst ablation in chick embryos. J Comp Neurol 202:47–57.

Parks TN, Collins P, Conlee JW (1983) Morphology and origin of axonal endings in nucleus laminaris of the chicken. J Comp Neurol 214:32–42.

Parks TN, Gill SS, Jackson H (1987) Experience-independent development of dendritic organization in the avian nucleus laminaris. J Comp Neurol 260:312.

Pasic TR, Moore DR, Rubel EW (1994) Effect of altered neuronal activity on cell size in the medial nucleus of the trapezoid body and ventral cochlear nucleus of the gerbil. J Comp Neurol 348:111–120.

Potashner SJ, Suneja SK, Benson CG (1997) Regulation of D-aspartate release and uptake in adult brain stem auditory nuclei after unilateral middle ear ossicle removal and cochlear ablation. Exp Neurol 148:222–235.

Powell TPS, Erulkar SD (1962) Transneuronal cell degeneration in the auditory relay nuclei of the cat. J Anat 96:268.

Puel JL, Uziel A (1987) Correlative development of cochlear action potential sensitivity, latency, and frequency selectivity. Brain Res 465:179–188.

Reichling DB, Kyrozis A, Wang J, Mac Dermott AB (1994) Mechanisms of GABA and glycine depolarization-induced calcium transients in rat dorsal horn neurons. J Physiol (Lond) 476:411–421.

Reuss S (2000) Introduction to the superior olivary complex. Microsc Res Technique 51:303–306.

Rietzel H-J, Friauf E (1998) Neuron types in the rat lateral superior olive and developmental changes in the complexity of their dendritic arbors. J Comp Neurol 390:20–40.

Rivera C, Voipio J, Payne JA, Ruusuvuori E, Lahtinen H, Lamsa K, Pirvola U, Saarma M, Kaila K (1999) The K^+/Cl^- co-transporter KCC2 renders GABA hyperpolarizing during neuronal maturation. Nature 397:251–255.

Robertson D (1996) Physiology and morphology of cells in the ventral nucleus of trapezoid body and rostral periolivary regions of the rat superior olivary complex studied in slices. Audit Neurosci 2:15–31.

Rogowski BA, Feng AS (1981) Normal postnatal development of medial superior olivary neurons in the albino rat: a Golgi and Nissl study. J Comp Neurol 196:85–97.

Roth B, Bruns V (1992a) Postnatal development of the rat organ of Corti. I. General morphology, basilar membrane, tectorial membrane and border cells. Anat Embryol 185:559–569.

Roth B, Bruns V (1992b) Postnatal development of the rat organ of corti. II. Hair cell receptors and their supporting elements. Anat Embryol 185:571–581.

Rubel EW (1978) Ontogeny of structure and function in the vertebrate auditory system. In: Jacobsen M (ed), Handbook of Sensory Physiology. New York: Springer-Verlag, pp. 135–237.

Rubel EW, Parks TN (1988) Organization and development of the avian brain-stem auditory system. In: Edelman GM, Gall WE, Cowan WM (eds), Auditory Function. Neurobiological Bases of Hearing. New York, Chichester, Brisbane, Toronto, Singapore: John Wiley, pp. 3–92.

Ruben RJ (1992) The ontogeny of human hearing. Acta Otolaryngol (Stockh) 112:192–196.

Ruben RJ, Rapin I (1980) Plasticity of the developing auditory system. Ann Otol Rhinol Laryngol 89:303–311.

Rübsamen R, Gutowski M, Langkau J, Dörrscheidt GJ (1994) Growth of central nervous system auditory and visual nuclei in the postnatal gerbil (Meriones unguiculatus). J Comp Neurol 346:289–305.

Russell FA, Moore DR (1995) Afferent reorganisation within the superior olivary complex of the gerbil: development and induction by neonatal, unilateral cochlear removal. J Comp Neurol 352:607–625.

Russell FA, Moore DR (1999) Effects of unilateral cochlear removal on dendrites in the gerbil medial superior olivary nucleus. Eur J Neurosci 11:1379–1390.

Sanes DH (1993) The development of synaptic function and integration in the central auditory system. J Neurosci 13:2627–2637.

Sanes DH, Chokshi P (1992) Glycinergic transmission influences the development of dendrite shape. NeuroReport 3:323–326.

Sanes DH, Hafidi A (1996) Glycinergic transmission regulates dendrite size in organotypic culture. J Neurobiol 31:503–511.

Sanes DH, Rubel EW (1988) The ontogeny of inhibition and excitation in the gerbil lateral superior olive. J Neurosci 8:682–700.

Sanes DH, Siverls V (1991) Development and specificity of inhibitory terminal arborizations in the central nervous system. J Neurobiol 8:837–854.

Sanes DH, Takács C (1993) Activity-dependent refinement of inhibitory connections. Eur J Neurosci 5:570–574.

Sanes DH, Walsh EJ (1998) The development of central auditory processing. In: Rubel EW, Popper AN, Fay RR (eds), Development of the Auditory System. New York: Springer-Verlag, pp. 271–314.

Sanes DH, Markowitz S, Bernstein J, Wardlow J (1992a) The influence of inhibitory afferents on the development of postsynaptic dendritic arbors. J Comp Neurol 321: 637–644.

Sanes DH, Song J, Tyson J (1992b) Refinement of dendritic arbors along the tonotopic axis of the gerbil lateral superior olive. Dev Brain Res 67:47–55.

Sanes DH, Reh TA, Harris WA (2000) Development of the Nervous System. San Diego: Academic Press.

Sato K, Kiyama H, Tohyama M (1992) Regional distribution of cells expressing glycine receptor α2 subunit mRNA in the rat brain. Brain Res 590:95–108.

Sato K, Kuriyama H, Altschuler RA (1995) Expression of glycine receptor subunits in the cochlear nucleus and superior olivary complex using non-radioactive in-situ hybridization. Hearing Res 91:7–18.

Saunders JC, Coles RB, Gates GR (1973) The development of auditory evoked responses in the cochlea and cochlear nucleus of the chick. Brain Res 63:59–74.

Saunders JC, Kaltenbach JA, Relkin EM (1983) The structural and functional development of the outer and middle ear. In: Romand R (ed) Development of Auditory and Vestibular Systems. New York: Academic Press, pp. 3–25.

Schofield BR, Cant NB (1991) Organization of the superior olivary complex in the guinea pig. I. Cytoarchitecture, cytochrome oxidase histochemistry, and dendritic morphology. J Comp Neurol 314:645–670.

Schwartz IR (1992) The superior olivary complex and lateral lemniscal nuclei. In: Webster DB, Popper AN, Fay RR (eds), The Mammalian Auditory Pathway: Neuroanatomy. New York: Springer-Verlag, pp. 117–167.

Shatz CJ (1990) Impulse activity and the patterning of connections during CNS development. Neuron 5:745–756.

Shatz CJ (1992) How are specific connections formed between thalamus and cortex? Curr Opin Neurobiol 2:78–82.

Shatz CJ (1996) Emergence of order in visual system development. Proc Natl Acad Sci USA 93:602–608.

Silverman MS, Clopton BM (1977) Plasticity of binaural interaction. I. Effect of early auditory deprivation. J Neurophysiol 40:1266–1274.

Smith DI, Kraus N (1987) Postnatal development of the auditory brainstem response (ABR) in the unanesthetized gerbil. Hear Res 27:157–164.

Smith ZDJ (1981) Organization and development of brain stem auditory nuclei of the chicken: dendritic development in n. laminaris. J Comp Neurol 203:309–333.

Smith ZDJ, Rubel EW (1979) Organization and development of brain stem auditory nuclei of the chicken: dendritic gradients in nucleus laminaris. J Comp Neurol 186: 213–239.

Spangler KM, Warr WB (1991) The Descending Auditory System. In: Altschuler RA,

Bobbin RP, Clopton BM, Hoffman DW (eds), Neurobiology of Hearing: The Central Auditory System. New York: Raven Press, pp. 27–45.

Suneja SK, Potashner SJ, Benson CG (1998) Plastic changes in glycine and GABA release and uptake in adult brain stem auditory nuclei after unilateral middle ear ossicle removal and cochlear ablation. Exp Neurol 151:273–288.

Taber-Pierce E (1967) Histogenesis of the dorsal and ventral cochlear nuclei in the mouse: an autoradiographic study. J Comp Neurol 131:27–54.

Taber-Pierce E (1973) Time of origin of neurons in the brain stem of the mouse. In: Ford DH (ed) Neurological Aspects of Maturation and Aging. Progress in Brain Res. Amsterdam: Elsevier, pp. 53–65.

Takahashi T, Momiyama A, Hirai K, Hishinuma F, Akagi H (1992) Functional correlation of fetal and adult forms of glycine receptors with developmental changes in inhibitory synaptic receptor channels. Neuron 9:1155–1161.

Tecott LH, Julius D (1999) A new wave of serotonin receptors. Curr Opin Neurobiol 3: 310–315.

Tessier-Lavigne M, Goodman CS (1996) The molecular biology of axon guidance. Science 274:1123–1133.

Thompson AM (1998) Heterogeneous projections of the cat posteroventral cochlear nucleus. J Comp Neurol 390:439–453.

Thompson AM, Thompson GC (1991) Posteroventral cochlear nucleus projections to olivocochlear neurons. J Comp Neurol 303:267–285.

Thoss VS, Kungel M, Friauf E, Hoyer D (1996) Presence of somatostatin sst₂ receptors in the developing rat auditory system. Dev Brain Res 97:269–278.

Tokimoto T, Osako S, Matsuura S (1977) Development of auditory evoked cortical and brain stem responses during the early postnatal period in the rat. Osaka City Med J 23:141–153.

Tolbert LP, Morest DK, Yurgelun-Todd DA (1982) The neuronal architecture of the anteroventral cochlear nucleus of the cat in the region of the cochlear nerve root: horseradish peroxidase labelling of identified cell types. Neuroscience 7:3031–3052.

Uziel A, Romand R, Marot M (1981) Development of cochlear potentials in rats. Audiology 20:89–100.

Walsh EJ, McGee J (1988) Rhythmic discharge properties of caudal cochlear nucleus neurons during postnatal development in cats. Hear Res 36:233–248.

Walsh EJ, McGee J, McFadden SL, Liberman MC (1998) Long-term effects of sectioning the olivocochlear bundle in neonatal cats. J Neurosci 18:3859–3869.

Warr WB (1972) Fiber degeneration following lesions in the multipolar and globular cell areas in the ventral cochlear nucleus of the cat. Brain Res 40:247–270.

Warr WB (1992) Organization of olivocochlear efferent systems in mammals. In: Webster DB, Popper AN, Fay RR (eds), Mammalian Auditory Pathway: Neuroanatomy. New York: Springer-Verlag, pp. 410–448.

Webster DB, Webster M (1977) Neonatal sound deprivation affects brainstem auditory nuclei. Arch Otolaryngol 103:392–396.

Weliky M, Katz LC (1997) Disruption of orientation tuning in visual cortex by artificially correlated neuronal activity. Nature 386:680–685.

Wenthold RJ (1991) Neurotransmitters of brainstem auditory nuclei. In: Altschuler RA, Bobbin RP, Clopton BM, Hoffman DW (eds), Neurobiology of Hearing: The Central Auditory System. New York: Raven Press, pp. 121–139.

Willard FH (1995) Development of the mammalian auditory hindbrain. In: Malhotra BS (ed), Advances in Neural Science. Greenwich CT: JAI Press, pp. 205–234.

Willard FH, Martin GF (1986) The development and migration of large multipolar neurons into the cochlear nucleus of the North American opossum. J Comp Neurol 248: 119–132.

Wong WT, Faulkner-Jones BE, Sanes JR, Wong ROL (2000) Rapid dendritic remodeling in the developing retina: dependence on neurotransmission and reciprocal regulation by Rac and Rho. J Neurosci 20:5024–5036.

Woolf NK, Ryan AF (1985) Ontogeny of neural discharge patterns in the ventral cochlear nucleus of the mongolian gerbil. Dev Brain Res 17:131–147.

Yin TCT (2002) Neural mechanisms of encoding binaural localization cues in the auditory brainstem. In: Oertel D, Fay RR, Popper AN (eds), Integrative Functions in the Mammalian Auditory Pathway. New York: Springer-Verlag, pp. 99–159.

Yu TW, Bargmann CI (2001) Dynamic regulation of axon guidance. Nat Neurosci 4: 1169–1176.

4

Plasticity of Binaural Systems

DAVID R. MOORE AND ANDREW J. KING

1. Introduction

The ability to localize a sound source in space relies on the detection and interpretation of spatial cues that arise from the interaction between sound waves and the head and external ears. The dominant cues for localization in the horizontal dimension are binaural cues: interaural time differences (ITDs) and interaural level differences (ILDs). Spectral localization cues are generated by the head and external ears and are utilized for resolving front–back confusion, localization in the vertical plane, and for localization using one ear alone (see Wightman and Kistler 1997a). To localize sound sources accurately and unambiguously, the central auditory system (CAS) must extract, process, and combine information over different frequency channels and from both ears to form an internal representation of these cues. Acoustical measurements in humans and other animals have shown that the spatial cue values available can vary quite markedly from one individual to another. The relationship between the cue values and sound location must therefore be learned on the basis of experience. Moreover, as the head grows, the monaural and binaural cue values that correspond to particular directions in space will change. The developing CAS must therefore adjust to the changing cues to maintain accurate localization. Plasticity, particularly of the pathways responsible for binaural processing, is thus a necessary requirement for the retention of normal function through the period of head growth (up to about 12 years in humans; see Clifton 1992).

This chapter addresses three issues related to these basic observations. The first is whether binaural plasticity contributes to other aspects of normal hearing. The second is whether, and to what extent, binaural plasticity is induced by abnormal auditory experience or by inputs from other sensory modalities. The third is how the neural systems serving binaural hearing are altered by either normal or abnormal experience. In this chapter, we define plasticity broadly to mean any change in the structure or function of the binaural auditory system induced by altered inputs, occurring either naturally as a consequence of head growth or as a result of the introduction of abnormal inputs. The "abnormal input" may be either clinically or experimentally induced, and can be reversible or irreversible.

1.1 Binocular Plasticity

A major impetus for the work described in this chapter came from the classic study of Wiesel and Hubel (1963), who first demonstrated in cats that one eye deprived of pattern vision from birth to the age of 2–3 months lost the ability to influence the firing of neurons in the visual cortex. This simple but clear and dramatic result led to an explosion of interest in the structural and functional effects of various types of visual deprivation (reviewed by Movshon and Van Sluyters 1981; Sherman and Spear 1982; Kind 1999; Thompson 2000), leading to the suggestion that a balanced and correlated input from the two eyes is necessary for the normal development and maintenance of central neural connections.

Binaural systems have several properties in common with binocular systems: there are two end organs, each consisting of a topographically arrayed receptor sheet. The receptors project centrally, via a relay nucleus, and topographically to synapse on target neurons that receive matching input from the other side. Almost all neurons in the target nuclei receive input from the two sides, and project on to other targets performing higher order functions. It is therefore unsurprising that auditory neuroscientists should have asked, on the basis of the highly interesting and clinically relevant visual data, whether developmental plasticity also occurs in binaural systems.

There are, of course, major differences between the two sensory modalities, and one that is of particular importance to this chapter is the relative difficulty of controlling input to the auditory system. For example, Hubel and Wiesel (1995) were able to create a large and easily reversible mismatch between the eyes simply by suturing closed the lids of one eye. In the auditory system, ear plugging has been widely used (Table 4.1) to create an analogous input mismatch between the ears. However, ear plugging potentiates bone-conducted sound (see Section 2.3), and long-term ear plugging may sometimes produce irreversible hearing loss (see Section 3.2).

1.2 Applied and Clinical Interest

Conductive hearing loss in children, produced by otitis media (OM), has been widely reported to result in a variety of cognitive (Roberts et al. 1991; Gravel and Wallace 1992; Mody et al. 1999) and social (Haggard et al. 1994) problems. Because OM reaches a peak of prevalence between the ages of 6–18 months (Haggard and Hughes 1991; Hogan et al. 1997), when language acquisition is also normally at a peak (Hasenstab 1987), it has sometimes been assumed that the early failure or delay in language learning leads to cognitive retardation and, thence, to social and behavioral problems. However, it is also possible that these problems may derive from poor auditory processing, including binaural pro-

TABLE 4.1. Manipulations of auditory input in studies of the central auditory system.

Category	Method	Species	Reference
Conductive hearing loss	Ear plug	Ferret	King et al. (1988)
		Human	Florentine (1976)
		Barn owl	Knudsen (1985)
	Canal ligation	Cat	Moore and Irvine (1981)
		Rat	Clerici and Coleman (1986)
		Gerbil	Tucci et al. (2001)
	Canal cauterization	Mouse	Webster (1983a)
	Ossicular disarticula-	Rat	Paterson and Hosea (1993)
	tion or removal	Gerbil	Cook et al. (2002)
	Reshape external ear	Ferret	Schnupp et al. (1998)
		Barn owl	Knudsen et al. (1994)
	Tympanic membrane puncture	Chicken	Tucci and Rubel (1985)
	Otosclerosis	Human	Hall and Grose (1993)
	Canal atresia	Human	Wilmington et al. (1994)
	Otitis media	Human	Moore et al. (1991)
Amplification	Hearing aid	Human	Gatehouse (1992)
Sound chamber	Pure tones	Cat	Moore and Aitkin (1975)
	Quiet	Rat	Batkin et al. (1970)
	Pulsed tones	Mouse	Sanes and Constantine-Paton (1985)
	Omnidirectional noise	Guinea pig	Withington-Wray et al. (1990b)
Sensorineural hearing loss	Cochlear removal	Cat	Reale et al. (1987)
		Chicken	Parks (1981)
		Ferret	Moore (1990b)
		Gerbil	Nordeen et al. (1983)
		Mouse	Trune (1982)
		Guinea pig	Palmer and King (1985)
	Partial cochlear le-	Cat	Rajan et al. (1993)
	sions	Guinea pig	Robertson and Irvine (1989)
	Ototoxicity	Cat	Matsushima et al. (1991)
		Guinea pig	Dodson et al. (1994)
	Ototoxicity and elec- trical stimulation	Cat	Matsushima et al. (1991)
	Oval window puncture	Chicken	Tucci and Rubel (1985)

cessing. For example, children who have an early history of recurrent OM perform more poorly in later life on tests of binaural hearing involving interaural phase differences (binaural unmasking; see Section 5.4) than do children lacking a known OM history (Moore et al. 1991; Pillsbury et al. 1991). In the high-OM children, binaural hearing continued to be compromised after the OM had resolved and peripheral sensitivity had been restored to normal. Poor binaural hearing would be expected to reduce the ability to detect speech in noisy environments (see also Hall et al. 1998). Thus, if children with recurrent OM have poor speech detection between and after, as well as during episodes of

OM, the cognitive and social sequelae mentioned at the beginning of this section may, in part, be due to maladaptation in the binaural auditory system.

Adults experiencing binaurally imbalanced hearing have also been reported to undergo gradual changes in hearing ability that are not simply due to the abnormal peripheral input. In some studies, a unilateral hearing loss, produced either clinically (e.g., by otosclerosis; Hall and Grose, 1993) or experimentally (by ear plugging; Florentine 1976; McPartland et al. 1997), has been found to lead to initial changes in binaural performance following imposition of the loss and/or to further changes following removal of the loss. In another series of studies (e.g., Silman et al 1984; Byrne et al. 1992; Gatehouse 1992), long-term, unilateral hearing aid use by bilaterally hearing-impaired listeners was found to produce changes in auditory performance, both in the aided ear and in the unaided ear. Binaural plasticity has been implicated in these results by findings that the unaided ear is affected only when the other ear receives amplified input, and by the lack of change found in patients receiving binaural amplification.

Clinical interest in the potential for CAS plasticity, and particularly in the neural pathways that underlie spatial hearing, is not restricted to the consequences of hearing disorders. The growing evidence for enhanced auditory localization abilities in blind subjects (Röder et al. 1999) raises the possibility of developing therapeutic strategies that capitalize on these cross-modal changes.

1.3 Basic Interest

The activity of most neurons in the CAS is influenced by acoustic stimulation of either ear. A prolonged, interaural imbalance in the input to these neurons, usually produced in experimental animals by a unilateral hearing loss, has been shown to result in the generation and rearrangement of presynaptic axons and their terminal processes, and in modifications to postsynaptic dendrites and intracellular processes (e.g., Rubel, Parks, and Zirpel, Chapter 2). Unsurprisingly, these cellular changes have their functional correlates, including increased excitation in some neurons, and a shift in the range of effective acoustic stimuli for both neural and behavioral responses to sound source location.

In addition to their obvious relation to human binaural plasticity, these results have been of interest to developmental neurobiologists. Auditory brain stem pathways underlying binaural interaction include some of the largest synapses in the central nervous system (the end bulbs and calyses of Held; Ryugo 1992; Schwartz 1992), which exhibit unique patterns of synaptic input. For example, the spatial segregation between the afferent supply from each cochlear nucleus (CN) to the dendrites of medial superior olive (MSO; or nucleus laminaris, NL[1]) neurons provides a particularly convenient assay for studying issues such as

---

1. Throughout this chapter the default reference will be to the mammalian auditory system with specific reference to avian auditory studies where, as in many cases, they have provided seminal data in the field. Unless otherwise noted, the homologies listed in the Abbreviations at the end of this chapter will be assumed.

neural competition, and the pre- and postsynaptic contributions to synaptic plasticity.

An issue of intense recent interest in neuroscience has been the discovery of the plasticity of sensory and motor maps in the cerebral cortex of adult animals (e.g., Kaas 1991; Gilbert 1993; Darian-Smith and Gilbert 1995). In the auditory cortex, a rearrangement of frequency representation has been shown to follow discrete lesions of the contralateral, but not of the ipsilateral cochlea (Rajan et al. 1993; Irvine and Rajan 1995). Although little is known about binaural plasticity in the adult cortex, these and related training data (e.g., Recanzone et al. 1993; Kilgard and Merzenich 1998) suggest the likelihood that both normal and abnormal binaural input will influence the representation of auditory space within the CAS (see Clarey et al. 1992). Moreover, it is possible that neural plasticity of the sort revealed by these experiments may underlie some of the psychoacoustic phenomena described above and, in more detail, in Section 5 of this chapter.

1.4 Organization and Scope of Chapter

This chapter is divided into sections reflecting the main methods (cellular, electrophysiological, behavioral) that have been used in experimental studies of binaural plasticity. These sections are preceded by a consideration of some more theoretical and general methodological issues, and followed by overviews of the mechanisms of binaural plasticity and some directions of present and future research. Each of the experimental sections is based around a table providing a comprehensive list of references. Although the chapter will attempt to list, in these tables, all known sources in the area, the review is limited to a discussion of the major findings and focuses on concepts rather than trying to present an exhaustive literature review.

The scientific limits of the presentation have been defined by the other chapters in this book and by the other volumes in this series. In particular, the chapter will not discuss in detail plasticity in invertebrate auditory systems (see Lake-Harlan, Chapter 6; Huber 1987), the normal development of binaural hearing (see Werner and Gray, Chapter 2; Sanes and Walsh, Chapter 6; Cant, Chapter 7 in Rubel et al. 1998), the influence of auditory input on the CN (see Rubel, Parks, and Zirpel, Chapter 2), or plasticity in the tonotopic organization of the CAS (see Weinberger, Chapter 4). However, it is inevitable that reference will be made to issues considered in more detail in those volumes and chapters to provide information necessary for an overview of binaural plasticity. Other recent reviews of this and related subjects are presented in Romand (1992), Rauschecker (1995), Gravel et al. (1996), King (1999), Altschuler et al. (2000), King et al. (2001a), Moore et al. (2001), and Knudsen (2002).

2. Manipulating Auditory Input

2.1 Neural Activity

In the normally functioning ear, neural activity is constantly generated in auditory nerve (AN) fibers by the release of excitatory transmitters from the base of cochlear inner hair cells (see Ruggero 1992). Activity is present in all fibers, in the absence of auditory stimulation, and neural responses to sound consist of an increase in the probability of discharge of the fibers (Kiang 1984). Thus, eliminating sound input does not lead to a cessation of neural input to the CAS. Conversely, the discharge level of most AN fibers reaches a maximum at relatively modest sound levels (Irvine 1986), so a reduction in the level of an intense sound may have little effect on the total neural input to the CAS. It follows that changes in auditory inputs may result in a modulation rather than an elimination of the ongoing activity in the AN.

2.2 Environmental and Conductive Manipulations

Several experimental methods have been used to produce a conductive hearing loss, or to manipulate the acoustic environment, in studies of central auditory processing. Several other variants of auditory input, occurring as a result, or in the treatment, of clinical hearing loss, have also been studied (Table 4.1). While all these conditions were assumed (and, in many cases, were measured) to result in a reduced input to the cochlea, it is unclear how the resulting modulation of AN activity would influence CAS neurons. Tucci and colleagues (2001) have recently shown that CAS neural activity is reduced by a conductive loss, both with and without deliberate acoustic stimulation. Data from several neural systems have shown that the *presence* of activity in afferent axons is necessary for the formation of normal connections (e.g., Stretavan et al. 1988; Shatz 1990) and for the maintenance of postsynaptic neurons (Born and Rubel 1988; Pasic et al. 1994; see Rubel, Parks, and Zirpel, Chapter 2). However, much less is known about the *level* of activity that is required for these functions. What does seem clear is that the *pattern* of activity in neighboring neurons is important.

Studies of the developing visual system have shown that the spontaneous activity of neighboring retinal ganglion cells is correlated, and that this activity appears to drive a segregation between the two eyes into eye-specific terminal zones within the lateral geniculate nucleus (Wong et al. 1993, 2000). These results suggest that, in the auditory system, correlated activity between the AN fibers deriving from each ear may be needed for the refinement of tonotopicity within the CAS, whereas correlated activity between the ears may be necessary for the development and refinement of binaural connections. Data relevant to this hypothesis are presented below. The general point is that the AN is the gateway to the CAS and changes in auditory input are only going to be as effective as the filtering properties of this gateway allow.

2.3 Practical Considerations

In addition to these theoretical issues, there are practical considerations in studies of variable auditory input. One is the level of control of the input. In long-term laboratory studies of animal hearing it is not possible to specify or control the exact level of sound reaching the tympanum because it is dependent on a variable auditory environment. Even in the most controlled chambers, there are usually sound level differences between different parts of the chamber. It is also impossible to prevent an animal from producing sound, or to insulate it totally from the external sounds produced by, for example, feeding, maternal grooming, and sibling behavior. Rather than controlling the environment, most studies (Table 4.1) have sought to reduce input by attenuating sounds passing through the outer or middle ear. However, ear canal occlusion potentiates bone conducted sound (Tonndorf 1972), thereby providing a low-frequency bias to the perceived spectrum as well as adding extra weight to uncontrollable self-vocalizations. Occlusion of the outer or middle ear (Knudsen et al. 1984a; Hartley and Moore 2003) also produces delays in sound transmission (see Section 4.1). Removal of one or more ossicles (Cook et al. 2002) reduces sound transmission and neural output from the cochlea in a highly stable way, even in the absence of controlled sound stimulation. However, that manipulation is irreversible. Studies on humans and, particularly, studies on clinical subjects, suffer from all these problems. In addition, the level of control is further diminished by factors including the limited range of different types of hearing loss (or amplification) available, time constraints, and the often fluctuating nature of conductive losses.

2.4 Sensorineural Manipulations

In an attempt to circumvent the above problems, some animal studies (Table 4.1) have eliminated peripheral input entirely by surgically or chemically ablating the cochlea. This procedure has the advantage of reproducibility and, as it abolishes activity in the AN and anteroventral CN (AVCN; Koerber et al. 1966; Born et al. 1991), there are fewer interpretational concerns than with other forms of input manipulation. A further point in favor of cochlear ablation is that it is radical. Pathways and processes that might be affected by more subtle forms of hearing loss may, initially, be identified using this method. There are, however, two major shortcomings with the method. The first is that mechanical lesions of the cochlea, at least in postpartum mammals (Kelley et al. 1995), produce irreversible damage to the sensory epithelium. In some instances, however, recovery studies are possible following experimentally induced elimination of peripheral activity. If, instead of ablating the cochlea, the drug tetrodotoxin (TTX) is applied to the round window, a short-term blockade of AN activity results (e.g., Born and Rubel 1988), as TTX blocks sodium channels in nerve membranes. Another important instance where recovery is possible is following hair cell regeneration in nonmammalian vertebrates (see Cotanche

1999; Bermingham-McDonogh and Rubel 2003). Finally, temporary threshold shift, induced by moderate level sound exposure, has been used to examine adult cortical plasticity (Calford et al. 1993; Rajan 1998). As these methods have been used in only a single study involving binaural systems (Pasic et al. 1994), the reader is referred to Chapter 2 for further discussion. Nevertheless, they are methods that might profitably be exploited in future studies of binaural plasticity.

The second shortcoming of cochlear ablation is that it fails to simulate any known form of hearing disorder. Clinically oriented critics may therefore argue that the method does not inform us about human auditory pathology. On the other hand, in contrast to most forms of clinical hearing loss, cochlear ablation does produce a stable (i.e., zero activity) and reproducible outcome. Moreover, unilateral ablation is the only reliable way of eliminating binaural inputs and therefore of revealing plasticity in monaural sound localization mechanisms. Experimentally induced hearing loss or environmental manipulation can inform clinical judgements by providing data that, in many instances, describe a "worst case" scenario. However, now that the effects of stable manipulations of input are better understood, carefully controlled experiments involving fluctuating input are beginning to be performed.

2.5 Age Dependence of Auditory Manipulations

Manipulations of sensory input generally have more marked and wider ranging effects on the central nervous system (CNS) of immature than of mature animals. While CNS sensory plasticity in adult animals is now firmly established (Wall and Egger 1971; Merzenich et al. 1983; Weinberger, Chapter 4), an important aspect of auditory system plasticity is how manipulations of input might affect AN activity and higher level processing, at different ages. Studies of the development of hearing (see Rubel et al. 1998) have shown that, in most mammalian species, function begins postnatally. There then follows a period during which the peripheral auditory system is less sensitive to sound than it subsequently becomes. Several other aspects of hearing (e.g., spatial processing, complex sound processing) appear to remain immature beyond the age at which sensitivity matures. Nevertheless, it is clear that, at least in terms of peripheral sound transmission and activation of AN fibers, there will be major differences in the effect of any form of auditory input manipulation between younger and older animals. For example, ear plugging will not produce a reduction in fiber activity before the age at which functional hearing begins (the "onset of hearing"), and its modulatory effect will steadily increase thereafter as sound transmission through the auditory periphery, spontaneous activity in the nerve, and the dynamic range of fibers all increase. In older animals, conductive loss can also produce different effects at different ages on CAS neural activity (Tucci et al. 2001). A sensorineural loss, such as cochlear ablation, will also produce different effects at different ages. Spontaneous activity of nerve fibers starts some time before the onset of hearing (e.g., Lippe 1992), and this will, presumably, be abolished from the time of the hearing loss. As the level of activity in the nerve normally increases during development, the difference in activity between the normal and the pathological periphery increases with age. For CAS neurons and connections dependent on this activity, there may be a threshold level of input required, with the threshold possibly also changing with age.

A related issue is, given a constant level of AN input, how the neurons of the CAS change their response to that input as a function of age. This issue is dealt with in detail later, and elsewhere in this book (Rubel, Parks, and Zirpel, Chapter 2; Friauf, Chapter 5). It should be noted here, however, that the same input manipulation can and does produce different consequences, over different time scales, in animals of different ages. For example, the population of CN neurons projecting to the ipsilateral inferior colliculus (IC) was, initially, thought to be unchanged in number following ablation of the contralateral cochlea (see Fig. 4.1) in "adolescent" ferrets (3–6 months old; Moore and Kowalchuk 1988). A subsequent investigation, using much longer survival after the cochlear ablation, found that this was not the case; ablation in these older animals could change the projection (Moore 1994). The cellular bases of central processing (e.g.,

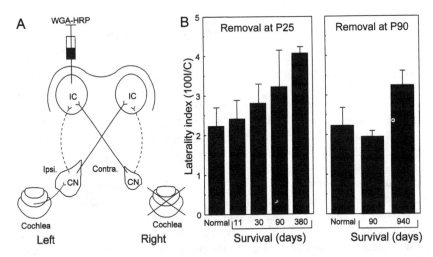

FIGURE 4.1. Plasticity of binaural connections in the ferret auditory brain stem and midbrain. (A) The retrograde tracer WGA-HRP was injected into the IC of ferrets that had one cochlea surgically destroyed at various ages and survived for various times after cochlear removal. In normally hearing ferrets, the direct projection from the CN to the contralateral (Contra.) IC is much larger than the ipsilateral (Ipsi.) projection. (B) Counts of retrogradely labeled neurons in each CN showed that the normal contralateral projection consists of about 50 times as many neurons as the ipsilateral projection. The Laterality Index, expressing the ratio of 100 I/C, increased gradually after cochlear removal at P25, becoming significant after 30 days survival, and continuing to increase beyond 90 days survival. Removal at P90 also increased the size of the ratio, but only if the ferret survived for more (perhaps considerably more) than a further 90 days. [Adapted from Moore and Kowalchuk (1988); Moore (1994).]

neurotransmitter, growth factor and receptor types and their actions; Tierney et al. 2001) have also recently been shown to change with age, at least during very early development. Thus, the central neural activation produced by a given level of peripheral activity can be age dependent (see Friauf, Chapter 5). It may therefore be a mistake to conclude, on the basis of a constant time scale of manipulation and its effect on animals of other ages, that there is an age-dependent difference in the magnitude or extent of the effect. The real difference may be in the time it requires to become manifest or in its mechanism of action.

2.6 Laterality of Manipulations

Most of the experimental studies listed in Table 4.1 have used unilateral ear manipulations. In many cases, the rationale has been that the unmanipulated side would provide a control for the manipulated side. For example, in studies of the CN following unilateral cochlear ablation (see Chapter 2), the CN on the side of the intact ear has generally been found to be unaffected by the removal of the contralateral cochlea (but see Jackson and Parks 1988; Dodson et al. 1994). However, in the studies that will concern us in this chapter, the manipulation has usually been performed unilaterally in order to create an imbalance of input between the ears and therefore to alter or eliminate binaural cues for sound location. In some cases (e.g., OM) the manipulation has been beyond the control of the researcher, but has been assumed (or measured), when bilateral, to be interaurally imbalanced. Finally, in studies focusing on either monaural or binaural processing, a bilateral manipulation has sometimes been used, usually as a control for the unilateral case. In several instances, bilateral manipulations of auditory input have been found to produce less marked neural consequences than unilateral manipulations.

Although it might be expected that total deafness would produce more marked neural consequences than unilateral deafness, this observation has been interpreted as particularly strong evidence for the importance of binaurally balanced input. However, in interpreting the results that follow, care must be taken in dissociating the results of a study employing a single technique, to study a single part of the CAS, from the overall effects of the auditory input manipulation. For this reason, the discussion that follows is divided into different levels of analysis, from the more reductionist to the more holistic, before an overview of binaural plasticity is attempted.

2.7 Nonacoustic Manipulations of Neural Activity

External events are often registered by the sense organs of more than one modality and then processed by multisensory areas of the brain. Indeed, the integration and coordination of information derived from different sensory systems is an important part of perception, attention and the control of movement (Calvert et al. 1998; Driver and Spence 1998). It is therefore not surprising that auditory performance and processing can be influenced by the presence or ab-

TABLE 4.2. Manipulations of visual input in studies of the central auditory system.

Category	Method	Species	Reference
Visual deprivation	Early blindness	Human	Spigelman (1976)
	Binocular eyelid suture	Ferret	King and Carlile (1993)
		Barn owl	Knudsen et al. (1991)
		Guinea pig	Withington (1992)
	Dark rearing	Guinea pig	Withington-Wray et al. (1990a)
	Neonatal eye enuclea-tion	Hamster	Izraeli et al. (2002)
Displacement of the visual field	Prisms	Barn owl	Knudsen and Knudsen (1990)
	Surgically induced strabismus	Ferret	King et al. (1988)

sence of other sensory stimuli. This seems to apply particularly to spatial processing, where auditory plasticity has been observed, particularly during development, in response to altered visual experience (see Table 4.2). As well as providing valuable insights into the extent of and mechanisms that underlie the plasticity of auditory spatial representations, and of binaural processing in particular, these studies have also shed light on the behavioral significance of experience-driven changes in auditory response properties.

2.8 Free Field and Closed Field Stimulation

In psychophysical and neurophysiological studies of spatial hearing, sounds can be delivered either from loudspeakers positioned around the subject's head in the free field or via headphones. The free-field approach, which provides the full complement of acoustic cues associated with a real sound source, is more suitable for demonstrating plasticity in neural representations of auditory space and in sound localization behavior. On the other hand, dichotic presentation of sounds over headphones has been used to reveal experience-driven changes in sensitivity to particular binaural cues. To an increasing extent, these traditional approaches for presenting auditory stimuli are being supplemented by the use of virtual acoustic space (VAS) stimuli, which are headphone signals that replicate all the spatial cues associated with a real sound source, including the spectral filtering carried out by the external ears (Wightman and Kistler 1997a; King et al. 2001a). By presenting VAS stimuli based on acoustical measurements made from different subjects, psychophysical studies have shown that humans localize sounds most accurately when listening through their own ears (Wenzel et al. 1993; Middlebrooks 1999), suggesting that the neural mechanisms responsible are plastic and calibrated by experience of the cues provided by an individual's own ears. This approach is likely to be useful for investigating neuronal correlates of these behavioral findings (Mrsic-Flogel et al. 2001) and for assessing the relative contributions of monaural and binaural inputs in CAS changes produced by abnormal sensory experience.

3. Anatomical Studies

The binaural auditory system consists of CAS neurons at and above the level of the superior olivary complex (SOC), together with the axons that interconnect them, and associated supporting cells. Axons of CN neurons that project out of the CN will also be included, for the purposes of this chapter, but plasticity occurring wholly within or peripheral to the CN is not discussed, as it is the subject of Chapter 2 in this volume. Additional material on cellular plasticity in the developing SOC is presented in Chapter 5. A detailed presentation of the cells and pathways mentioned in this section may be found in other volumes of this series (normal, adult mammals—Webster et al. 1992; normal, adult birds—Dooling et al. 2000; normal, developing animals—Rubel et al. 1998).

3.1 Cochlear Nucleus Axons

Axons of mammalian AVCN spherical bushy cells synapse on neurons in the ipsilateral MSO and lateral superior olive (LSO) before projecting contralaterally, to synapse in the other MSO, and rostrally, to synapse in the nuclei of the lateral lemniscus (NLL) and the inferior colliculus (IC). AVCN globular bushy cell axons provide the major excitatory input to the neurons in the contralateral medial nucleus of the trapezoid body (MNTB), and these neurons send a dominant inhibitory input, deriving from the contralateral ear, to neurons in the LSO (Fig. 4.2; Cant 1991; Schwartz 1992). Neurons in the dorsal (DCN) and posteroventral (PVCN) CN project directly to the midbrain, sending a major projection that terminates in the contralateral IC, and a much more minor projection to the ipsilateral IC (Fig. 4.1; Oliver and Shneiderman 1991; Oliver and Huerta 1992).

Studies of CN projections to the SOC following input manipulations (Table 4.3) have shown that CN axons are capable of forming functional connections in inappropriate target zones (Fig. 4.3). Kitzes and colleagues (1995) and Russell and Moore (1995) performed unilateral cochlear ablations on postnatal day (P2–P14) gerbils. Axons from the VCN on the side of the intact ear were traced either with a carbocyanine dye (DiI) or with horseradish peroxidase (HRP). Normally, axons deriving from the left or right VCN terminate almost exclusively on the left or right dendrites, respectively, of neurons in both MSOs (Fig. 4.2). Neonatal (P2–P5) cochlear ablation induced, within 3–5 days, the formation of new VCN terminal zones in the region of MSO dendrites that are normally innervated by the VCN on the side of the ablated ear (Fig. 4.4). Terminals in the LSO contralateral to the intact ear were distributed throughout the nucleus, in contrast to their normal concentration in the hilar region (Fig. 4.4). Most dramatically, labeled calycine endings were found in the MNTB ipsilateral

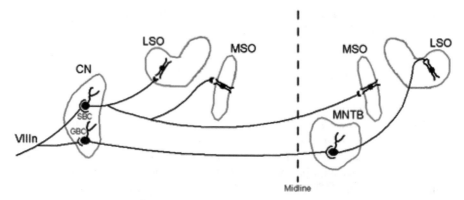

FIGURE 4.2. Organization of binaural connections in the auditory brain stem. Major connections between the eighth nerve, the AVCN, and the principal nuclei of the superior olivary complex are shown, based on studies in several species (see text). Note that the projection from the CN to the bipolar neurons of the MSO is bilateral, with the inputs from each side terminating on the homolateral dendrites. The projection to the LSO is also bilateral, but the dominant contralateral influence is via inhibitory neurons in the MNTB. The MNTB receives only contralateral input from the AVCN.

to the VCN in which the tracer was inserted, whereas these endings are, normally, found only in the contralateral MNTB (Figs. 4.3 and 4.5). These de novo connections were, apparently, produced by sprouting of existing axons, as there was no exuberant growth in normally developing gerbils of the same age range, and the new connections could sometimes be seen as buds from existing axons when the trapezoid body (TB) axons were examined at high magnification. The induced calyx terminals in the ipsilateral MNTB formed apparently normal, asymmetric synapses on their targets, and those synapses were functionally excitatory (Kitzes et al. 1995).

Two constraints have been found to limit the ability of VCN axons to make these aberrant connections. The first is that the animals had to be very young. Cochlear ablation in slightly older (P10–P14) gerbils did not result in any induced terminals (Russell and Moore 1995). The second is that total deafferentation of the SOC may not induce the connections. Rubel et al. (1981) found that, in the chicken, unilateral cochlear ablation induced axons from neurons in the nucleus magnocellularis (NM) on the side of the intact ear to form aberrant connections on neuron dendrites on the "wrong" side of NL. This result is comparable with that found in the gerbil (Fig. 4.3). However, section of the crossed dorsal cochlear tract (XDCT) in the chicken, a procedure that removes all afferent input from the ventral dendrites of the NL neurons, did not induce any sprouting of connections from the ipsilateral NM. Because ventral NL dendrites atrophy more rapidly and completely after XDCT lesions than after cochlear ablation (see Section 3.2), Rubel and colleagues suggested that the failure

TABLE 4.3. Anatomical plasticity of binaural systems.

System	Method	Species	Reference
CN to CN	Intracellular HRP	Chicken	Jackson and Parks (1988)
CN to SOC	Anterograde Fink-Heimer	Chicken	Rubel et al. (1981)
	Anterograde degeneration	Chinchilla	Morest and Bohne (1983)
	Anterograde DiI and HRP	Gerbil	Kitzes et al. (1995) Russell and Moore (1995)
	Silver staining	Chinchilla	Morest et al. (1997)
	EM	Gerbil	Russell and Moore (1995)
CN to IC	Anterograde Fink-Heimer	Gerbil	Moore and Kitzes (1985)
	Retrograde HRP	Gerbil	Nordeen et al. (1983)
	Retrograde HRP	Ferret	Moore and Kowalchuk (1988) Moore et al. (1989) Moore (1990a) Moore (1994)
SOC to SOC	Intracellular HRP	Gerbil	Sanes and Takacs (1993)
SOC to IC	Retrograde HRP	Gerbil	Nordeen et al. (1983)
	Retrograde HRP	Ferret	Moore et al. (1995)
IC to IC	Extracellular biocytin	Gerbil	Hafidi et al. (1995)
	Retrograde biotinylated dextran amine	Barn owl	Feldman and Knudsen (1997)
	Anterograde biocytin	Barn owl	DeBello et al. (2001)
AI to IC	Anterograde Fink-Heimer	Rat	Land et al. (1984)
CN neurons	Nissl	Guinea pig	Dodson et al. (1994)
		Rat	Coleman and O'Connor (1979)
		Rat	Dyson et al. (1991)
SOC neurons	Nissl	Cat	Powell and Erulkar (1962) Schwartz and Higa (1982) Conlee et al. (1984)
	Nissl	Mouse	Webster (1983a,b)
	Nissl	Ferret	Moore (1992)
	Nissl	Gerbil	Pasic et al. (1994)
	Golgi	Rat	Feng and Rogowski (1980)
	Golgi	Chicken	Parks (1981) Smith et al. (1983)
	Golgi	Gerbil	Sanes and Chokshi (1992) Sanes et al. (1992)
	EM	Chicken	Dietch and Rubel (1984, (1989a,b)
	EM	Cat	Schwartz and Higa (1982)
IC neurons	Nissl	Mouse	Webster (1983a,b)
	Golgi	Rat	Killackey and Ryugo (1977)
	Cytochrome oxidase	Rat	Paterson and Hosea (1993)
AI neurons	Golgi	Rabbit	McMullen and Glaser (1988) McMullen et al. (1988)
	Golgi	Mouse	Gyllensten et al. (1966) Ryugo et al. (1975)

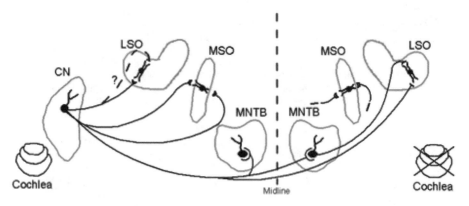

FIGURE 4.3. Plasticity of binaural connections in the gerbil auditory brain stem. Following contralateral cochlear removal very early in life, AVCN axons form terminal sprouts to innervate two, formerly inappropriate SOC targets—the heterolateral dendrites of neurons in both MSOs and the principal neurons of the ipsilateral MNTB. In addition, there is a greatly increased, direct projection to the contralateral LSO. [Based on Kitzes et al. (1995); Russell and Moore (1995).]

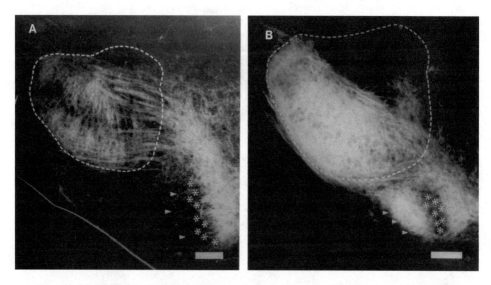

FIGURE 4.4. Plasticity of auditory afferents in the neonatal gerbil LSO and MSO. Small crystals of DiI were placed in the right AVCN of gerbil pup brains that had been fixed following sacrifice at P12–P14. (A) In control animals, labeled axons invade the border regions of the lateral limb of the left LSO (outlined with *dashed lines*), and arborize extensively in the neuropil containing the medial (homolateral) dendrites of neurons in the left MSO. Note the absence of arbors lateral to the somata (*asterisks*) of the left MSO (*arrowheads*). (B) In gerbils that had the right cochlea removed at P2–P5, the axons arborized more extensively in the LSO and formed inappropriate terminals lateral to the MSO. [See also Figs. 4.2 and 4.3. Scale bars: 0.1 mm. These micrographs from Russell and Moore (1995).]

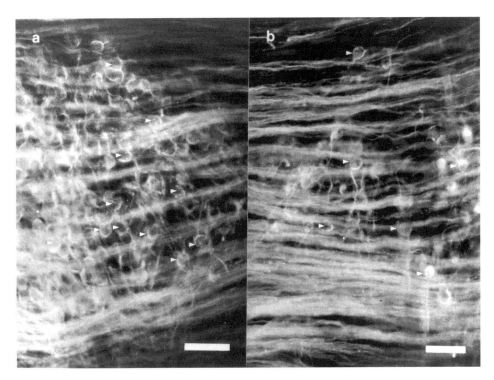

FIGURE 4.5. Plasticity of auditory afferents in the neonatal gerbil MNTB. Methods were as described in Fig. 4.4. Afferents from the right AVCN terminate in giant (calyces of Held) synapses on principal neurons in the MNTB. (A) In control animals, calyces (*arrowheads*) are found only in the left MNTB (see Fig. 4.2). (B) In gerbils that had the right cochlea removed at P2–P5, calyces of apparently identical form and function (Kitzes et al. 1995) have also formed in the right MNTB. [See Fig. 4.3. Scale bars: 0.1 mm. These micrographs from Russell and Moore (1995).]

to observe axon sprouting in the former case may be related to the loss of the postsynaptic target.

The functional consequences of unilateral cochlear ablation are discussed in detail in Section 4.3, but it is worth considering here how the induced VCN connections in the SOC might affect auditory processing. Increased afferent convergence onto MSO neurons might be expected to result in stronger excitatory responses to stimulation of the intact ear and, depending on the specificity of the connections, to broader frequency tuning. The novel input to the ipsilateral MNTB has been shown to result in sound-evoked excitation (Kitzes et al. 1995), and that excitation would, presumably, be conveyed as an inhibitory input to the ipsilateral LSO. The direct effect of the ablation on the LSO ipsilateral to the intact ear is not known, mainly because of the massive number of labeled fibers running through the nucleus after the placement of a DiI crystal in the nearby VCN. Finally, the increased direct projection from the intact side to the contralateral LSO might turn the normally inhibitory influence of the contralateral ear on that nucleus into an excitatory influence, thus further boosting the overall level of excitation produced by the intact ear. Alternatively, the novel projections might make inhibitory connections with neurons of the MSOs and contralateral LSO. This possibility, requiring investigation by immunocytochemistry, EM and/or recording, is intriguing, as it addresses questions concerning the influence of afferents and target cells on the formation of excitatory and inhibitory synapses.

Cochlear lesions in older gerbils (Russell and Moore 2002) and chinchillas (Morest et al. 1997) produce longer-term degeneration of some CN axon terminals in the MSO. Degenerating fine and medium-sized axons were first seen after 14 days survival. After 1–20 months, the remaining terminals containing round vesicles (presumed to be excitatory) were smaller, and less densely populated with vesicles, where they synapsed onto the dendrites of MSO neurons on the side of the deafened ear.

3.2 The Superior Olivary Complex

A number of studies have investigated the effect of conductive and sensorineural hearing loss, and of XDCT section, on the morphology and ultrastructure of SOC neurons (Table 4.3). Nissl studies have shown that, as in the CN (see Chapter 2), neurons in some SOC nuclei die following cochlear ablation in infancy, and the mean soma area of the remaining neurons is smaller than normal. However, unlike the CN, neurons in different divisions of the SOC are not affected equally by cochlear removal. For example, significant neuron loss has been found in the ipsilateral, but not in the contralateral LSO of the ferret, following P5 cochlear ablation (Moore 1992), and significant neuron shrinkage has been found in the contralateral, but not in the ipsilateral MNTB following ablation in older (P32–P40) gerbils (Pasic et al. 1994). Neuron numbers and areas in the MSO do not appear to be affected by unilateral ablation (D.R. Moore and S.L. Pallas, unpublished data). Unilateral cochlear ablation in adult cats (Powell and Erulkar 1962) also produced neuron shrinkage in the ipsilateral LSO and the contralateral MNTB, but no neuron loss.

Dendrites of MSO and LSO neurons have been found to be affected by hearing loss and deafferentation. In the chicken, otocyst ablation on the third day of incubation (E3) resulted, by E17, in a >40% reduction of mean dendrite length on the side of NL neurons normally innervated by fibers deriving from the ablated ear (Parks 1981). XDCT section (at posthatch day P10) reduced the length of dennervated dendrites by 10% within 1 hour, and by >60% 16 days after the surgery (Deitch and Rubel 1984). In the gerbil LSO, cochlear ablation at P7 produced a complex pattern of dendritic elongation and branching that depended on the laterality of the nucleus examined, relative to the deafened ear, and the position of the parent neuron along the tonotopic axis of the nucleus (Sanes et al. 1992).

Two findings from these morphological studies of the SOC seem particularly noteworthy. While cochlear ablation in infancy generally produces an atrophy of neurons throughout the auditory brain stem, neuron somata and dendrites in the LSO contralateral to the ablated cochlea are, apparently, *larger* than those in a normal animal. Sanes and his colleagues have attributed this response to a withdrawal of the inhibitory input to the neurons from the ablated side. This suggestion has received support from an experiment in which strychnine, a glycine receptor antagonist, has been applied in the vicinity of the LSO for 18 days from P3 (Sanes and Chokshi 1992). Inhibitory transmission blockade and contralateral cochlear ablation produced similar effects on LSO neurons. Thus, both excitatory and inhibitory connections can contribute to the response of binaural neurons to modifications of acoustic input.

The second noteworthy finding comes from a study by Pasic et al. (1994). Unilateral cochlear ablation, or application of TTX to the round window, led to neuron shrinkage in the VCN and MNTB. However, shrinkage in the VCN *preceded* that in the MNTB. Gerbils that survived for 24 hours after the ablation showed a 10–20% reduction in neuron soma size in the spherical and globular bushy cell areas of the VCN, but no reduction in size of the principal cells of the MNTB. After 48 hours survival, both VCN and MNTB neurons were reduced in size by 10–20%. These results confirmed that the maintenance of target cell metabolism, assumed to underlie soma size, is dependent on afferent activity (see Chapters 2 and 5). In addition, they suggested that a slower process or processes, possibly axonally transported trophic substances, released from afferent fibers, may also contribute to this maintenance. In terms of binaural plasticity, the results show that neural changes contingent on auditory input can occur over different time courses that may depend on, among other factors, the length and transport characteristics of the axons connecting the periphery to the central target. Where these characteristics are believed to differ between the two sides (e.g., in "delay lines" to the MSO; see also Section 4.2), the maintenance of the target may be dependent on the interaural symmetry of auditory input.

Conductive hearing loss has been reported to reduce neuron size in the ipsilateral VCN, and in the contralateral MNTB and IC following ear canal atresia in neonatal mice (P4; Webster 1983a), but not in older animals (P45; Webster 1983b) (see Rubel, Parks, and Zirpel, Chapter 2). Other forms of unilateral conductive loss in other species have, however, produced a mixture of results. In some studies, rearing with a conductive loss failed to produce any neuron shrinkage in the VCN (Tucci and Rubel 1985; Moore et al. 1989). In other studies, changes in the length of MSO dendrites receiving input from the deprived side have been reported (Feng and Rogowski 1980; Smith et al. 1983). The interpretation of these results is confounded by the finding of different outcomes from apparently identical manipulations of auditory input. One explanation is that at least some forms of conductive loss inadvertently produce a sensorineural loss by promoting infection and/or inflammation in the middle and inner ear (e.g., Sterritt and Robertson 1964). Since most of these anatomical studies have not used any functional assessment of the auditory periphery, and since, in those studies that have used such an assessment, more or less subtle, long-term and apparently irreversible effects of the "conductive loss" have been noted, this explanation is difficult to refute. Nevertheless, physiological and behavioral results, reviewed below, strongly suggest that reversible conductive hearing loss does have longer term, functional effects on binaural systems. Because all functional effects have a structural correlate, it is therefore possible that some or all of the cellular changes reported to have followed from a conductive hearing loss have, in fact, been caused by that loss. In any case, the value of combining a functional assessment of the auditory periphery with cellular studies of conductive hearing loss should be clear.

Whatever the cause of the loss and shrinkage of neurons in the SOC, it is necessary to consider what implications those cellular changes might have for binaural processing. Reductions in neuron numbers presumably limit the ability of nuclei to relay information to higher structures. If there is an imbalance between the two sides of the brain, as there is in the LSO and MNTB after neonatal cochlear ablation, the appropriate integration of information at higher levels will be compromised. However, neurons seem to be lost only following total deafness or activity blockade early in life, before the onset of hearing (Tierney et al. 1997), and these manipulations would, in themselves, profoundly affect integration. Unilateral dendritic shrinkage of neurons in the MSO, following more subtle forms of hearing loss, would be expected to unbalance the delicate integrative function that these neurons are thought to possess (e.g., Yin and Chan 1990). Unfortunately, there is a lack of ultrastructural data on synaptic distribution and densities on MSO neurons after unilateral hearing loss. Nevertheless, even if the number and size of input synapses remained constant, the relationship of those synapses to the impulse generation site of the neurons and the passive electrical properties of the dendrites would undoubtedly be affected by the shrinkage. This would change the temporal integration of the neurons that is thought to be pivotal to the "delay line" (Jeffress 1948) concept of interaural phase coding (see Irvine 1992).

3.3 Input to the Inferior Colliculus

The direct projection from the CN to the IC (Fig. 4.1) has been the subject of several studies of experimental hearing loss (Table 4.3). The main advantage of this projection is that it is experimentally tractable. As outlined above, the system consists of a crossed component, from all three divisions of the CN to the contralateral central nucleus of the IC (ICC), and an uncrossed component, again from the three divisions of the CN, to the ipsilateral ICC. In terms of the number of CN neurons retrogradely labeled by an injection of a tracer in the IC, the system is highly asymmetric. In the ferret, for example, about 50 times as many CN neurons from each division project contralaterally as project ipsilaterally (Moore 1988). Nordeen et al. (1983) originally showed that, following neonatal, unilateral cochlear ablation, the number of ipsilaterally projecting neurons increases substantially—by up to 100% (Fig. 4.6). In contrast to the other effects of cochlear ablation described above, the increase in the CN to ipsilateral IC projection (CN-ICi) occurs slowly. No effect was found until several weeks after the ablation in P24 ferrets, then increases in the CN-ICi gradually accrued for at least several months after the ablation (Moore 1994; Fig. 4.1). The CN-ICi has also been found to remain plastic relatively late in life. Cochlear ablation at P90 was, following three months survival, originally thought not to affect CN-ICi (Moore and Kowalchuk 1988). However, following a much longer survival (2.5 years) the projection was, again, found to be significantly larger than that in a normal ferret. It is unknown whether this projection can be modified by very long survival following cochlear ablation in adulthood. Nordeen et al. (1983) did not find any change in adult gerbils that survived for 5–13 months after the ablation. The projection is not altered by bilateral cochlear ablation (at P24; Moore 1990a), but is changed by unilateral ear plugging from P24 (Moore et al. 1989).

Two conclusions may be drawn from these results. The first is that specific neural pathways that may contribute to binaural hearing are affected by modi-

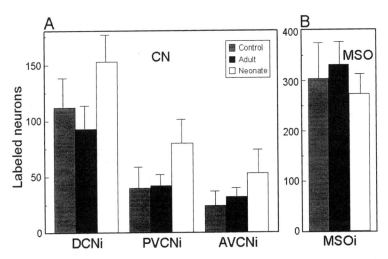

FIGURE 4.6. Contrasting effects of cochlear removal on two binaural pathways in the gerbil auditory brain stem. HRP was injected into the IC of adult gerbils that had the contralateral cochlea removed at P0–P2 (Neonate) or in adulthood (see Fig. 4.1A). (A) Following neonatal cochlear removal, the mean number of neurons retrogradely labeled in each division of the CN on the unoperated side (ipsilateral to the injected IC) was about twice that of the control (unoperated) or adult operated animals. (B) In contrast, in the same animals, the number of retrogradely labeled neurons in the MSO ipsilateral to the injected IC was unaffected by cochlear removal in either age group. [Adapted from Nordeen, Killackey and Kitzes (1983).]

fications of auditory input after the onset of hearing. It follows that some cellular events, the reorganization of the CN-ICi, together with the SOC neuron dendritic changes discussed above, are candidates for modification by auditory experience, whereas other events, including neuron death in the brain stem and the reactive sprouting of VCN axons, occurring only after input modifications before the onset of hearing, are not candidates. The second conclusion is that, in contrast to all the other changes mentioned above, the CN-ICi changes slowly; over a time scale of weeks to months, rather than minutes to hours. This time scale suggests a dependence on long-term input modification, as might occur in various forms of hearing loss and amplification. It also suggests a different cellular mechanism from those that have been found to underlie the rapid responses of CN neurons to input blockade (see Chapter 2). In the CN-ICi, changes may involve axon sprouting and synaptogenesis, mechanisms found to be associated with plasticity in the adult visual cortex occurring over similar time scales (Darian-Smith and Gilbert 1994). In an anterograde study of the CN-ICi (Moore and Kitzes 1985) it was found that, following cochlear ablation in the neonatal gerbil, terminal arbors of CN neurons distributed very widely in the IC, suggesting the possibility that the apparent increase in the number of CN neurons labeled in retrograde studies was, in fact, the result of a greater uptake of tracer through terminal sprouts. Recently, Gabriele and colleagues (2000) have shown an asymmetry between the two ICs and the two dorsal nuclei of the lateral lemniscus (DNLL) in the uptake of tracer injected into the commissure of Probst following unilateral cochlear ablation in neonatal rats. Again, the distribution of both axonal terminal fields and their parent somata are influenced by asymmetric withdrawal of afferent input in infancy. It will be of great interest to discover if similar cellular events underlie experience dependent plasticity in binaural systems later in life.

Projections from the SOC to the IC following neonatal cochlear ablation have also been examined in several studies (Table 4.3). In contrast to the direct projection to the IC from the CN, the number of MSO neurons projecting to the ipsilateral IC has been found to be unaffected by unilateral ablation of either ear (Nordeen et al. 1983; Fig. 4.6). Similarly, neurons in the LSO that normally project in about equal numbers to either IC continue to do so after unilateral ablation (Moore et al. 1995). Although negative, these findings allow the evaluation of certain hypotheses concerning the effects of auditory input. An obvious but important one is that cochlear ablation does not change all brain stem pathways. Thus, the changes that have been found are not due to some general effect of ablation on brain stem neurons. Nor are they simply the result of degeneration, leading to an imbalance of afferent input between pathways deriving from the two ears since, as detailed previously, one LSO is atrophied and the other is hypertrophied following unilateral cochlear ablation in infancy. It may therefore be that there is something special about primary deafferentation. Alternatively, there may be a "threshold" effect, where a certain level of activity reduction (relative to the other input[s]?), or a certain amount of atrophy, is required to trigger changes in the other input[s]. It is also possible that the failure to find changes in the SOC to IC projections was influenced by the other changes produced lower down the system, in the CN to SOC projections. Sprouting of VCN axon terminals to provide additional input to SOC neurons might provide sufficient support, for example, to the LSO on the side of the ablation, to allow it to "compete" with the other LSO for target space in the IC. Most of these ideas are speculative, but they are all testable within the auditory brain stem.

The functional implications of the changes in the CN-ICi are unclear. Because retrograde studies have shown it to be a minor projection, compared with the much more numerous contralateral projection, its influence on spatial hearing may be subtle. On the other hand, the limited anterograde evidence suggests that cochlear removal has a profound effect on the distribution of the projection within the target IC. If the exuberant terminals in the ipsilateral IC have a functional parallel, the influence of the intact ear on that IC might increase substantially relative to its influence on the contralateral IC. This anatomical rewiring, which is also observed, to some extent, after plugging one ear in infancy (Moore et al. 1989), may influence the processing of acoustic localization cues at higher levels of the auditory system and contribute to changes in sound localization and related behavior that have been reported following a unilateral hearing loss (see Section 5.1).

3.4 Other Areas and Connections

There have been several reports of other cellular changes in the auditory system following cochlear ablation that may be relevant to binaural plasticity (Table 4.3). One particularly interesting pathway is a novel, functional, and topographically appropriate one formed between the two NMs in the chicken following either unilateral or bilateral ablation (Jackson and Parks 1988; Lippe et al. 1992). This pathway is described elsewhere in this volume (see Chapter 2). Equivalent de novo pathways have not previously been sought in species other than the chicken, but a CN–CN connection has been demonstrated in the normal cat (Cant and Gaston 1986). If conductive hearing loss or other manipulations of the ears leads to an asymmetric, enhanced connection between the CNs, this could compromise the balance of binaural processing at higher levels of the auditory system.

An anatomical reorganization of projections within the different subdivisions of the IC has been observed in the barn owl after raising juvenile birds with either a unilateral conductive hearing loss (Gold and Knudsen 2001) or with prismatic spectacles that displace the animal's visual field laterally (Feldman and Knudsen 1997; DeBello et al. 2001). For localization in the azimuthal plane, mammals rely on ITDs at low frequencies and ILDs at high frequencies (Middlebrooks and Green 1991). Barn owls differ, however, in being able to use both binaural cues for localization over the same frequency range. Because of the asymmetry in their external ears, horizontal localization in this species depends principally on ITDs, whereas vertical localization is based on ILDs

(Knudsen and Konishi 1979; Moiseff and Konishi 1981). In the midbrain, ITD information is transmitted via topographic projections from the ICC to the external nucleus of the IC (ICX) and thence to the optic tectum (OT). Altered sensory experience during development has been shown to bring about a systematic remodelling of the projection from the ICC to the ICX, which appears to underlie the adaptive changes in the representation of auditory space that have been observed using recording techniques (Section 4).

Altered sensory inputs can also lead to changes in neuron morphology in the auditory cortex (AC; Table 4.3). Two months after unilateral deafening in neonatal rabbits, AC spiny pyramidal neurons were found to have fewer basal dendritic spines than those in normal-hearing animals, although no changes were found in soma size or in the number and length of dendrites (McMullen and Glaser 1988). In contrast, nonspiny, nonpyramidal neurons had longer than normal dendrites, with at least some dendrites showing signs of abnormal growth (McMullen et al. 1988). These and other, more recent results (Pallas et al. 1999) show that auditory input manipulations can produce demonstrable cellular effects in neurons several synapses removed from the site of the manipulation. Changes in neuron morphology have also been reported as far apart as the CN (Dyson et al. 1991) and AC (Gyllensten et al. 1966; Ryugo et al. 1975) in visually deprived rodents. These studies suggest that all levels of the CAS may be affected by at least some forms of sensory manipulation. Moreover, they show that morphological specializations (dendritic spines), thought to be major sites of synaptic input to AC pyramidal neurons, are sensitive to peripheral manipulations. These are some of the very few studies (see also Deitch and Rubel 1989a,b; Friauf, Chapter 5) that have searched for cellular changes in the CAS beyond the level of the CN. They point the way to possible candidates for the influence of binaurally imbalanced hearing on the CAS and it will be of interest to determine how such changes might affect binaural interaction, for example, in the AC.

4. Physiological Studies

Physiological investigations of binaural plasticity (Table 4.4) may be divided into three groups that, to some extent, reflect historical influences as well as scientific considerations. Following the lead of Wiesel and Hubel (1963; see Section 1.1), and other pioneers of visual system plasticity (e.g., Blakemore and Cooper 1970), auditory neuroscientists began by rearing animals in controlled acoustic environments, or with a unilateral conductive hearing loss. The expectation of these experiments was that, as neurons in the visual cortex were found to change their response properties in line with visual experience, so auditory neurons would become exclusively responsive to, for example, the tone frequency that was experienced, or the ear that was preferentially stimulated during rearing. Some of the methodological problems with these experiments are discussed above (see Sections 2.3 and 3.2), and the experiments involving monaural

TABLE 4.4. Physiological plasticity of binaural systems.

Stimulus method/ input manipulation	Recording method	Nucleus	Species	Reference
Closed field/ear occlusion	Single unit	IC	Rat	Clopton and Silverman (1977, 1978)
				Silverman and Clopton (1977)
			Cat	Moore and Irvine (1981)
			Barn owl	Mogdans and Knudsen (1993)
				Gold and Knudsen (2000b)
		OT	Barn owl	Mogdans and Knudsen (1992)
				Gold and Knudsen (2000a)
		AC	Cat	Brugge et al. (1985)
	ABR	—	Rhesus monkey	Doyle and Webster (1991)
		—	Guinea pig	Laska et al. (1992)
Closed field/pinna removal	Multiunit	OT	Barn owl	Knudsen et al. (1994)
Closed field/cochlear ablation	Single unit	IC	Gerbil	Nordeen et al. (1983)
				Kitzes (1984)
				Kitzes and Semple (1985)
				Moore and Kitzes (1986)
			Ferret	Moore et al. (1993)
		AC	Cat	Reale et al. (1987)
	ABR	IC, AC	Guinea pig	Popelár et al. (1994)
Closed field/visual deprivation	Single unit	OT	Barn owl	Knudsen et al. (1991)
Closed field/prisms	Single unit	IC	Barn owl	Brainard and Knudsen (1993)
				Feldman and Knudsen (1998a)
				Gutfreund et al. (2002)
		OT	Barn owl	Knudsen and Brainard (1991)
				Brainard and Knudsen (1995, 1998)
				Knudsen (1998)
				Linkenhoker and Knudsen (2002)
Electrical stimulation/ cochlear ablation	Single unit	IC	Cat	Snyder et al. (1990, 1991)
Free field/ear occlusion	Single unit	IC	Big brown bat	Jen and Sun (1990)
		OT	Barn owl	Knudsen (1983, 1985)
				Gold and Knudsen (1999)
		SC	Ferret	King et al. (1988, 2000)
	2-DG	IC	Rat	Clerici and Coleman (1986)
Free field/reshape external ear	Single unit	SC	Ferret	Schnupp et al. (1998)
Free field/cochlear ablation	Single unit	SC	Ferret	King et al. (1994)

TABLE 4.4. *Continued*

Stimulus method/ input manipulation	Recording method	Nucleus	Species	Reference
	ABR	—	Ferret	Moore (1993)
	2-DG	CN, MSO	Chicken	Lippe et al. (1980)
		SOC, IC	Chicken	Heil and Scheich (1986)
		—	Guinea pig	Sasaki et al. (1980)
Free field/cochlear lesion	Evoked potential	IC	Chinchilla	Salvi et al. (1990)
Free field/omnidirectional noise	Multiunit	SC	Guinea pig	Withington-Wray et al. (1990a)
Free field/visual deprivation	Single unit	SC	Ferret	King and Carlile (1993)
	Multiunit	SC	Guinea pig	Withington (1992) Withington-Wray et al. (1990a)
	Single unit	Field AES in cortex	Cat	Korte and Rauschecker (1993)
Free field/surgically induced strabismus	Single unit	SC	Ferret	King et al. (1988)
Free field/partial SC lesion	Single unit	SC	Ferret	King et al. (1998)
Free field/receptor modulation	Single unit	SC	Ferret	Schnupp et al. (1995)

occlusion or other manipulations of the outer ear are detailed below (Sections 4.1 and 4.2).

The second group of experiments involved unilateral cochlear ablation. Recognizing the difficulties of controlling auditory experience and the likely greater effects of deafferentation (Levi-Montalcini 1949; see Section 3, and Rubel et al., Chapter 2), it was believed that unilateral deafening would both reveal processing that could be changed by auditory input, and provide fundamental data on the effect of various afferent influences on neuronal responses to sound. The extent to which this method has succeeded in those aims is discussed in Section 4.3.

Finally, the discovery of topographically aligned maps of visual and auditory space in the OT of the barn owl (Knudsen and Konishi 1978) and the SC of mammals (Palmer and King 1982), and their likely contributions to sound localization (Cohen and Knudsen 1999), led to questions concerning the role of visual as well as auditory experience in the formation of the maps, reviewed in Section 4.4, and in the acquisition of accurate localization (see Section 5.1). Physiological and imaging studies in humans have also attempted to reveal changes in auditory processing that may be relevant to the effects of early blindness on the accuracy of sound localization in humans (Sections 4.4 and 5.5).

4.1 Ear Occlusion

Occlusion of the ear canal, usually achieved in animal studies by plugging, ligature, or cauterization, produces a level reduction and a time delay for air-borne sound reaching the tympanic membrane. Where the occlusion is unilateral, this creates an interaural level difference (ILD) and an interaural time difference (ITD) that are added to the well-known ILD and ITD produced, for a laterally positioned sound, by the influence on the acoustic field of the head and external ears. The binaural cues associated with a particular point in space are therefore abnormal; a sound on the mid-sagittal plane, for example, will appear to be displaced toward the unoccluded ear.

Unilateral ear occlusion produces a number of other, less obvious changes in the perception of sounds. Although most forms of occlusion result in a hearing loss that is relatively flat across the audible spectrum, they will invariably introduce some spectral distortion, with high frequencies typically attenuated more than low frequencies (e.g., Moore et al. 1989). Thus, ILDs and ITDs will be affected in a frequency-dependent manner, as clearly demonstrated in recent studies in which barn owls were raised with an acoustic filtering device fitted to one ear (Gold and Knudsen 1999). As outlined above, ear canal occlusion potentiates bone-conducted sounds, particularly those of low frequency (Tonndorf 1972). Self-produced sounds and intense, external sounds will therefore receive further spectral shaping through this mechanism. Finally, an important consideration in the interpretation of the results from ear occlusion experiments is the extent to which binaural cues are altered or eliminated altogether. Indeed, because of the shadowing effect of the head and the attenuating effect of the occlusion, low-level and high-frequency sounds may not be detected at all if they are positioned on the side of the occluded ear. As psychophysical studies using VAS stimuli (Wightman and Kistler 1997b) have shown, however, some authors may have incorrectly assumed that their ear-plugged subjects were using monaural cues alone to localize sounds.

Silverman and Clopton (1977; Table 4.4) performed the first physiological studies of the effects of ear occlusion on the response of neurons in the rat IC (Fig. 4.7). Animals received either unilateral or bilateral ear canal ligation, starting at P10–P60, and lasting for 3–5 months. The ligation was removed and click stimuli were delivered to either or both ears. Sensitivity to ILDs was determined by comparing the mean peak response to stimulation of the contralateral ear with that produced by binaural stimulation. In normal-hearing animals, this comparison revealed a marked binaural suppression for ILDs that were more positive at the ear ipsilateral to the recording site. For animals unilaterally ligated from P10, the binaural suppression was reduced when recordings were

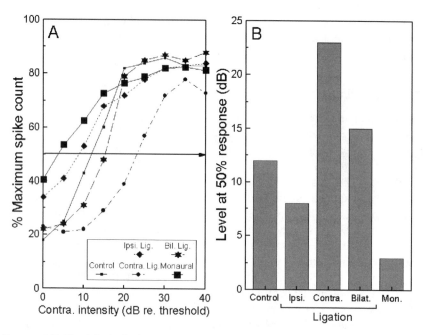

FIGURE 4.7. Physiological plasticity in the rat IC. Rats were reared, from P10, with a ligature of neither, either, or both ear canals. After 3–5 months, the ligature was opened and single-unit recordings were made in the IC. **(A)** For each unit, the response to contralateral stimulation (0–40 dB re. contra. threshold) was compared with the response to binaural stimulation of varying ILD (ipsi. +40 to −40 dB re. contra.). All responses were normalized to the maximum response for that unit and mean % maximum spike counts were derived for each group. In control animals, the Monaural, contralateral response (*filled squares*) increased monotonically with stimulus level, reaching saturation at about 30 dB above threshold. Adding the ipsilateral stimulus (Control—small squares) suppressed the contralateral response at ILDs that favored the ipsi. ear (low contra. intensities). In rats that had been reared with a bilateral ligation (Bil. Lig.), the ipsilateral suppression was near the control level. In rats that had been reared with a ligature of one ear canal, the degree of suppression was higher than normal in the IC contralateral to the formerly ligated canal (Contra. Lig.), and lower than normal in the IC ipsilateral to that canal (Ipsi. Lig.). **(B)** The sound level at the contralateral ear for which the mean response of the IC neurons in each group was half-maximal shows the dramatic difference between the effects of Ipsi., Contra. and Bil. ligation. These results provide evidence for experience-dependent plasticity of responses to ILD in the rat auditory system. [Adapted from Silverman and Clopton (1977).]

made on the side of the previously ligated ear, and increased when recordings were made on the side of the unligated ear. Bilateral ligation produced little change in binaural suppression. The extent of the changes in binaural suppression following unilateral ligation was dependent on the age of the animal at the time the ligation was first imposed. The largest changes occurred with ligation from P10, with reduced change from P30 and little or no change from P60.

A key issue in these experiments, as in all experiments involving ear occlusion, was the peripheral contribution to the observed, central changes. Silverman and Clopton measured the cochlear microphonic (CM) from a sample of previously ligated ears and found only a small ($\sim$5 dB) loss of sensitivity in those ears. In contrast, the shift in binaural suppression produced by the ligation, comparing the extreme ipsilateral and contralateral recording conditions (Fig. 4.7), was in the order of 20 dB. The conclusion from these comparisons was that the difference between the CM and unit data showed that most of the observed change in binaural suppression was due to CAS plasticity.

Two other studies used similar techniques to examine the effect of unilateral ear occlusion during infancy on the responses of cat CAS neurons to ILDs. In the first of these studies (Moore and Irvine 1981), cats were reared for about 3 months, from just after birth or as adults, with a unilateral ligature of the ear canal. Single neurons in the IC were tested for sensitivity to ILDs following removal of the ligature. Both infant and adult ligated animals were found to have a dearth of ILD sensitive neurons, compared with normal controls. The insensitivity reflected a diminished inhibitory input from the nonligated ear. In the second study (Brugge et al. 1985), cats were reared from soon after birth (P1–P4) for 8–14 months with a unilateral ear canal atresia, induced by cutting and tying the canal. Single neurons in the AC ipsilateral or contralateral to the surgically opened atretic ear were assessed for sensitivity to ILDs. In contrast to the results of Moore and Irvine, the form of the ILD sensitivity functions of these neurons was normal and, in apparent agreement with Silverman and Clopton, the ILD sensitivity was shifted in favor of higher sound levels at the previously atretic (occluded) ear. However, based on monaural response thresholds to stimulation of the atretic ear, Brugge and colleagues concluded that the shift in ILD sensitivity could be entirely accounted for by the residual hearing loss (20–30 dB) in the atretic ear.

Because of the conflicting outcomes of these studies, it is difficult to draw firm conclusions about the ILD response of mammalian CAS neurons to ear occlusion. The finding of a considerable number of ILD sensitive neurons in the cat AC, compared with the earlier difficulty of finding these neurons in the IC, suggests that there may have been a sampling difference between the studies. Indeed, one of the major problems in single unit studies is the diversity of response types in the CAS, and the definition of inclusion criteria for neurons recorded in different experiments. A second problem, as discussed in the preceding paragraphs, is the extent to which residual peripheral insensitivity, produced by the occlusion, can account for the CAS results. In two of the above studies, opposite conclusions were reached following an assessment of the contribution of this factor.

Clearer evidence for plasticity of spatial processing has come from studies of the role of binaural experience in the maturation and maintenance of maps of auditory space in the midbrain. In particular, the presence of superimposed maps of visual and auditory space in the mammalian SC and its avian homologue, the OT, has allowed the alignment between the visual and auditory receptive fields of individual bimodal neurons, or of different neurons recorded at corresponding points within each map, to be used as a measure of adaptive plasticity in the auditory representation.

Knudsen (1983, 1985) found that, in barn owls reared with one ear plugged, the visual and auditory maps were aligned when recordings were made with the earplug in place (Fig. 4.8). Thus, the auditory space map in the OT appears to adjust appropriately to the altered binaural cues produced by the plug so as to maintain the alignment with the visual map. Dichotic studies, carried out in the absence of the earplug, have revealed that the basis for this adjustment involves a shift in the ITD and ILD tuning of OT neurons toward the side of the unplugged ear (Mogdans and Knudsen 1992). These findings demonstrate binaural plasticity in this pathway, as the shifts in ITD and ILD sensitivity were in the opposite direction to those expected had there been any residual peripheral insensitivity due to debris or other changes in the external ear. They also explain why removal of the earplug moved the auditory receptive fields, mapped with free-field stimuli, out of alignment with the visual receptive fields and toward the side of the previously plugged ear. As outlined in Section 4.1, an alternative form of monaural occlusion, an acoustic filtering device, has been used more recently to show that the emergence of a normal representation of auditory space is based on frequency-specific adjustments in tuning to the binaural cues (Gold and Knudsen 2000a).

The capacity of owls to adapt to, and recover from, a long period of monaural occlusion appears to be developmentally regulated (Knudsen 1985). Thus, the auditory spatial tuning of OT neurons in owls reared with a plug in one ear from 30 to 50 days after hatching returned to normal and realigned with the visual map if plug removal occurred before about 200 days of age. After that age, the misalignment between the auditory and visual maps found immediately after plug removal persisted for at least 1 year. Moreover, plugging in adult owls did not produce an adaptive shift in the auditory space map in the OT.

The site of the plasticity seems not to be within the OT, but at or below the level of the ICX, the other known auditory space mapped nucleus in the barn owl. Abnormal auditory experience produced similar changes in ITD tuning in the ICX to those previously demonstrated in the OT (Gold and Knudsen 2000b), and a frequency-specific change in the pattern of connections from the ICC to the ICX (Gold and Knudsen 2001). ILD plasticity has been found at the first site of interaural level comparison in the owl, the posterior division of the ventral nucleus of the lateral lemniscus, VLVp (Mogdans and Knudsen 1994). Interestingly, the degree of the plasticity found at that level was not large enough to

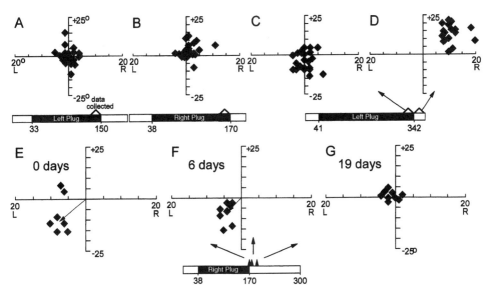

FIGURE 4.8. Plasticity of spatial tuning in the barn owl optic tectum. Single-unit recordings have shown that tectal neurons respond most strongly to sounds and lights presented from distinct regions of space and that the visual and auditory receptive field (RF) centers are aligned. When an earplug is introduced to one ear of an adult owl, the normal binaural cues underlying the auditory RF are disrupted, and a misalignment results between the auditory and visual RF. The alignment of the RFs has been used to study the effects of auditory experience on the representation of auditory space in the tectum. Visual RFs are unaffected by ear plugging. (**A,B**) Two owls were raised with plugs in one ear from P33 to P150 (**A**) and from P38 to P170 (**B**), as represented by the auditory history bars at the bottom of each figure. Data were collected shortly before the plug was removed and the auditory RF center of each unit was plotted relative to the visual RF center of the same unit. L, Left; R, right; +, above; −, below. Plug rearing produced auditory and visual RFs that were in alignment while the plug remained in place. (**C,D**) In another owl, longer rearing (to P342) also produced auditory/visual alignment with the plug in but, when the plug was removed (**D**), the fields became misaligned to a degree that was predictable from the displacement normally produced by plug insertion. (**E–G**) Recordings made from a further owl immediately after (0 days, **E**) plug removal at P170 showed the same displacement of the auditory RFs as had been observed in the older owl (**D**). However, subsequent recordings made 6 (**F**) and 19 (**G**) days after plug removal revealed a gradual realignment of the auditory and visual RFs, as shown by the median vectors (*arrows*) in each figure. Auditory experience can therefore change the registration of auditory and visual representations in the owl tectum. [Adapted from Knudsen (1985).]

account for the changes found in the OT, indicating that processing changes at another site above the level of the VLVp, possibly the ICX, also contribute to the tectal plasticity.

Plasticity of the auditory space map in the mammalian SC has also been shown using single-unit recording in adult ferrets following unilateral ear plugging during infancy (King et al. 1988, 2000). While the earplugs were in place, most neurons had spatially tuned receptive fields, and the map of auditory space was close to normal and in approximate register with the visual map (Fig. 4.9). In contrast, a comparable period of monaural occlusion in adult ferrets did not lead to adaptive changes in auditory spatial tuning. Thus, removal of the earplug in these animals immediately led to normal alignment between the auditory and visual maps of space. In these respects, plasticity in the mammalian SC resembled that in the barn owl OT. However, the mechanisms underlying generation of the space map, and sound localization in general, in the barn owl differ in several important ways from those in mammals. Of relevance to the issue of binaural plasticity is the evidence that the mammalian map is synthesized from a combination of monaural spectral cues and ILDs, whereas the owl map appears

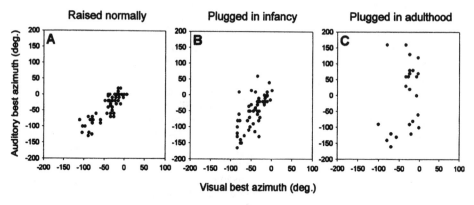

FIGURE 4.9. Effects of chronic monaural occlusion on the registration of the auditory and visual maps of space in the ferret superior colliculus (SC). Each panel shows the relationship between the representations of visual azimuth in the superficial layers and auditory azimuth in the deeper layers of the SC. (A) Data from normal, adult ferrets. (B) Data from adult ferrets that had been raised with the ear ipsilateral to the recording site occluded. (C) Data from adult ferrets that had the ipsilateral ear plugged for a comparable period, this time beginning when they were at least 6 months old. The data shown in (B) and (C) were obtained with the earplug still in place. As the superficial-layer visual map showed a high degree of topographic order in each case, the increased scatter in the relationship between the two maps in the plugged animals is indicative of poorer topographic order in the auditory representation. These comparisons indicate some adaptation to the altered cues in the ferrets that were raised with one ear occluded, but not in the ferrets that were plugged as adults. [Adapted from King et al. (1988, 2000).]

to be based purely on binaural cues. Indeed, in the SC of guinea pigs (Palmer and King 1985) and ferrets (King et al. 1994), a map of auditory space was retained, at low sound levels, after deafening of the ipsilateral ear. At higher sound levels, topographic order in the representation of auditory space in these animals was lost, with many of the neurons responding equally well to sounds presented from a wide range of directions. These data suggest that the normal spatial tuning at high levels involves an interaction between contralaterally evoked excitation and ipsilateral inhibition, and the majority of neurons in the cat SC have been found to respond in this way to dichotically presented stimuli (Wise and Irvine 1983).

In contrast to the wealth of data available for the barn owl, the sensitivity of SC neurons to monaural and binaural localization cues has not been examined in ear-plugged mammals. However, following plug removal in ferrets that were reared with one ear plugged, the alignment between the visual and auditory maps of space actually improved (King et al. 1988). This raises the possibility that the adaptation observed in the auditory space map may involve changes in the sensitivity of SC neurons to monaural spectral cues, rather than a systematic adjustment in their binaural tuning.

4.2 Other Acoustic Cue Manipulations

Monaural occlusion changes the magnitude of the binaural cues available. An alternative strategy for altering the localization cue values corresponding to different sound directions is surgically to modify or remove the structures of the external ear. In barn owls, bilateral removal of the facial ruff and preaural flap alters the spatial pattern of both binaural cues. This procedure resulted in a compensatory change in the tuning of OT neurons to ILDs and especially ITDs (Knudsen et al. 1994). Interestingly, and in contrast to the effects of ear plugging, these adaptive changes, which were sufficient to reestablish a partial correspondence between the auditory and visual representations, were observed following external ear modifications in adult as well as juvenile owls.

Acoustical measurements have shown that removal of the pinna and concha of the external ear in ferrets degrades both monaural and binaural spectral localization cues (Carlile and King 1994). The filtering of sounds by the external ear normally results in frequency-dependent gains in signal transmission that are distributed asymmetrically with respect to the interaural axis. Removal of the pinna and concha eliminated these features and, when performed acutely in adult ferrets, led to an ambiguous representation of auditory space in the SC, with many neurons tuned to two different directions, one on either side of the interaural axis. This ambiguous map of space failed to develop, however, in ferrets in which the pinna and concha were surgically excised before the onset of hearing (Schnupp et al. 1998). Auditory localization behavior was also impaired in these animals (Parsons et al. 1999), suggesting that the provision of impoverished spectral localization cues during development somehow limited the ability of the system to utilize the binaural cues that were still present.

These studies illustrate that relatively subtle changes in peripheral cues can influence the formation of binaural interaction in the CAS. An important issue for further research will be the extent to which other manipulations (e.g., acoustic trauma; Salvi et al. 1990) influence binaural plasticity and, more generally, what the limits are on those influences. The results reviewed here (and in Section 5) suggest that plasticity in some binaural systems may be dynamically driven throughout life by a wide range of auditory experiences.

4.3 Cochlear Ablation

The early results of ear occlusion studies and the dramatic increase in the number of afferents deriving from the CN on the side of the intact ear (see Sections 3.1 and 3.3) led to the expectation that unilateral cochlear ablation might increase the physiological responsiveness of CAS neurons to stimulation of the intact ear. This turned out to be the case, but what was unexpected was the degree of the increase. In the gerbil IC (Table 4.4), for example, about 40% of the neurons are normally excited by appropriate acoustic stimulation of the ipsilateral ear. Following neonatal (P2; Kitzes 1984) ablation of the contralateral cochlea, this proportion increased significantly, to 86%. In addition to increasing the proportion of IC neurons excited by stimulation of the ipsilateral, intact ear, neonatal cochlear ablation was found to increase the sensitivity and response strength of single neurons to stimulation of the intact ear (Kitzes and Semple 1985; Fig. 4.10). Qualitatively similar results were obtained in the cat AC (Reale et al. 1987) and in the ferret IC (Moore et al. 1993), showing that the effect was generalized across species and levels of the auditory system.

The co-occurrence of increased activity in the IC and increased afferent projections to the IC (see Section 3.3) initially suggested that both the distribution and the local density of functional excitatory synapses in the IC had been increased by the ablation. However, more recent data suggest another possibility, that the increased excitation may be due, at least in part, to a loss of inhibition, to an "unmasking" of excitation. Popelár et al. (1994) compared IC and AC compound evoked responses ipsilateral and contralateral to the stimulated ear at various times (from 6 hours to 32 days) after sisomicin-induced unilateral cochlear lesions in adult guinea pigs. Increased ipsilateral response amplitudes and decreased thresholds were found within 1–6 days of the lesions, with no changes contralateral to the stimulated ear. McAlpine et al. (1997), using microelectrodes, showed that the responses of IC neurons to stimulation of the intact ear could be greatly increased by ablating the contralateral ear in adult ferrets. In addition, they showed that a dramatically increased responsiveness occurred rapidly, within the course of a single day. Similar results were obtained in the ferret AC (Moore et al. 1997). Further experiments on the adult gerbil IC (Mossop et al. 2000) show that the increased responsiveness occurs within minutes. Because neural connections are not thought able to reform within this time frame, these experiments suggest that the increase is due to functional switching. Presumably, inactivation of neural pathways deriving from the con-

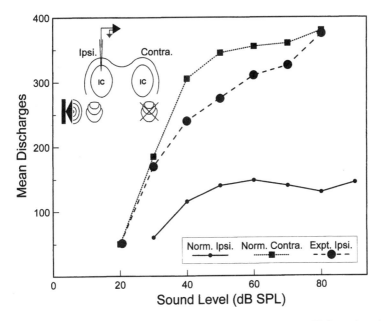

FIGURE 4.10. Effects of cochlear removal on gerbil IC physiology. Unilateral cochlear removal was performed at P0–P2. Several months later, microelectrode recordings were made from the IC ipsilateral to the intact ear, while that ear was acoustically stimulated (*inset*) and the mean response of all the recorded units was plotted as a function of sound pressure level. In normal gerbils, ipsilateral stimulation excites only 40% of neurons in the ipsilateral IC (Kitzes 1984), and those that are excited have high thresholds, narrow dynamic ranges and low peak discharge levels (Norm. Ipsi.). In contrast, contralateral stimulation produces more widespread excitation, and single neurons respond more sensitively and more strongly (Norm. Contra.) than they do to ipsilateral stimulation. Following cochlear removal in infancy, ipsilateral stimulation also produces widespread excitation (Kitzes 1984) that, at the single-unit level (Expt. Ipsi.), resembles the response to normal, contralateral stimulation. [Adapted from Kitzes and Semple (1985).]

tralateral, deafened ear removes the input from higher level inhibitory circuits that normally suppress the responses of IC and AC neurons to stimulation of the ipsilateral ear.

Another possibility is that rapid changes in the expression of neurotransmitters, their receptors, or signalling pathways could contribute to the observed change in activity. Caspary and his colleagues (1990; Milbrandt et al. 2000) have shown that hearing loss produced by aging or acoustic trauma downregulates GABAergic systems in the IC, and acoustic trauma and ototoxicity have been associated with increased responsiveness of central auditory neurons, analogous to the effects of cochlear ablation (Salvi et al. 2000). It has been proposed (e.g., Salvi et al. 2000) that this increased responsiveness may explain or contribute to tinnitus in hearing impaired individuals. Evidence for changes in the strength of central synapses following neonatal cochlear ablation has been produced by Sanes and his colleagues (Kotak and Sanes 1996, 1997; Vale and Sanes 2000). Whole cell responses to electrical stimulation of input pathways were made from brain slices of gerbils that survived for 1–6 days following ablation at P7. In the LSO, deafening the contralateral ear (normally providing inhibitory input to LSO neurons) reduced the number and strength of evoked inhibitory postsynaptic potentials (IPSPs), while deafening the (excitatory) ipsilateral ear reduced the efficacy of excitatory PSPs. In the IC, bilateral ablation further reduced IPSPs evoked from stimulation of the lateral lemniscus, as well as postsynaptic chloride homeostasis. Together, these studies demonstrate that deafening produces changes in both excitatory and inhibitory synaptic transmission. The IC results indicate that these changes can accumulate at higher levels of the system.

CAS plasticity following bilateral, ototoxic deafening in infancy and unilateral, chronic electrical stimulation has been reported by Snyder and colleagues (1990, 1991). In these experiments, the spatial extent and temporal properties of ICC single- and multiunit excitation produced by electrical stimulation of one cochlea were compared between adult cats that had been deafened during the first 16 postnatal days, and then implanted at 8–16 weeks with an active electrical stimulation device ("stimulated" cats), cats that had been deafened and implanted in infancy, but not chronically stimulated ("unstimulated" cats), and cats that were unilaterally deafened and implanted in adulthood ("control" cats). Unstimulated cats yielded very similar results to control cats, showing that the distribution of excitation in the ICC and the response pattern and latency of single units were not dependent on acoustically evoked activity from P16. In the stimulated group, however, the spatial extent of excitation within the ICC was significantly broader, the response latencies shorter, and the incidence of inhibitory and "late" responses greater than in either the unstimulated or control groups.

Although the data from all these studies provide clear evidence for physiological plasticity in binaural systems, there are inconsistencies and unresolved questions. Several reports have noted that neural responses ipsilateral to the intact ear resemble normal contralateral responses, but sound-evoked responses of neurons contralateral to the intact ear do not appear to be affected by the ablation. In contrast to these studies, plasticity in the cochlear implant experiments was found in the ICC contralateral to the stimulated ear. Thus, despite the suggestion (Snyder et al. 1990) that the mechanisms underlying the phenomena might be common (competitive interactions between the ears; see Section 6), the results suggest otherwise. Another question concerns the type, timing, and degree of damage to the cochlea. The most common form of cochlear lesioning in mammals has been complete surgical ablation, including removal of the spiral ganglion cell bodies. However, in several studies, ototoxic drugs have been used. These tend to have a variable effect on the cochlea, and they rarely produce complete hair cell loss after a single administration. Repeated doses must therefore be given, usually over a period of several days. This can be an interpretive complication, especially in developmental studies, as susceptibility of the CAS to deafening is known to change rapidly and dramatically during the neonatal period (Hashisaki and Rubel 1989; Moore 1990b; Tierney et al. 1997). Another difficulty with ototoxic drugs is that they may have an action, either direct or indirect, on the spiral ganglion cells, and this action can follow a time course that is different from their action on the hair cells. Some studies have commented on the effects of partial cochlear lesions. In one such study, Kitzes (1984) found that an incomplete lesion, that nevertheless destroyed much of the organ of Corti and at least some ganglion cells, did not produce the profound physiological changes he showed to follow from complete ablation. In contrast, puncture of the oval window in chickens, without any further apparent cochlear involvement, has been found to abolish activity in the AN and lead to degeneration of CN neurons (Tucci and Rubel 1985; Tucci et al. 1987). One explanation for this apparent discrepancy is that some of the effects of partial lesions may be transitory, and that restoration of input can preserve or reverse the immediate effects of peripheral manipulations. For example, some of the cellular effects produced by activity blockade with TTX are reversible (Pasic et al. 1994) and AN activity in cats is abolished in the short term, but not necessarily in the longer term, by neomycin ototoxicity (Snyder et al. 1990).

Age at the time of deafening has also been considered important, with most studies ablating the cochlea early in infancy. Nordeen and her colleagues (1983) presented qualitative evidence suggesting that long-term cochlear ablation in adulthood did not affect sound-evoked excitation. While more recent work has questioned this finding, the studies of McAlpine and colleagues (1997) presented evidence for quantitatively greater effects of deafening very early in life. Finally, there is evidence that responsiveness in the AC increases more than that in the IC (Popelár et al. 1994), and that the greater effects of early deafening occur at later ages in the AC than in the IC (Moore et al. 1997).

It is unknown at present whether the anatomical and physiological changes induced by a unilateral cochlear ablation contribute to plasticity of spatial processing. The level-dependent spatial tuning observed in the SC of adult guinea pigs within a few hours of ablating the ipsilateral cochlea (Palmer and King 1985) resembles that found in adult ferrets that had been deafened in infancy (King et al. 1994). In both cases, topographic order in the representation was present only at sound levels close to neuronal thresholds. Although the suprathreshold spatial response profiles were more variable in the ferrets, there was no obvious indication that early deafening had systematically altered the spatial tuning of these neurons. It might be expected from the studies outlined above, however, that any changes would be manifest primarily in the SC ipsilateral to the intact ear. In both the guinea pig and ferret experiments, recordings were restricted to the contralateral SC.

4.4 Auditory Consequences of Modifying Visual Experience

In the mammalian SC and avian OT, the registration of different sensory maps allows both unimodal and multimodal stimuli originating from the same location in space to direct orienting movements (Stein and Meredith 1993). Aligning maps of visual and auditory space involves matching specific monaural and binaural cue values to positions on the retina. Studies in barn owls and in several mammalian species have shown that this is an activity-dependent process, in which visual experience is used to shape the auditory spatial tuning of tectal neurons, so that map alignment is preserved in spite of growth-related changes in the relative geometry of the sense organs.

Plasticity of auditory spatial processing has been demonstrated either by depriving animals of a normal visual input or by systematically altering the relationship between the visual and auditory maps. Abnormalities in the organization of the map of auditory space have been reported in barn owls (Knudsen et al. 1991), ferrets (King and Carlile 1993), and guinea pigs (Withington 1992) that were deprived of patterned visual experience during development by binocular eyelid suture. Elimination of all visual experience by dark rearing guinea pigs (Withington-Wray et al. 1990a) and cats (Wallace and Stein 2000) was found to be more disruptive. In this instance, the auditory neurons in the SC remained broadly tuned and topographic order in the representation failed to emerge by the time the recordings were made. Although these findings are consistent with a role for visual signals in refining the auditory space map, their functional significance remains uncertain in view of the highly variable results that have been reported in behavioral studies of auditory localization in blind humans and other animals (see Sections 5.1 and 5.5).

Because of their limited ocular mobility, a mismatch between the auditory and visual maps can be achieved in barn owls by mounting on the animal's head prismatic spectacles that displace the visual field to one side. In adult owls that were reared wearing prisms, the auditory receptive fields of OT neurons were shifted laterally so that they became aligned with the optically displaced visual receptive fields (Knudsen and Brainard 1991). This adaptive change in the map of auditory space, which results from a shift in ITD tuning, was, in the initial studies carried out by Knudsen and colleagues, found to take place only if the prisms were introduced within a so-called sensitive period of development that ends at approximately 200 days of age (Brainard and Knudsen 1995). In fact, the capacity for ITD tuning to shift in response to the prisms, and particularly to shift back to normal values following their removal, seems to depend not only on the age of the animal, but also on the environment in which it was raised. Thus, in owls that were housed individually in small cages, recovery of normal ITD tuning was observed only if the prisms were removed during the sensitive period, whereas this extended into adulthood in birds that were kept with other individuals in large aviaries (Brainard and Knudsen 1998). Moreover, shifted ITD maps can be induced by prism experience in adult barn owls that were previously exposed to the same prisms in infancy followed by a period of normal visual experience (Knudsen 1998). Another factor affecting the capacity for adaptive binaural plasticity at different ages appears to be the magnitude of the change required. Whereas juvenile owls could adapt to a single prismatic shift of 23°, older, normally raised owls could do so only when the visual field was shifted incrementally in small steps (Linkenhoker and Knudsen 2002).

The vision-dependent plasticity of the auditory space map in the barn owl's OT can be explained by changes that take place earlier in the ITD pathway (Fig. 4.11). Equivalent changes in the ITD map were found in the ICX of prism-reared owls, whereas ITD tuning in the ICC was unaltered (Brainard and Knudsen 1993). As outlined in Section 3.4, the altered ITD tuning of neurons in the ICX appears to result from the growth of topographically ordered novel projections from the ICC. In addition, the existing excitatory connections, which would otherwise confer normal ITD tuning, are inhibited via GABA$_A$ receptors (Zheng and Knudsen 1999). The continued existence of both sets of ICC–ICX connections presumably accounts for the plasticity of the auditory space map in prism-reared adult birds.

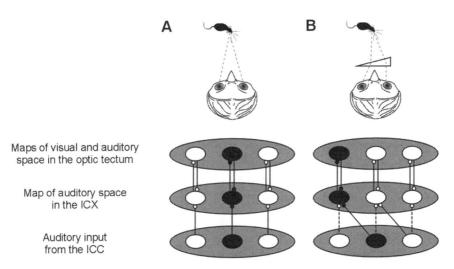

Maps of visual and auditory space in the optic tectum

Map of auditory space in the ICX

Auditory input from the ICC

FIGURE 4.11. Vision shapes the development of ITD tuning in the midbrain of the barn owl. **(A)** The owl's ICX contains a map of auditory space, which is derived from topographic projections that combine spatial information across different frequency channels in the ICC. The ICX map of auditory space is then conveyed to the optic tectum, where it is superimposed on a map of visual space. These connections are thought to develop in an experience-independent manner. From around 40 days after hatching, the auditory space maps in both the optic tectum and the ICX are refined by visual experience, thereby ensuring that the two maps remain aligned despite individual differences in the size and shape of the head and ears. **(B)** Evidence for this is provided by the effects of chronically shifting the visual field in young owls by mounting prisms in front of their eyes. The same stimulus in (A) now activates a different set of neurons in the optic tectum. The auditory space maps in the ICX and tectum gradually shift by an equivalent amount in prism-reared owls, thereby reestablishing the alignment with the optically displaced visual map in the tectum. This involves the growth of novel projections from the ICC to the ICX (*unbroken lines*); the original connections remain in place but are suppressed (*broken lines*). Recent evidence suggests that this remodeling may be brought about by a retrograde signal, delivered via a topographically organized projection from the optic tectum to the ICX.

Because of the likelihood of compensatory eye movements, prisms have not been used in mammals as a means of altering the relationship between auditory localization cue values and directions in visual space. Instead, the visual field has been shifted relative to the head by removal of the medial rectus muscle in order to deviate the eye laterally. Induction of this exotropic strabismus in juvenile ferrets led to a corresponding shift in the auditory representation in the contralateral SC, so that the maps of visual and auditory space remained in register (King et al. 1988). In contrast, a complete inversion of the visual field, produced by section of all six extraocular muscles and rotation of the eye, resulted in a less precise auditory representation, rather than an adaptive realignment. The range of visual–auditory misalignments that lead to compensatory changes in auditory spatial tuning therefore appears to be limited.

As with the effects of prism rearing in barn owls, these results point to a guiding role for vision in the plasticity of auditory spatial processing. The interpretation of experiments in which a change in eye position is induced surgically is complicated by the fact that the auditory responses of SC neurons recorded in awake mammals can be modulated whenever the direction of gaze is altered (Jay and Sparks 1984; Hartline et al. 1995). However, early eye rotation had much less effect on the maturing auditory representation in ferrets that were visually deprived at the same time (King and Carlile 1995). The superficial layers of the SC receive inputs from the retina and visual cortex that are in topographic register with the underlying auditory representation in the intermediate and deep layers. Partial lesions of the superficial layers in neonatal ferrets disrupted the auditory representation in the deeper layers, indicating a role for vision (King et al. 1998). Moreover, because the auditory responses were affected only in the region of the deeper SC layers lying beneath the lesion, this study suggests that a topographically organized visual input from the superficial layers (Doubell et al. 2000, 2003) is required for the normal development of the auditory space map. Recent studies in prism-reared barn owls have also concluded that the auditory receptive fields of neurons in the OT are shaped according to a visually based template (Hyde and Knudsen 2002).

The growing evidence that neuronal sensitivity to acoustic localization cues is matched to a map of visual space that matures at an earlier stage of development is consistent with a Hebbian mechanism of synaptic plasticity (King 1999). Although there is no direct evidence that temporal correlation of converging signals from the eyes and ears leads to a selective strengthening or weakening of auditory inputs, two studies have implicated N-methyl-D-aspartate (NMDA) receptors in auditory map development and plasticity. Schnupp and colleagues (1995) examined the effect of chronically applying the NMDA receptor antagonists MK801 or AP5 to the SC of ferrets during the postnatal period over which topographic order in the auditory representation normally emerges. The drugs were delivered from a slow release polymer (Elvax) placed on the surface of the SC from P25 to P70. Single-unit recording in the SC, after removal of the Elvax, showed that the map of auditory space was degraded, relative to animals reared with drug-free implants. In contrast, no changes in auditory topography were found after chronic application of MK801 in older animals. Thus, activation of NMDA receptors in the SC during the period after the onset of hearing was found to be necessary for the elaboration of the auditory space map. This result is consistent with age-dependent changes in the expression of NMDA receptors (King 1999) and in the relative contribution of those receptors to the synaptic currents of SC neurons (Hestrin 1992; Shi et al. 1997). It also supports the concept of a sensitive developmental period during which changes in auditory spatial tuning can be induced by abnormal experience.

A different approach was taken by Feldman and Knudsen (1998a) who used iontophoretic application of glutamate receptor antagonists to show that, in comparison to the normal responses of ICX or OT neurons, the altered ITD tuning produced by prism experience transiently showed a greater dependence on NMDA receptors. It is interesting to note that changes in NMDA/AMPA current ratios have also been implicated in the activity-dependent emergence of functional glutamatergic synapses during early development (Durand et al. 1996; Wu et al. 1996), suggesting that similar mechanisms of synaptic plasticity can be induced at later stages in response to sensory experience.

The great majority of physiological studies that have examined vision-dependent plasticity of auditory spatial processing have been carried out on the midbrain. There is some evidence, however, that changes also take place at higher levels. The adaptive nature of plasticity in the forebrain, where there is no evidence for a topographic representation of auditory space, is potentially harder to assess than in the space-mapped structures of the midbrain. Nevertheless, Miller and Knudsen (1999) found that prism rearing induced similar shifts in ITD tuning in the barn owl's OT and in the auditory archistriatum, a forebrain region involved in auditory spatial memory. Moreover, early blindness resulted in an increase in the number of auditory neurons and sharper spatial tuning in the cat's anterior ectosylvian sulcus (Korte and Rauschecker 1993).

4.5 Other Methodological Approaches for Measuring Central Auditory Plasticity

In recent years there has been a great deal of interest in the new imaging methods that enable simultaneous, if indirect, observation of activity in neuron populations. Using functional magnetic resonance imaging (fMRI), Scheffler and colleagues (1998) have shown that unilateral acoustic stimulation produces substantially greater activation in the AC contralateral to the stimulated ear than in the ipsilateral AC of normal-hearing adult humans. Long-term monaurally deaf patients were found to have a more laterally balanced AC activation pattern, consistent with the animal physiology described in Section 4.3. In one patient with a sudden onset deafness produced by cochlear nerve resection, fMRI at 1 week, 5 weeks, and 1 year after surgery suggested a progressive change in the laterality of AC activation to stimulation of the hearing ear (Bilecen et al. 2000), but further data are needed to substantiate this finding.

The older method of 2-deoxyglucose (2-DG) radiography has provided data on the effects of both cochlear ablation and ear occlusion on populations of CAS neurons (Table 4.3). In the chicken NL, Lippe et al. (1980) showed that 2-DG could be used selectively to mark the presynaptic processes that normally provide excitatory input to the dendrites on each side of the NL bipolar neurons. Following unilateral cochlear ablation, the processes corresponding to the deafened side remained unlabeled, indicating a withdrawal of presynaptic activity from that side. Heil and Scheich (1986) extended this analysis to other areas of the chicken CAS and varied the survival time following either unilateral or bilateral cochlear ablation. They failed to find any change in the symmetry or distribution of 2-DG labeling with variation in age, survival time, or laterality of the lesion. In contrast, Sasaki et al. (1980) found that unilateral cochlear ablation in adult guinea pigs led, acutely (1–18 hours), to the predicted imbalance of label between the two sides, with the IC particularly reduced in label contralateral to the ablation. Following longer survival (10–48 days), the level of 2-DG uptake in the contralateral IC and MGB was restored to normal, suggesting increased activity in those nuclei relative to the levels seen shortly after the ablation.

Following unilateral ligation of the ear canal between P12–P17, Clerici and Coleman (1986) found that the region of maximum activation of the rat IC produced by high-frequency sounds shifted from a medioventral to a lateral position in the ICC. The authors interpreted this result in terms of ligation-induced changes in stimulus representation in the cochlea during a period of elevated central susceptibility to experiential factors. Working with gerbils, Stuermer and Scheich (2000) found that ear canal atresia from before the time of hearing (P9) to P27 increased 2-DG uptake in the AC ipsilateral to the open ear, and that the atresia affected the tonotopicity of the AC. Taken together, the 2-DG data present a somewhat confusing array of apparently contradictory observations. However, the weight of evidence suggests that acute cochlear ablation silences the AVCN and TB, leading to a loss of uptake in those pathways normally activated only by the deafened ear, but not in pathways higher in the system that continue to receive some excitatory input from the intact ear. Long-term conductive loss in one ear leads to increased uptake levels higher in the contralateral CAS, where the normally hearing ear gains increased synaptic influence, as happens following longer-term deafening.

Functional imaging and event-related potential studies have revealed changes in the cortical distribution of sound-evoked activity in blind humans compared to normal-sighted subjects (Kujala et al. 2000). Of particular interest to the issue of auditory plasticity is the possibility that areas of occipital cortex that are normally involved in visual processing may be activated during auditory localization tasks in these subjects (see Section 5.5).

5. Behavioral Studies

Until recently, behavioral research on binaural plasticity has been based mostly on the use of two species, the barn owl and the human (Table 4.4). The use of the barn owl has been justified by the animal's exquisite sound localization accuracy (e.g., Konishi 1993; see Section 5.1) and by the obvious and considerable benefits of carrying out neurophysiological and behavioral studies in the same species. In view of this, it is unclear why mammals have not been more frequently used in psychophysical research, particularly if the neurobiological basis of hearing disorders in humans is to be investigated. Most common laboratory animals have highly accurate sound localization that, in the case of carnivores and primates, may approach or equal the accuracy of the barn owl (Fay 1988).

5.1 Plasticity of Binaural Hearing in Animals

Naturalistic observations of the ability of barn owls to catch prey in the dark led to the laboratory finding that the owl turns its head quickly and accurately toward a sound source (Konishi 1993). The head turning response, although common among animals, was particularly useful for studying barn owl localization, as the eyes in this species are nearly immobile. A head turn is therefore necessary to foveate an acoustic stimulus. By measuring these movements, using infrared light reflected by a mirror mounted on the owl's head, sound localization accuracy was quantified (Knudsen et al. 1984a). In normal owls, minimum audible angles of 3.6° and 2.2° have been measured in azimuth and elevation, respectively, using noise stimuli (Knudsen and Konishi 1979).

Unilateral ear plugging resulted in less accurate head orienting responses, indicating their dependence on binaural processing. However, young owls (<2 months of age) were found to adapt to the abnormal cue values produced by long-term monaural occlusion and recovered accurate sound localization despite the presence of the earplug (Fig. 4.12; Knudsen et al. 1984a). Older animals were unable to make this adjustment. When the plug was removed from the owls that had adapted to the altered cues, the orientation responses were initially displaced to the side of the previously plugged ear, but then recovered in accuracy so long as the owls were <7 months old at the time of plug removal (Knudsen et al. 1984b). Interestingly, this recovery in performance was not observed if the owls were visually deprived when the plug was removed, suggesting a role for vision in the behavioral adaptation to altered binaural cues (Knudsen and Knudsen 1985).

Prism rearing also affected the accuracy of auditory head orienting responses, which were shifted into alignment with the optically displaced visual field (Knudsen and Knudsen 1990). Although this adjustment ensured that orienting responses to visual and auditory targets remained in correspondence, the owls actually localized less accurately because the head was turned to one side of the sound source. These changes were observed only if the prisms were introduced in juvenile owls (<7 months old), whereas the accuracy of the head orienting responses in prism-reared owls was found to shift back to normal following prism removal at any age (Brainard and Knudsen 1998). The results of this study illustrate the powerful influence of the visual system over the development of spatial hearing in this species and, like the effects of long-term monaural

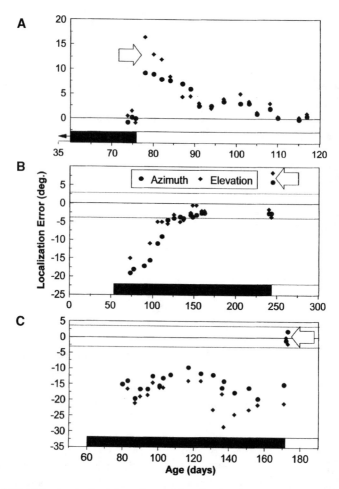

FIGURE 4.12. Behavioral plasticity of sound localization in the barn owl. Owls were trained and tested on a head orientation task to a variable noise stimulus. In normal owls, the average localization error was 1.4° ± 0.8° (SEM) in azimuth and 1.2° ± 0.9° in elevation. Errors <2SD of these values (*horizontal lines*) satisfied the criterion for normal localization. (**A**) Following right ear plug rearing from P35 to P76 an owl had, on initial testing with the plug in place, errors within the normal range. When the plug was removed, the owl systematically mislocalized in both azimuth and elevation (open arrow), before returning to normal during the next 2 weeks. (**B**) An owl that had normal hearing before receiving a right ear plug at P55 initially mislocalized toward the side of the unplugged ear. With 3 months of continuous plugging, errors gradually decreased to normal. When the plug was removed (at P244), the owl mislocalized toward the side of the previously plugged ear (open arrow). (**C**) An owl receiving a plug for the first time at P60 did not adjust to the altered cues; mislocalization continued until the plug was removed at P175, when performance immediately returned to normal (open arrow). Abnormal auditory experience beginning before 8 weeks of age can therefore lead to a systematic and long-lasting change in sound localization. After 8 weeks, abnormal cues do not change sound localization, but a return to normal cues (i.e., unplugging) continues to lead to recovery until owls reach 38–42 weeks. [Adapted from Knudsen et al. (1984a,b).]

occlusion, are consistent with the changes produced by abnormal sensory experience in the tuning of neurons in the midbrain to binaural cues (Sections 4.1 and 4.4).

A consistent finding of these studies is a different time course for the acquisition phase, during which the auditory system can undergo large and rapid adaptations in response to altered auditory or visual cues, and the recovery phase for the restoration of normal orienting responses after earplug or prism removal. As Knudsen et al. (1984b) pointed out, the capacity of the auditory system to recover well beyond the acquisition stage may indicate an innate expectation of what normal adult cue values should be. These experiments also show that the period of plasticity depends not only on the age of the animal, but also on the environment in which it was raised (Brainard and Knudsen 1998), a finding that has important implications for studies of auditory system plasticity in mammals.

The role of auditory experience in the development and maintenance of behavioral tasks that rely on binaural hearing has been examined in ferrets. By training adult animals to approach the location of a sound source, King and colleagues (2000, 2001b) found that plugging one ear disrupted this ability, particularly on the side of the plugged ear. However, like barn owls, ferrets that were raised with one ear occluded could perform this task almost as well as normal controls, indicating that they had learned to reinterpret the acoustic cues corresponding to different directions in space (Fig. 4.13). Subsequent removal and reinsertion of the earplug resulted in much smaller changes in localization accuracy than those observed when adult ferrets were first plugged. The basis for this adaptation may therefore involve learning to utilize monaural spectral cues and to ignore the abnormal binaural cues. Sound localization accuracy also improved after long-term monaural occlusion in adult ferrets. Indeed, rapid improvements in performance could be observed within the first few days of plugging. The accuracy of both approach to target responses and head orienting movements recovered rapidly after plugging one ear in both sighted and binocularly deprived ferrets, indicating that this form of adult plasticity is not dependent on instructive signals from the visual system (Kacelnik et al., submitted for publication).

The effects of occluding one ear in ferrets at different ages have also been examined on the magnitude of a free-field version of the binaural masking level difference (MLD; Moore et al. 1999). An animal model of this task was developed because the MLD is probably the most commonly used test of binaural hearing in humans (see Section 5.4). MLD levels of around 10 dB were recorded in normal adults. Much lower levels were found with one ear occluded, even after 3–12 months of ear plugging in infancy or adulthood. The MLD was also impaired after the earplugs were removed and took several months to reach normal levels. Because pure-tone sensitivity in the previously plugged ear was normal, this presumably reflects changes in binaural processing that were induced by the hearing loss. In contrast to the localization experiments, however, these animals showed little sign of adaptation to the altered cue values produced by the earplug.

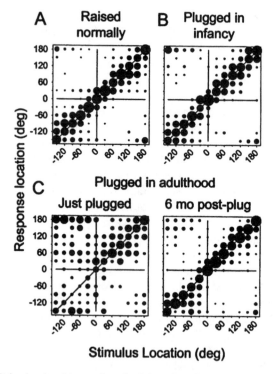

FIGURE 4.13. Behavioral evidence for plasticity of auditory localization in mammals. Stimulus–response plots showing the combined data of 3 normal, adult ferrets (**A**) and 3 ferrets that had been raised and tested with the left ear occluded with a plug that produced 30–50 dB attenuation (**B**). These plots illustrate the distribution of responses (ordinate) as a function of stimulus location (abscissa). The size of the dots indicates, for a given speaker angle, the proportion of responses made to different spout locations. The juvenile plugged ferrets could localize sound as accurately as the controls. (**C**) Data from 3 normally reared adult ferrets immediately (Just plugged) and 6 months after (6 mo post-plug) occlusion of the left ear. Prior to plugging, these animals performed as well as the normals shown in (**A**). Following insertion of the earplug, they initially localized poorly, but showed a marked improvement when they were next tested 6 months later. [Adapted from King et al. (2001b).]

As expected from their role in resolving the ambiguities that are present in binaural cues, disruption of spectral localization cues by pinna removal caused adult chinchillas to make more elevation and front-back errors (Heffner et al. 1996). This was also the case in ferrets in which the pinna and concha were removed bilaterally in infancy (Parsons et al. 1999). In fact, these animals showed more general deficits in their ability to localize broadband sounds within the horizontal plane, a result that accords with the disruptive effects of pinna removal on the development of the map of auditory space in the SC (Schnupp et al. 1998), although these deficits recovered to some degree with repeated testing.

The other peripheral manipulation that has been used to probe the capacity of the auditory system to adapt to altered sensory inputs is visual deprivation. Individual differences in the age of onset, degree and duration of visual loss have undoubtedly contributed to variable findings reported from investigations of spatial hearing in blind humans (see Section 5.5). Animal studies therefore provide an opportunity for examining the consequences of more controlled and reproducible forms of sensory deprivation. However, due to differences in experimental design and in the species used, we still do not have a clear-cut answer to the question of whether and how a loss of vision affects the development of auditory localization. Barn owls raised with their eyelids sutured together made less precise auditory orienting responses, particularly in elevation, than normal birds, a result that accorded with changes to the map of auditory space in the OT (Knudsen et al. 1991). In contrast, visually deprived cats (Rauschecker and Kniepert 1994) and ferrets (King and Parsons 1999) showed improvements in horizontal localization (except for frontal locations) and, in the case of the ferrets, no change in elevation performance (Parsons and King, submitted for publication).

In addition to experience-driven plasticity, it is also of considerable interest to determine whether compensatory changes in spatial hearing can result from central damage. Numerous studies have shown that lesions of the AC lead to localization deficits in the contralateral hemifield. Where reported, these deficits appeared to be permanent (Jenkins and Merzenich 1984), although some recovery was observed with subsequent testing in ferrets after reversibly inactivating the AC using Elvax implants that chronically released the $GABA_A$ agonist muscimol (Smith et al., 2004). Clear evidence for plasticity of sound localization has been found in adult barn owls in which small lesions were made in the ICX or OT, the two midbrain structures that contain a map of auditory space (Wagner 1993). These lesions degraded head-turning accuracy for sound directions corresponding to the region of the space map that had been damaged. In most animals, the behavioral performance recovered over time and, in two of five cases, the recovery was complete. This finding is, in some senses, congruent with other findings of plasticity in adult sensory systems, and extends those findings into the spatial domain. The cellular basis for this recovery remains unclear, but might involve either collateral sprouting of intrinsic connections to establish new functional domains within the deafferented tissue, as is the case in other examples of adult plasticity (e.g., Darian-Smith and Gilbert 1994), or the owl's forebrain taking over this aspect of its auditory localization behavior.

5.2 Adaptation to Altered Auditory-Localization Cues in Humans

Studies of binaural interaction in human listeners with pathological hearing (Table 4.4) have, in contrast to the animal experiments cited above, produced evidence that is often difficult to compare and, where comparisons are possible, frequently contradictory. Many of the difficulties with this literature are undoubtedly attributable, as Durlach et al. (1981) suggested, to differences in impairments and general experimental procedures between the studies. In addition, it seems likely that binaural plasticity also played a (generally unrecognized) role in promoting the confusion through effects that psychoacousticians have variously labeled practice, reorientation, adaptation, acclimatization, deprivation, and perceptual learning. Despite these variables, Durlach and colleagues were able to draw certain conclusions from their review of the older literature. Localization and lateralization performance is (1) degraded more by unilateral deafness and binaural asymmetry than by bilateral hearing loss, (2) degraded more by middle ear and auditory nerve disorders than by cochlear disorders, and (3) not easily predicted on the basis of audiograms. Similar conclusions have been reached by a variety of other techniques described in this and the following sections.

More recent studies indicate that the neural processing of acoustic localization cues is altered in subjects with binaural asymmetries resulting from conductive or sensorineural hearing losses. Wilmington et al. (1994) examined the effects on binaural hearing of an extreme form of congenital conductive hearing loss (atresia) resulting from abnormalities in the external or middle ear. The listeners in their study had unilateral hearing losses that ranged from 30 to 70 dB. They were tested on various binaural tasks before and after corrective surgery (at ages 6–33 years) that restored sensitivity in the impaired ear to within 10 dB of the normal range. Surgery resulted in an immediate improvement in performance on all tasks. However, the extent of the improvement varied considerably between the tasks and between subjects. For example, because of the magnitude of the hearing loss, ITD thresholds could not be measured before surgery, but assumed normal levels for nearly all listeners when they were next measured 4 weeks afterwards. In contrast, the sound localization ability of these patients varied greatly before surgery, with a few individuals making errors that were only slightly outside the normal range. Moreover, in most cases, performance in this task showed only minor improvement even months after surgery.

Häusler et al. (1983) tested minimum audible angles (MAA) and the minimum discriminable ILD and ITD in a large sample of normally hearing and hearing-impaired listeners. Among listeners with long-standing conductive impairments, those with mild losses of not more than 25 dB hearing level (HL) performed nearly normally, whereas those with larger (>35 dB HL) losses performed poorly on most or all of the tests. In spite of wide variations in performance, this study also found that some subjects with only one functional ear had MAAs within the normal range. A similar result was found by Slattery and Middlebrooks (1994), who reported that three (out of five) subjects with unilateral congenital deafness were able to judge the location of broadband sounds more accurately than control subjects wearing an earplug for up to 24 hours. These findings strongly suggest that at least some monaural human listeners can learn to localize reasonably well in the horizontal plane. Whether this reflects a change in the way monaural spectral cues are processed in the brain will require future animal studies.

In an attempt to study binaural plasticity without the complication of the often variable nature of clinical conditions, some human studies have employed experimental hearing impairments in normal-hearing adult subjects (Table 4.5). These impairments were usually unilateral ear occlusions, but included other means for altering ILD and ITD. In an early study, Bauer and colleagues (1966) found that monaural occlusion resulted in localization errors toward the side of the unplugged ear. The degree of this shift was measured for approximately 70 hours of continuous plugging and was found to decline (i.e., improve) somewhat over that time. The rate of improvement was much more rapid (about 5 hours), and complete, if training (results feedback) was provided during plug wearing. When the plug was removed, normal performance accuracy resumed.

In another study that used headphones, Florentine (1976) tested thresholds, interaural equal loudness matches, and centering judgments before, during, and after the insertion of an earplug. She found that equal loudness matches, based on alternating presentation of tones to the two ears, required an increase in the level of the tone presented to the plugged ear that approximated the degree of attenuation produced by the plug. The degree to which the sound level to the plugged ear had to be increased did not change markedly during up to 20 days of continuous plug wearing. On initial plugging, median plane centering, based on the ILD required for a dichotically presented tone to be perceived in the median plane of the subject's head, behaved in the same way as the equal loudness match, requiring an increase in the level of the tone to the plugged ear to match the attenuation of the plug. However, during the next 4 days, the sound level in the plugged ear required for centering reduced (in three of four listeners), for low-frequency tones, to near the level that had been required for centering before the plug was inserted (Fig. 4.14). Removal of the plug, after many days of continuous wearing, led almost immediately to balanced sound levels for equal loudness. Centering, in contrast, required at least seven days to return to preplug ILDs.

More recent studies of binaural adaptation to ear inserts have suggested more subtle and complicated plasticity. McPartland et al. (1997) found little adaptation to unilateral ear plugs worn constantly for a week. Two listeners (of six tested) did produce a small (3 dB), statistically significant adaptation during plugging, but normal binaural hearing resumed immediately on withdrawal of the plug. Instead of introducing a binaural asymmetry, Hofman and colleagues (1998) fitted both external ears with inserts that altered the pattern of spectral cues corresponding to different directions in space. As expected, elevation judgments in these subjects were initially disrupted, but recovered gradually over a period of several weeks. One of the most striking findings in this study was that, following removal of the inserts, the subjects were immediately able to localize as accurately as they did before the experiment was started. In other words, they were apparently able to use two different sets of spectral cue values to determine the vertical coordinates of a sound source.

Together, these results suggest that adult human listeners do show some capacity to learn to use abnormal spatial cue values in order to localize sound. Whether

TABLE 4.5. Behavioral plasticity of binaural systems.

Ear manipulation/hearing loss	Stimulus method/spectrum	Task	Test method	Species	Reference
Ear plug	Free field vocalizations	Localization	Approach response	Guinea pig	Clements and Kelly (1978)
Ear plug	Free field BBN	Localization	Approach response	Ferret	King et al. (2000, 2001b)
Reshape external ear	Free field BBN	Localization	Approach response	Ferret	Parsons et al. (1999)
Visual deprivation	Free field BBN	MAA, Localization	Approach response	Ferret	King and Parsons (1999)
Visual deprivation	Free field BBN	MAA, Localization	Conditioned suppression, head orienting	Hamster	Izraeli et al. (2002)
Visual deprivation	Free field tones	Localization	Approach response	Cat	Rauschecker and Kniepert (1994)
Ear plug	Free field BBN	Localization	Head turn	Barn owl	Knudsen et al. (1984a,b); Knudsen and Knudsen (1985)
Visual deprivation	Free field BBN	Localization	Head turn	Barn owl	Knudsen et al. (1991); Knudsen and Knudsen (1990)
Prisms	Free field BBN	Localization	Head turn	Barn owl	Brainard and Knudsen (1998)
Brain (tectal) lesions	Free field BBN	Localization	Head turn	Barn owl	Wagner (1993)
Unilateral CL	Dichotic 500Hz in NBN	MLD	3AFC, adaptive	Human	Hall and Derlacki (1986); Hall and Grose (1993)
Unilateral and bilateral SNL	Free field Speech	Audiogram	Word recognition	Human	Hood (1984)
Bilateral SNL	Free field Dichotic Speech	Speech perception	Word recognition	Human	Jerger et al. (1993)
Hearing aid Bilateral SNL	Free field Whistle and drum	Localization	Pointing	Human	Beggs and Foreman (1980)
Hearing aid Bilateral SNL	Free field Speech	Localization Speech perception	2AFC Word recognition	Human	Sebkova and Bamford (1981)
Ear plug	Free field BBN, NBN	Localization	Pointing	Human	Bauer et al. (1966)
Ear plug Unilateral CL, SNL	Dichotic Tones	BLB Centering	Subject controlled adjustment	Human	Florentine (1976)
Pseudophone	Free field	Localization	Pointing	Human	Held (1955)
Unilateral and bilateral CL and SNL	Free field	MAA, ITD, ILD	2AFC, adaptive	Human	Häusler et al. (1983)
Unilateral noise	Dichotic Noise			Human	
Unilateral SNL	Dichotic Tones	BLB	Subject controlled adjustment	Human	Jerger and Harford (1966)

TABLE 4.5. *Continued*

Ear manipulation/ hearing loss	Stimulus method/ spectrum	Task	Test method	Species	Reference
Hearing aid Bilateral SNL	Dichotic Speech	Speech perception	Word recognition	Human	Gelfand et al. (1987)
Hearing aid Bilateral SNL	Dichotic Speech	Speech perception	Word recognition	Human	Silman et al. (1984)
Hearing aid Asymmetric SNL	Dichotic Speech	Speech perception	Word recognition	Human	Silverman and Emmer (1993)
Unilateral and bilateral CL and SNL	Free field Tones, NBN	Localization	Pointing	Human	Nordlund (1964)
Congenital, unilateral atresia of the ear canal	Free field Dichotic Clicks, 500Hz in NBN	BLB, ITD, Localization, MLD	4AFC, adaptive; naming; adjustment	Human	Wilmington et al. (1994)
OM	Dichotic 500Hz in NBN	MLD	2AFC, adaptive	Human	Moore et al. (1991)
OM	Dichotic 500Hz in NBN	MLD	3AFC, adaptive	Human	Pillsbury et al. (1991)
OM	Dichotic Clicks, 500Hz in NBN	BLS, ILD, ITD, MLD	2AFC, constant; 3AFC, adaptive	Human	Stephenson et al. (1995)
Bilateral SNL MS	Dichotic NBN	IC, ILD, ITD, MLD	2AFC, constant	Human	Gabriel et al. (1992)
Ear plug	Free field NBN	Localization	Head turn	Human	Slattery and Middlebrooks (1994)
Early blindness	Free field BBN	MAA, Localization	2AFC, adaptive; reaching	Human	Ashmead et al. (1998)
Early blindness	Free field Tones	Localization	Button press	Human	Spigelman (1976)
Early and late blindness	Free field Tones	Localization	Naming; pointing	Human	Wanet and Veraart (1985)
Early blindness	Free field BBN	Localization	Button press	Human	Röder et al. (1999)
Early blindness	Free field BBN	Localization	Pointing	Human	Lessard et al. (1998)
				Human	Zwiers et al. (2001)
				Human	Lewald (2002)

Hearing loss: CL, Conductive; MS, multiple sclerosis; SNL, sensorineural; OM, otitis media.

Stimulus: BBN, Broad band noise; NBN, narrow band noise.

Task: BLB, Binaural loudness balance; BLS, binaural loudness summation; IC, interaural correlation; ILD, interaural level difference; ITD, interaural time difference; MAA, minimum audible angle; MLD, masking level difference.

Test: nAFC, n alternative, forced choice.

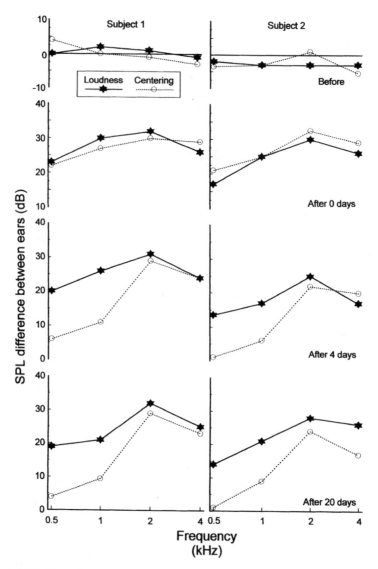

FIGURE 4.14. Adaptation to ear plugging in human binaural hearing. Normally hearing adult listeners were tested on an interaural loudness task, requiring matching of tones presented successively to each ear, and on a centering task, requiring a midline judgment for tones presented simultaneously to both ears. Both equal loudness and centering judgments were made for ILDs close to zero. When a plug was placed in one ear, performance in both tasks was initially (0 days) offset by 15–35 dB, toward the plugged ear, corresponding to the attenuation produced by the plug. However, after 4 days of plugging, the ILD required for a match on the centering task had, for the lower frequencies, returned to near zero, while the loudness match ILD retained the offset it had acquired at the time of plugging. Binaural adaptation to the plug was maintained for as long as the plug remained in the ear. Following removal of the plug, an ILD favoring the unplugged ear had initially to be set to achieve centering. This offset also returned to normal over the ensuing days. These results suggest that the adult binaural system retains sufficient plasticity to adjust to at least some changes in auditory experience. [Adapted from Florentine (1976).]

this represents plasticity of binaural interactions, however, remains unclear. An alternative interpretation was offered by Knudsen (1984), who suggested that the listeners in the ear-plugging studies had learned to ignore misleading cues and attend selectively to the cues that were unaffected by the plugging. This would explain why, in some of the studies described in the preceding paragraphs, normal localization or lateralization accuracy was observed as soon as the earplug was removed. Had there been a reinterpretation of the relationship between binaural cue values and directions in space, the restoration of a balanced binaural input should have led to an after-effect toward the side of the previously plugged ear.

One possibility, supported by the plasticity observed under conditions where only those cues are available for localization (Slattery and Middlebrooks 1994; Hofman et al. 1998), is that adaptation to a unilateral conductive hearing loss may involve learning to place greater emphasis on spectral cues. King and colleagues (2000) came to the same conclusion on the basis of the effects of chronically plugging, and unplugging, one ear in ferrets, although these animals continued to rely to some extent on binaural cues for localization in the horizontal plane. Similarly, other studies (e.g., Held 1955) have indicated that changes in the association between binaural cues and locations in space are possible in adult humans, provided that the extent of those changes is limited.

An alternative strategy for investigating the plasticity of human spatial hearing has been to use VAS techniques to manipulate the cues available, so that they either corresponded to sound directions that were different from normal or were abnormally large (Shinn-Cunningham 2001). Short-term training revealed some adaptation even within a single testing session, although this adaptation was incomplete (Shinn-Cunningham et al. 1998). Plasticity on this time scale is clearly important for recalibrating auditory localization as subjects encounter different acoustic environments.

Spatial processing plasticity is not restricted to situations where subjects have to localize sounds using cues that have been altered, either experimentally or as a consequence of a hearing disorder. Like many other sensory abilities, the ability to localize sounds using normal, unmodified spatial cues can improve with practice (Butler 1987; Recanzone et al. 1998; Wright and Fitzgerald 2001). Examination of the specificity of these improvements has the potential to provide valuable insights into the way in which spatial information is processed within the human CAS.

5.3 Monaural Hearing Aids

Human patients with bilateral sensorineural hearing losses have often been fitted with monaural hearing aids. Silman et al. (1984) reported that long-term (4–5 years) wearing of a monaural aid in such patients led to a decrement in speech recognition scores in the *unaided* ear relative to the aided ear of the same patients or either ear of those who had been binaurally aided. Although they termed this a "deprivation" effect, a subsequent study (Gelfand et al. 1987) found no decrement in patients with similar hearing losses who had never worn hearing aids. Thus, the performance decrement was contingent on the other ear being aided. However, several other studies (e.g., Hood 1984; Silverman and Emmer 1993) have suggested that it is the interaural asymmetry of input, rather than unilateral amplification, that leads to this phenomenon; listeners with asymmetric hearing impairments can also show superior speech discrimination in the better hearing ear relative to the worse hearing ear, even where these two ears, in different subjects, are matched for HL.

Although these results suggest a form of binaural plasticity in adult humans, it is unclear whether this is mediated through the sorts of relatively simple neural mechanisms discussed in this chapter, or through some form of "higher level" cognitive learning (Silman 1993). For example, it may be that the greater speech information available through the better ear, whether amplified or less impaired, produces a rearrangement of the neural code for speech that is fundamentally different from that produced in the tonotopic arrangement of the primary auditory cortex by restricted cochlear lesions (e.g., Irvine and Rajan 1995). More recent research has also questioned whether the unaided ear of binaurally impaired, but unilaterally aided listeners, is affected by the aid. Gatehouse (1992) found that, while speech identification in noise was changed in the aided ear, it was unaffected in the unaided ear. Similarly, Robinson and Gatehouse (1994) failed to find any effect of aiding on intensity discrimination in the unaided ear. In these studies, changes were restricted to the aided ear and there was, therefore, no need to invoke binaural plasticity.

5.4 Binaural Unmasking and Conductive Hearing Loss

A common binaural test in both the clinic and the laboratory is the masking level difference (MLD), a measure of the spatial separation of target and masking sounds based mainly on interaural phase comparisons (Moore 1997). Hall and Grose (1993) presented MLD data from adult patients undergoing surgery for the alleviation of otosclerosis. Prior to surgery, the patients had a unilateral, flat hearing loss of 25–40 dB and their mean MLD (8.6 dB) was about half that of a normally hearing, control group (17.9 dB). After surgery, the hearing loss was reduced to 0–15 dB across the same frequency range. Although the mean MLD became larger (13.9 dB) by 1 month after surgery, it was still significantly smaller than that of the control group. By 1 year after surgery the mean MLD had increased further (to 15.8 dB) and was not significantly different from that of the controls, despite the same small, persistent loss that had been seen just after surgery. These results show that binaural hearing, as measured by the MLD, is degraded by a unilateral hearing loss. They also demonstrate dynamic binaural plasticity; the MLD was initially poorer than normal, then improved during the year after surgery when thresholds were stable.

As outlined above, OM in childhood has also been found to lead to a reduced MLD. In two retrospective studies (Moore et al. 1991; Pillsbury et al. 1991), children with a history of OM were tested between the ages of 5 and 13 years and were found to have significantly smaller MLDs than an age-matched, control sample. Pillsbury and colleagues tested children before, 1 month after, and 3 months after the surgical placement of ventilation tubes in the tympanic membrane. As for the otosclerosis patients described above, pure-tone thresholds were normal, or near normal, 1 month after surgery, whereas MLDs remained smaller than normal in some listeners. MLDs were still depressed, to about the same extent, 3 months after surgery. A correlation between the MLD data and the symmetry of hearing loss before surgery showed that interaural asymmetry was a better predictor of postsurgery MLD performance than hearing loss per se; the asymmetric cases performed more poorly.

Unfortunately, Pillsbury and colleagues (1991) did not present data from longer postsurgery recovery periods, and it is therefore impossible to compare the results of that study with those of the adult, otosclerosis study performed in the same laboratory. However, the data of Moore and colleagues (1991) suggest that recovery from the effects of the abnormal binaural cues created by OM may be very slow. As children have a peak of OM incidence during the second year (Hogan et al. 1997), lower than normal MLDs some years later would suggest a binaural disadvantage that was more persistent than that reported in the former otosclerotic adults by Hall and Grose (1993). Thus, one developmental effect of unilateral or asymmetric conductive hearing loss may be to extend the period during which binaural hearing problems are experienced after restoration of peripheral input. However, more recent evidence has suggested that the effects of OM in childhood are not permanent. Hall and colleagues (1995) studied children who had received tympanostomy tubes for persistent OM for 4 years following the surgery. MLDs improved slowly and reached control levels between 2–3 years after surgery. The children originally tested by Moore and colleagues (1991) when they were 6–12 years old were retested some 6 years later and found to have control level MLDs (Hogan et al. 1996).

In their study of binaural hearing in patients before and after corrective surgery for congenital conductive hearing losses (atresias), Wilmington et al. (1994) found that MLDs improved substantially following surgery, although in many cases not to the values that were measured in normal subjects. The MLDs of a few of these subjects increased slightly between tests carried out at 4 weeks and at 6 months (or more) after surgery. Overall, however, there was no significant improvement over this time period. In view of the results of the studies described above, it seems likely that larger MLDs would have been measured eventually. On the other hand, the large and continuous asymmetries experienced from birth by these individuals may have more seriously compromised their binaural abilities than in other forms of conductive hearing loss.

5.5 Blindness in Humans

Many reports have suggested that early blindness in humans may give rise to superior performance in auditory or tactile tasks (Rauschecker 1995). In the context of auditory localization, this is complicated by the animal literature showing that instructive signals provided by the visual system help to shape the development of auditory spatial tuning in midbrain neurons (Section 4.4). Indeed, as in the animal behavioral studies (Section 5.1), measurements of spatial hearing in blind humans have produced conflicting results, with some authors reporting impairments in performance (e.g., Spigelman 1976; Wanet and Veraart 1985; Zwiers et al 2001) and others that sound localization abilities were better than in sighted controls (e.g. Röder et al. 1999; Ashmead et al. 1998; Lessard et al. 1998). Much of this variation may reflect differences in the stimuli used, the responses measured, and the positions tested, as well as intersubject differences in performance and in the etiology of the blindness. Nevertheless, there is some consistency between the animal and human literature that blind subjects may be superior to the normal sighted when localizing sounds to one side of the head. As pointed out in a recent study by Izraeli et al. (2002), it remains to be seen whether these behavioral differences really reflect altered auditory acuity or the effects of other factors, such as attention and practice.

6. Overview and Future Directions

It may be noted from Tables 4.1–4.4 that studies of binaural plasticity are, with the exception of some human psychoacoustic reports, all less than 30 years old. Despite the immaturity of this field, some important contributions have been made, and a synthesis of the major highlights is presented below (Section 6.1). Although there is now a substantial body of data establishing the phenomenon of binaural plasticity, at levels ranging from cell biology to human behavior, there is much still to be learned, and listed below (Sections 6.2–6.4) are some of the areas and specific questions that it would now seem appropriate, tractable, and timely to address.

6.1 Overview

Despite some reservations (e.g., Brugge et al. 1985), there is now overwhelming evidence in support of the existence of binaural plasticity. The plasticity may be induced in developing and mature animals and humans by unilateral and bilateral sensorineural or conductive hearing loss. Binaural plasticity may also be observed following unilateral sound amplification, and it seems highly likely that some form of neural plasticity underlies the natural response to growth-related changes in head size and other natural and artificial changes of interaural cues. Finally, as most clearly demonstrated in studies of the barn owl's auditory system, adjustments in binaural processing may underlie crossmodal effects on spatial hearing.

The existence of binaural plasticity in the human auditory system has implications for education, pediatrics, otolaryngology, and geriatrics. OM undoubtedly reduces a child's ability to detect target sounds in noisy environments (e.g., nursery, schoolroom), and a certain amount of neural adjustment to the abnormal binaural cues produced in most cases of OM may produce continuing difficulties after the peripheral sensitivity has returned to normal. It would seem important that teachers are made aware of this possibility and that they endeavor to identify and pay special attention to children who have, or have had, persistent OM. The several examples of heightened binaural plasticity in early life raise issues concerning the medical management of OM in infancy. Although the condition generally resolves itself in the fullness of time, binaural plasticity may result in impaired hearing that, in the 15% of children having at least unilateral OM for more than half of the first 5 years of life (Hogan et al. 1997; Hogan and Moore 2003), might carry over between OM episodes, producing chronic and long-lasting difficulties. The decision whether, for example, to insert tympanostomy tubes, should be guided by an understanding of the potential contribution of binaural plasticity, as well as the most detailed practicable audiological history of the case. For bilateral OM, the binaural plasticity literature would indicate bilateral treatment. Conductive hearing loss later in life can also produce neural plasticity that may, on the one hand, help the listener to adjust to a mild bilateral asymmetry and, on the other hand, impede progress following remediation of the hearing loss. Finally, in sensorineural loss, peripheral inactivity may lead to central degenerative and generative effects that could, if unilateral or asymmetric, change perception through the normally or better hearing ear. Where amplification is prescribed, attempts should be made to provide spectral balancing between the ears.

Studies of binaural plasticity have made important contributions to our understanding of the functional organization of the CAS and have also provided models, and some data, on which more general issues of neurobiology may be fruitfully explored. Indeed, the sound localization pathway leading to the construction of a computational map of auditory space in the SC/OT has become one of the most useful systems for investigating the functional significance of sensory experience in shaping the developing brain (King 1999). The bilateral CN inputs to the SOC nuclei provide the other principal system on which many studies of binaural plasticity have been based. In the chicken NL (Deitch and Rubel 1989a,b) and gerbil MSO (Russell and Moore, 1999, 2002), substantial progress has been made in understanding the dendritic response to deafferentation, and studies on the gerbil MNTB (and the chicken NM, see Chapter 2) have shown that the unique form of calycine synapses can be reproduced by CNS axon sprouts (Kitzes et al. 1995; Russell and Moore 1995).

In terms of CAS processing, the on going dependence of neurons on the level and type of input they receive from the peripheral auditory system, as well as on the targets they innervate, is a major theme of binaural plasticity. Interestingly, this theme has emerged coincidentally with another thrust of auditory neuroscience that has emphasized the sensitivity of CAS neurons to dynamically modulated stimuli (e.g., Spitzer and Semple 1991; Takahashi and Keller 1992; Batra et al. 1993), and the dynamic nature of responses to stationary stimuli (e.g., Wagner 1990; Sanes et al. 1998). Nevertheless, activity-dependent plasticity and dynamic changes in neuronal response properties generally operate over different time courses and, presumably, utilize different mechanisms.

In addition to the direct contribution made by studies of binaural plasticity in understanding normal and pathological hearing, there has been an important, indirect contribution relating to the points made in the preceding paragraphs. It is often assumed that the structure and connections of neurons, their response properties, and the auditory percepts they underlie are, particularly in adults, static entities that may be affected by factors (such as tissue processing, anesthetics, and practice) that the experimenter should endeavor to control. However, neural plasticity of the type described in this volume implies individual variability, and is a part of the normal function of the brain. In other words, variations in the anatomical and physiological characteristics of neurons and in the perceptual performance of individual listeners, including use-dependent changes observed following repeated presentation of the same acoustic stimuli, may be regarded as symptoms not of poor experimental control, but of the slowly changing, intrinsic response of the living nervous system.

6.2 Mechanisms of Binaural Plasticity

Studies of the effects of cochlear removal on neurons in the chicken NM have documented many of the cellular and molecular processes involved in the rapid degenerative response of those neurons (see Rubel, Parks, and Zirpel, Chapter 2). Few comparable studies have been performed of the effects of alterations of auditory input on neurons in the binaural systems of the CAS, but considerable data are available on the occurrence, time course, and functional consequences of degenerative and generative phenomena in binaural systems (see Sections 3–5 of this chapter).

The most obvious additional process occurring in binaural plasticity is what has here been called "generation"—the ability of neurons to make new, functional connections following a modification of sensory inputs. For example, a loss or change in the pattern of activity produced by one ear leads to an expansion in the projection from the CN to the IC on the unmanipulated side, either through axonal sprouting or some other form of reactive synaptogenesis (see Section 3.3). In owls, both unilateral ear plugging and prism rearing can induce axon sprouting and synaptogenesis in the projection from the ICC to the ICX (Section 3.4). Coupled with the suppression, by GABAergic inhibition, of the normal anatomical projection, these changes appear to underlie the behaviorally adaptive shifts in ITD tuning of ICX and OT neurons that are induced by these sensory manipulations (Sections 4 and 5).

Further exploration of the cellular processes responsible for binaural plasticity should be a priority for future research. This will require the use of both morphological and intracellular recording techniques to investigate how individual neurons and the synaptic connections they make are affected by input manipulations. The nature of the trigger signal(s) that lead to degenerative and generative changes in binaural neurons need to be identified. In the case of the circuits that lead to the synthesis of a map of auditory space in the midbrain, it is now well established that inputs from the visual system can shape the maturation of auditory neurons. Progress has recently been made in identifying the source of these visual signals and (in owls) the site of crossmodal interaction (Sections 3.4 and 4.4). However, the basis for this "instructed learning" (Knudsen 2002) remains unclear.

Most forms of neural plasticity have been demonstrated by observing the consequences of experimental or clinical changes in sensory inputs. It has been argued that the cellular mechanisms involved may be essentially the same as those involved in the initial formation of the neural circuits (Feldman and Knudsen 1998b). Further work is needed to confirm this and to determine why and how synaptogenesis can be reinduced at later stages of development.

One of the main differences between binaural plasticity and the degenerative response of CN neurons is their relation to the age of the animal. In mammals, CN neurons die following cochlear removal before the onset of hearing, but do not appear to do so after that time (Tierney et al. 1997; Mostafapour et al. 2000). Thus, the mechanisms triggering or activating that cell death change early during development. In contrast, several aspects of binaural hearing (e.g., the connections between the CN and the IC, the physiology of cortical neurons, and the acclimatization to hearing aids and earplugs) continue to remain vulnerable to change well into or throughout adult life. Moreover, although many of the changes documented in binaural hearing occur quickly, there are examples of plasticity occurring over time scales of weeks to months, or even years.

Some of the possible mechanisms underlying plasticity in mature animals have been discussed above (Sections 3.3 and 5.1), and further overviews of early and late neural plasticity are available elsewhere (Purves and Lichtman 1985; Steward 1989; Kaas 1991; Weinberger, Chapter 4). It would appear that, while early plasticity can involve changes in long-range connections and innervation of normally inappropriate targets, in addition to the regressive phenomena seen in the CN, later plasticity is limited to short-range axonal growth and synaptogenesis and to unmasking of existing connections. These principles would suggest that later plasticity is likely to be more limited than earlier plasticity, both in degree and in the extent to which new functions can be performed by reinnervated targets. In general, this is what is found when comparing earlier with later binaural plasticity.

6.3 Behavior and Biology

Work on binaural plasticity in the barn owl (Sections 4 and 5) has demonstrated the power of combining behavioral and physiological approaches in a single species. The use of combined behavioral and, particularly, less invasive biological methods, permitting repeated monitoring over days to months, should provide important insights into understanding the effects of input manipulations on binaural hearing. Progress in the psychoacoustics of binaural impairments in humans has been notoriously slow (e.g., Gabriel et al. 1992), and acute physiological experiments, using stereotyped and simple stimuli to study a diversity of neuronal response patterns, have usually provided data that are difficult to relate to real hearing problems. Some suggestions for future research in this area follow.

Most of the physiological and anatomical studies of binaural plasticity have focused on the brain stem and midbrain. Future studies will certainly need to address the contributions of the thalamus and the cortex to the behavioral changes that result from manipulations of sensory inputs. For example, the possibility discussed in Section 5.1 that adaptive adjustments in auditory localization behavior by humans and other mammals following monaural occlusion may involve learning to make better use of the unaffected acoustic cues, rather than binaural plasticity per se, needs to be investigated. The physiological basis of use-dependent plasticity should also be extended to other forms of perceptual learning in adult animals, including training and acclimatization.

Evidence that the response properties of cortical neurons can be altered by behavioral training or after conditioning with various stimuli has come mainly from acute recordings in anesthetized animals (see Weinberger, Chapter 4). Given that the responses of neurons recorded under anesthesia may not fully reflect those of the awake animal (Lu et al. 2001), it seems certain that a growing number of recording studies will be carried out in the future on awake animals. For issues such as binaural plasticity and other forms of learning, these neuronal recordings should, ideally, be made from animals that are actively engaged in behavioral tasks. At the same time, less invasive techniques, including magnetoencephalography, event-related potential recording, and functional magnetic resonance imaging, which can also be used in human subjects, are likely to provide valuable information about the brain areas involved and the time course of physiological changes that accompany behavioral plasticity.

6.4 Clinical Issues

There would seem to be two priorities for future work. One is the acquisition of further knowledge about the short- and long-term auditory consequences of different forms of hearing loss and of other sensory abnormalities. The other is the dissemination of knowledge within the clinical community and its implementation into clinical practice. From the knowledge we already have it seems clear that one of the therapeutic goals should be to create a spectrally balanced input between the ears (see Section 6.1). A second goal should be, for therapies of proven efficacy, to provide the earliest possible intervention. Early intervention is likely to reduce the degenerative consequences of a sensorineural hearing loss, and may promote generative changes that might underlie compensatory improvements in perception.

With future advances in imaging and other recording techniques, it will hopefully be possible to develop better objective tests of binaural hearing in human listeners. Coupled with more extensive behavioral testing, including cognitive tests, this will allow a fuller assessment of how binaural processing is affected by hearing loss. At least some of the variations in localization performance that have been reported in the clinical literature likely reflect differences in patient age and etiology as well as the testing paradigms used. Future questions include whether the degree of hearing loss is important for the expression of binaural plasticity and whether blind subjects really do develop better auditory spatial acuity. Finally, the controversial issue of whether and how training can be used to improve the perceptual abilities of patients with either peripheral impairments or central processing disorders seems certain to be a topic of future research in this area.

7. Summary

Studies of binaural plasticity have been inspired by classical demonstrations of plasticity in other neural systems, by clinical observations and considerations, by a basic desire to understand the way in which the auditory system responds to variable input, and by the belief that binaural systems offer particular advantages for addressing general neurobiological issues. The starting point of this quest is the observation that sound localization remains accurate as head size and, hence, the binaural cues for localization, change during normal development. Binaural plasticity can also be induced by clinical and experimental perturbations of auditory input, including unilateral sensorineural (e.g., Meniére's disease, cochlear ablation) and conductive (e.g., middle ear disease, ear plugging) hearing loss, and unilateral sound amplification, and by abnormal visual experience.

Research into binaural plasticity has used a wide variety of techniques from the cellular to the behavioral levels. Early unilateral hearing loss has been found to result in rearrangements of connections between the axons of cochlear nucleus (CN) neurons and their targets in the superior olivary complex (SOC) and the inferior colliculus (IC). In the SOC, axons from the CN on the side of the normal ear have been found to sprout to form new terminals in novel territories. In the IC, axon terminals from the same CN occupy more territory than usual. Postsynaptic neurons in the SOC, IC, and higher levels may die or retract dendrites in response to hearing loss. Physiological responses include increased excitatory discharges to stimulation of the normal ear, changes in the sensitivity or response range to interaural level differences (ILDs), and compensatory adjustments of neuronal spatial tuning. Conductive hearing loss can reduce or alter sound localization performance, reduce binaural unmasking, and alter the ILD required for the midline centering of a tone. Unilateral sound amplification with a hearing aid can change auditory discrimination in the aided or in the unaided ear. Visual cues play a role in shaping the development of auditory spatial tuning in the midbrain; studies in the barn owl have shown that this provides one of the clearest examples of binaural plasticity. The effects of a loss of vision on binaural hearing are more controversial, but most studies in blind mammals have provided evidence for compensatory improvements in auditory localization.

These studies have contributed to our understanding of both normal and abnormal auditory function by emphasizing the dynamic nature of binaural hearing. In the future, we would expect to make further advances using, among other techniques, labeling of individual axons, intracellular recording, chronic recordings from awake, behaving animals, functional neuroimaging, and novel tests of binaural hearing.

Acknowledgments. The research in our laboratory in Oxford has been generously funded by the Medical Research Council (U.K.), the Wellcome Trust, Defeating Deafness (the Hearing Research Trust), the Oxford McDonnell-Pew Centre for Cognitive Neuroscience, the National Health Service, the E.P. Abraham Research Fund, the John Ellerman Foundation, and the National Organization for Hearing Research (U.S.).

Abbreviations

Nuclei in avian species are shown in parentheses. Mammalian homologies are indicated in *italics* after the explanation of the avian abbreviation.

2-DG	2-deoxyglucose
AC	auditory cortex
AN	auditory nerve
AVCN	anterior ventral cochlear nucleus
CAS	central auditory system
CM	cochlear microphonic
CN	cochlear nucleus
CN-ICi	cochlear nucleus projection to the ipsilateral IC
CNS	central nervous system
DCN	dorsal cochlear nucleus
DiI	1,1',Dioctadecyl-3,3,3',3'-tetramethylindocarbocyanine perchlorate
EI	excitatory–inhibitory
EM	electron microscopy
HL	hearing level
IC	inferior colliculus
ICC	central nucleus of the IC
ICX	external nucleus of the IC
ILD	interaural level difference
ITD	interaural time difference
LSO	lateral superior olivary nucleus
MAA	minimum audible angle
MLD	(binaural) masking level difference
MNTB	medial nucleus of the trapezoid body
MSO	medial superior olivary nucleus
NMDA	*N*-methyl-D-aspartate
(NL)	nucleus laminaris—*MSO*
NLL	nucleus of the lateral lemniscus
(NM)	nucleus magnocellularis—*AVCN*

OM	otitis media
(OT)	optic tectum—*SC*
P	postnatal day
PVCN	posterior ventral cochlear nucleus
RF	receptive field
SC	superior colliculus
SOC	superior olivary complex
TB	trapezoid body
TTX	tetrodotoxin
VAS	virtual acoustic space
VCN	ventral cochlear nucleus
(VLVp)	posterior division of the ventral NLL—*dorsal NLL* or possibly LSO
(XDCT)	crossed dorsal cochlear tract—*TB*

References

Altschuler RA (2000) Molecular mechanisms in central auditory function and plasticity. Hear Res 147:1–302.

Ashmead DH, Wall RS, Ebinger KA, Eaton SB, Snook-Hill MM, Yang X (1998) Spatial hearing in children with visual disabilities. Perception 27:105–122.

Batkin S, Groth H, Watson JR, Ansberry M (1970) Effects of auditory deprivation on the development of auditory sensitivity in adult rats. EEG Clin Neurophysiol 28:351–359.

Batra R, Kuwada S, Stanford TR (1993) High-frequency neurons in the inferior colliculus that are sensitive to interaural delays of amplitude-modulated tones: evidence for dual binaural influences. J Neurophysiol 70:64–80.

Bauer RW, Matuzsa JL, Blackmer RF (1966) Noise localization after unilateral attenuation. J Acoust Soc Am 40:441–444.

Beggs WD, Foreman DL (1980) Sound localization and early binaural experience in the deaf. Br J Audiol 14:41–48.

Bermingham-McDonogh O, Rubel EW (2003) Hair cell regeneration: winging our way towards a sound future. Curr Opin Neurobiol 13:119–126.

Bilecen D, Seifritz E, Radü EW, Schmid N, Wetzel S, Probst R, Scheffler K (2000) Cortical reorganization after acute unilateral hearing loss traced by fMRI. Neurology 54:765–767.

Blakemore C, Cooper GF (1970) Development of the brain depends on the visual environment. Nature 228:477–478.

Born DE, Rubel EW (1988) Afferent influences on brain stem auditory nuclei of the chicken: presynaptic action potentials regulate protein synthesis in nucleus magnocellularis neurons. J Neurosci 8:901–919.

Born DE, Durham D, Rubel EW (1991) Afferent influences on brain stem auditory nuclei of the chick: nucleus magnocellularis neuronal activity following cochlea removal. Brain Res 557:37–47.

Brainard MS, Knudsen EI (1993) Experience-dependent plasticity in the inferior colliculus: a site for visual calibration of the neural representation of auditory space in the barn owl. J Neurosci 13:4589–4608.

Brainard MS, Knudsen EI (1995) Dynamics of visually guided auditory plasticity in the optic tectum of the barn owl. J Neurophysiol 73:595–614.

Brainard MS, Knudsen EI (1998) Sensitive periods for visual calibration of the auditory space map in the barn owl optic tectum. J Neurosci 18:3929–3942.

Brugge JF, Orman SS, Coleman JR, Chan JCK, Phillips DP (1985) Binaural interactions in cortical area AI of cats reared with unilateral atresia of the external ear canal. Hear Res 20:275–287.

Butler RA (1987) An analysis of the monaural displacement of sound in space. Percept Psychophys 41:1–7.

Byrne D, Noble W, LePage B (1992) Effects of long-term bilateral and unilateral fitting of different hearing aid types on the ability to locate sounds. J Am Acad Audiol 3: 369–382.

Calford MB, Rajan R, Irvine DRF (1993) Rapid changes in the frequency tuning of neurons in cat auditory cortex resulting from pure-tone-induced temporary threshold shift. Neuroscience 55:953–964.

Calvert GA, Brammer MJ, Iversen SD (1998) Crossmodal identification. Trends Cogn Sci 2:247–253.

Cant NB (1991) Projections to the lateral and medial superior olivary nuclei from the spherical and globular bushy cells of the anteroventral cochlear nucleus. In: Altschuler RA, Bobbin RP, Clopton BM, Hoffman DW (eds), Neurobiology of Hearing: The Central Auditory System. New York: Raven Press, pp. 99–120.

Cant NB, Gaston KC (1986) Pathways connecting the right and left cochlear nuclei. J Comp Neurol 212:313–326.

Carlile S, King AJ (1994) Monaural and binaural spectrum level cues in the ferret: acoustics and the neural representation of auditory space. J Neurophysiol 71:785–801.

Caspary DM, Raza A, Lawhorn Armour BA, Pippin J, Arneric SP (1990) Immunocytochemical and neurochemical evidence for age-related loss of GABA in the inferior colliculus: implications for neural presbycusis. J Neurosci 10:2363–2372.

Clarey JC, Barone P, Imig TJ (1992) Physiology of the thalamus and cortex. In: Popper AN, Fay RR (eds), The Mammalian Auditory Pathway: Neurophysiology. New York: Springer-Verlag, pp. 232–334.

Clements M, Kelly JB (1978) Auditory spatial responses of young guinea pigs (*Cavia porcellus*) during and after ear blocking. J Comp Physiol Psychol 92:34–44.

Clerici WJ, Coleman JR (1986) Resting and high-frequency evoked 2-deoxyglucose uptake in the rat inferior colliculus: developmental changes and effects of short-term conduction blockade. Brain Res 392:127–137.

Clifton RK (1992) The development of spatial hearing in human infants. In: Werner LA, Rubel EW (eds) Developmental Psychoacoustics. Washington DC: American Psychological Association, pp. 135–157.

Clopton BM, Silverman MS (1977) Plasticity of binaural interaction. II. Critical period and changes in midline response. J Neurophysiol 40:1275–1280.

Clopton BM, Silverman MS (1978) Changes in latency and duration of neural responding following developmental auditory deprivation. Exp Brain Res 32:39–47.

Cohen YE, Knudsen EI (1999) Maps versus clusters: different representations of auditory space in the midbrain and forebrain. Trends Neurosci 22:128–135.

Coleman JR, O'Connor P (1979) Effects of monaural and binaural sound deprivation on cell development in the anteroventral cochlear nucleus of rats. Exp Neurol 64:553–566.

Conlee JW, Parks TN, Romero C, Creel DJ (1984) Auditory brain stem anomalies in albino cats: II. Neuronal atrophy in the superior olive. J Comp Neurol 225:141–148.

Cook RD, Hung TY, Miller RL, Smith DW, Tucci DL (2002) Effects of conductive hearing loss on auditory nerve activity in gerbil. Hear Res 164:127–137.

Cotanche DA (1999) Structural recovery from sound and aminoglycoside damage in the avian cochlea. Audiol Neurootol 4:271–285.

Darian-Smith C, Gilbert CD (1994) Axonal sprouting accompanies functional reorganization in adult cat striate cortex. Nature 368:737–740.

Darian-Smith C, Gilbert CD (1995) Topographic reorganization in the striate cortex of the adult cat and monkey is cortically mediated. J Neurosci 15:1631–1647.

DeBello WM, Feldman DE, Knudsen EI (2001) Adaptive axonal remodeling in the midbrain auditory space map. J Neurosci 21:3161–3174.

Deitch JS, Rubel EW (1984) Afferent influences on brain stem auditory nuclei of the chicken: time course and specificity of dendritic atrophy following deafferentation. J Comp Neurol 229:66–79.

Deitch JS, Rubel EW (1989a) Rapid changes in ultrastructure during deafferentation-induced dendritic atrophy. J Comp Neurol 281:234–258.

Deitch JS, Rubel EW (1989b) Changes in neuronal cell bodies in N. laminaris during deafferentation-induced dendritic atrophy. J Comp Neurol 281:259–268.

Dodson HC, Bannister LH, Douek EE (1994) Effects of unilateral deafening on the cochlear nucleus of the guinea pig at different ages. Dev Brain Res 80:261–267.

Dooling RJ, Popper AN, Fay RR, eds (2000) Comparative Hearing: Birds and Reptiles. New York: Springer-Verlag.

Doubell TP, Baron J, Skaliora I, King AJ (2000) Topographical projection from the superior colliculus to the nucleus of the brachium of the inferior colliculus in the ferret: convergence of visual and auditory information. Eur J Neurosci 12:4290–4308.

Doubell TP, Baron J, Skaliora I, King AJ (2003) Functional connectivity between the superficial and deeper layers of the superior colliculus: implications for sensorimotor integration. J Neurosci 23:6596–6607.

Doyle WJ, Webster DB (1991) Neonatal conductive hearing loss does not compromise brain stem auditory function and structure in rhesus monkeys. Hear Res 54:145–151.

Driver J, Spence C (1998) Attention and the crossmodal construction of space. Trends Cogn Sci 2:254–262.

Durand GM, Kovalchuk Y, Konnerth A (1996) Long-term potentiation and functional synapse induction in developing hippocampus. Nature 381:71–75.

Durlach NI, Thompson CL, Colburn HS (1981) Binaural interaction of impaired listeners. A review of past research. Audiology 20:181–211.

Dyson SE, Warton SS, Cockman B (1991) Volumetric and histological changes in the cochlear nuclei of visually deprived rats: a possible morphological basis for intermodal sensory compensation. J Comp Neurol 307:39–48.

Fay RR (1988) Hearing in Vertebrates: A Psychophysics Databook. Winnetka, IL: Hill-Fay Associates.

Feldman DE, Knudsen EI (1997) An anatomical basis for visual calibration of the auditory space map in the barn owl's midbrain. J Neurosci 17:6820–6837.

Feldman DE, Knudsen EI (1998a) Pharmacological specialization of learned auditory responses in the inferior colliculus of the barn owl. J Neurosci 18:3073–3087.

Feldman DE, Knudsen EI (1998b) Experience-dependent plasticity and the maturation of glutamatergic synapses. Neuron 20:1067–1071.

Feng AS, Rogowski BA (1980) Effects of monaural and binaural occlusion on the morphology of neurons in the medial superior olivary nucleus of the rat. Brain Res 189: 530–534.

Florentine M (1976) Relation between lateralization and loudness in asymmetrical hearing losses. J Am Audiol Soc 1:243–251.

Gabriel KJ, Koehnke J, Colburn HS (1992) Frequency dependence of binaural performance in listeners with impaired binaural hearing. J Acoust Soc Am 91:336–347.

Gabriele ML, Brunso-Bechtold JK, Henkel CK (2000) Plasticity in the development of afferent patterns in the inferior colliculus of the rat after unilateral cochlear ablation. J Neurosci 20:6939–6949.

Gatehouse S (1992) The time course and magnitude of perceptual acclimatization to frequency responses: evidence from monaural fitting of hearing aids. J Acoust Soc Am 92:1258–1268.

Gelfand SA, Silman S, Ross L (1987) Long-term effects of monaural, binaural and no amplification in subjects with bilateral hearing loss. Scand Audiol 16:201–207.

Gilbert CD (1992) Horizontal integration and cortical dynamics. Neuron 9:1–13.

Gilbert CD (1993) Rapid dynamic changes in adult cerebral cortex. Curr Opin Neurobiol 3:100–103.

Gold JI, Knudsen EI (1999) Hearing impairment induces frequency-specific adjustments in auditory spatial tuning in the optic tectum of young owls. Neurophysiol 82:2197–2209.

Gold JI, Knudsen EI (2000a) Abnormal auditory experience induces frequency-specific adjustments in unit tuning for binaural localization cues in the optic tectum of juvenile owls. J Neurosci 20:862–877.

Gold JI, Knudsen EI (2000b) A site of auditory experience-dependent plasticity in the neural representation of auditory space in the barn owl's inferior colliculus. J Neurosci 20:3469–3486.

Gold JI, Knudsen EI (2001) Adaptive adjustment of connectivity in the inferior colliculus revealed by focal pharmacological inactivation. J Neurophysiol 85:1575–1584.

Gravel JS, Wallace IF (1992) Listening and language at 4 years of age: effects of early otitis media. J Speech Hear Res 35:588–595.

Gravel JS, Wallace IF, Ruben RJ (1996) Auditory consequences of early mild hearing loss associated with otitis media. Acta Otolaryngol 116:219–221.

Guillery RW (1988) Competition in the development of the visual pathways. In: Parnavelas JG, Stern CD, Stirling RV (eds), The Making of the Nervous System. New York: Oxford University Press, pp. 356–379.

Gutfreund Y, Zheng W, Knudsen EI (2002) Gated visual input to the central auditory system. Science 297:1556–1559.

Gyllensten L, Malmfors T, Norrlin ML (1966) Growth alteration in the auditory cortex of visually deprived mice. J Comp Neurol 126:463–469.

Hafidi A, Sanes DH, Hillman DE (1995) Regeneration of the auditory midbrain intercommissural projection in organotypic culture. J Neurosci 15:1298–1307.

Haggard MP, Hughes EA (1991) Screening Children's Hearing: A Review of the Literature and the Implications for Otitis Media. London: HMSO.

Haggard MP, Birkin JA, Browning GG, Gatehouse S, Lewis S (1994) Behavior problems in otitis media. Pediatr Infect Dis J 13:S43–50.

Hall JW, Derlacki EL (1986) Effect of conductive hearing loss and middle ear surgery on binaural hearing. Ann Otol Rhinol Laryngol 95:525–530.

Hall JW, Grose JH (1993) Short-term and long-term effects on the masking level difference following middle ear surgery. J Am Acad Audiol 4:307–312.

Hall JW, Grose JH, Pillsbury HC (1995) Long-term effects of chronic otitis media on binaural hearing in children. Arch Otolaryngol Head Neck Surg 121:847–852.

Hall JW, Grose JH, Dev MB, Drake AF, Pillsbury HC (1998) The effect of otitis media with effusion on complex masking tasks in children. Arch Otolaryngol Head Neck Surg 124:892–896.

Hartley DEH, Moore DR (2003) Effects of conductive hearing loss on temporal aspects of sound transmission through the ear. Hear Res 177:53–60.

Hartline PH, Pandey Vimal RL, King AJ, Kurylo DD, Northmore DPM (1995) Effects of eye position on auditory localization and neural representation of space in superior colliculus of cats. Exp Brain Res 104:402–408.

Hasenstab MS (1987) Language Learning and Otitis Media. London: Taylor and Francis.

Hashisaki GT, Rubel EW (1989) Effects of unilateral cochlea removal on anteroventral cochlear nucleus neurons in developing gerbils. J Comp Neurol 283:465–473.

Häusler R, Colburn S, Marr E (1983) Sound localization in subjects with impaired hearing. Acta Otolaryngol (Stockh) (Suppl) 400:1–62.

Heffner RS, Koay G, Heffner HE (1996) Sound localization in chinchillas, III: Effect of pinna removal. Hear Res 99:13–21.

Heil P, Scheich H (1986) Effects of unilateral and bilateral cochlea removal on 2-deoxyglucose patterns in the chick auditory system. J Comp Neurol 252:279–301.

Held R (1955) Shifts in binaural localization after prolonged exposure to atypical combinations of stimuli. Am J Psychol 68:526–548.

Hestrin S (1992) Developmental regulation of NMDA receptor-mediated synaptic currents at a central synapse. Nature 357:686–689.

Hofman PM, Van Riswick JGA, Van Opstal JA (1998) Relearning sound localization with new ears. Nature Neurosci 1:417–421.

Hogan SC, Moore DR (2003) Impaired binaural hearing in children produced by a threshold level of middle ear disease. J Assoc Res Otolaryngol 4:123–129.

Hogan SC, Meyer SE, Moore DR (1996) Binaural unmasking returns to normal in teenagers who had otitis media in infancy. Audiol Neurootol 1:104–111.

Hogan SC, Stratford KJ, Moore DR (1997) Duration and recurrence of otitis media with effusion in children from birth to 3 years: prospective study using monthly otoscopy and tympanometry. Bri Med J 314:350–353.

Hood JD (1984) Speech discrimination in bilateral and unilateral hearing loss due to Meniere's disease. Br J Audiol 18:173–177.

Hubel DH, Wiesel TN (1965) Binocular interaction in striate cortex of kittens reared with artificial squint. J Neurophysiol 28:1041–1059.

Huber F (1987) Plasticity in the auditory system of crickets: phonotaxis with one ear and neuronal reorganization within the auditory pathway. J Comp Physiol A 161:583–604.

Hyde PS, Knudsen EI (2002) The optic tectum controls visually guided adaptive plasticity in the owl's auditory space map. Nature 415:73–76.

Irvine DRF (1986) The Auditory Brainstem. Progress in Sensory Physiology, Vol. 7 (Ottoson D, ed-in-chief). Berlin: Springer-Verlag.

Irvine DRF (1992) Physiology of the auditory brain stem. In: Popper AN, Fay RR (eds), The Mammalian Auditory Pathway: Neurophysiology. New York: Springer-Verlag, pp. 153–231.

Irvine DRF, Rajan R (1995) Plasticity in the mature auditory system. In: Manley GA, Klump GM, Köppl C et al. (eds), Advances in Hearing Research. Singapore: World Scientific.

Izraeli R, Koay G, Lamish M, Heicklen-Klein AJ, Heffner HE, Heffner RS, Wollberg Z (2002) Cross-modal neuroplasticity in neonatally enucleated hamsters: structure, electrophysiology and behaviour. Eur J Neurosci 15:693–712.

Jackson H, Parks TN (1988) Induction of aberrant functional afferents to the chick cochlear nucleus. J Comp Neurol 271:106–114.

Jay MF, Sparks DL (1984) Auditory receptive fields in primate superior colliculus shift with changes in eye position. Nature 309:345–347.

Jeffress LA (1948) A place theory of sound localization. J Comp Physiol Psychol 41: 35–39.

Jen PHS, Sun XD (1990) Influence of monaural plugging on postnatal development of auditory spatial sensitivity of inferior collicular neurons of the big brown bat, *Eptesicus fuscus*. Chin J Physiol 33:231–246.

Jenkins WM, Merzenich MM (1984) Role of cat primary auditory cortex for sound-localization behavior. J Neurophysiol 52:819–847.

Jerger JF, Harford ER (1966) Alternate and simultaneous binaural balancing of pure tones. J Speech Hear Res 3:15–30.

Jerger J, Silman S, Lew HL, Chmiel R (1993) Case studies in binaural interference: converging evidence from behavioral and electrophysiologic measures. J Am Acad Audiol 4:122–131.

Kaas JH (1991) Plasticity of sensory and motor maps in adult mammals. Annu Rev Neurosci 14:137–167.

Kelley MW, Talreja DR, Corwin JT (1995) Replacement of hair cells after laser microbeam irradiation in cultured organs of corti from embryonic and neonatal mice. J Neurosci 15:3013–3026.

Kiang NYS (1984) Peripheral neural processing of auditory information. In: Darian-Smith I (ed), Handbook of Physiology, Section 1, Volume III, Part 2. Bethesda, MD: American Physiological Society, pp. 639–674.

Kilgard MP, Merzenich MM (1998) Plasticity of temporal information processing in the primary auditory cortex. Nat Neurosci 1:727–731.

Killackey HP, Ryugo DK (1977) Effects of neonatal auditory system damage on the structure of the inferior colliculus of the rat. Anat Rec 187:624.

Kind PC (1999) Cortical plasticity: is it time for a change? Curr Biol 9:R640–643.

King AJ (1999) Sensory experience and the formation of a computational map of auditory space in the brain. BioEssays 21:900–911.

King AJ, Carlile S (1993) Changes induced in the representation of auditory space in the superior colliculus by rearing ferrets with binocular eyelid suture. Exp Brain Res 94:444–455.

King AJ, Carlile S (1995) Neural coding for auditory space. In: Gazzaniga MS (ed.), The Cognitive Neurosciences. Cambridge, MA: The MIT Press, pp. 279–293.

King AJ, Parsons CH (1999) Improved auditory spatial acuity in visually deprived ferrets. Eur J Neurosci 11:3945–3956.

King AJ, Hutchings ME, Moore DR, Blakemore C (1988) Developmental plasticity in the visual and auditory representations in the mammalian superior colliculus. Nature 332:73–76.

King AJ, Moore DR, Hutchings ME (1994) Topographic representation of auditory space in the superior colliculus of adult ferrets after monaural deafening in infancy. J Neurophysiol 71:182–194.

King AJ, Schnupp JWH, Thompson ID (1998) Signals from the superficial layers of the superior colliculus enable the development of the auditory space map in the deeper layers. J Neurosci 18:9394–9408.

King AJ, Parsons CH, Moore DR (2000) Plasticity in the neural coding of auditory space in the mammalian brain. Proc Natl Acad Sci USA 97:11821–11828.

King AJ, Schnupp JWH, Doubell TP (2001a) The shape of ears to come: dynamic coding of auditory space. Trends Cog Sci 5:261–270.

King AJ, Kacelnik O, Mrsic-Flogel TD, Schnupp JWH, Parsons CH, Moore DR (2001b) How plastic is spatial hearing? Audiol Neurootol 6:182–186.

Kitzes LM (1984) Some physiological consequences of neonatal cochlear destruction in the inferior colliculus of the gerbil, *Meriones unguiculatus*. Brain Res 306:171–178.

Kitzes LM, Semple MN (1985) Single-unit responses in the inferior colliculus: effects of neonatal unilateral cochlear ablation. J Neurophysiol 53:1483–1500.

Kitzes LM, Kageyama GH, Semple MN, Kil J (1995) Development of ectopic projections from the ventral cochlear nucleus to the superior olivary complex induced by neonatal ablation of the contralateral cochlea. J Comp Neurol 353:341–363.

Knudsen EI (1982) Auditory and visual maps of space in the optic tectum of the owl. J Neurosci 2:1177–1194.

Knudsen EI (1983) Early auditory experience aligns the auditory map of space in the optic tectum of the barn owl. Science 222:939–942.

Knudsen EI (1984) The role of auditory experience in the development and maintenance of sound localization. Trends Neurosci 7:326–330.

Knudsen EI (1985) Experience alters the spatial tuning of auditory units in the optic tectum during a sensitive period in the barn owl. J Neurosci 5:3094–3109.

Knudsen EI (1998) Capacity for plasticity in the adult owl auditory system expanded by juvenile experience. Science 279:1531–1533.

Knudsen EI (2002) Instructed learning in the auditory localization pathway of the barn owl. Nature 417:322–328.

Knudsen EI, Brainard MS (1991) Visual instruction of the neural map of auditory space in the developing optic tectum. Science 253:85–87.

Knudsen EI, Knudsen PF (1985) Vision guides the adjustment of auditory localization in young barn owls. Science 230:545–548.

Knudsen EI, Knudsen PF (1990) Sensitive and critical periods for visual calibration of sound localization by barn owls. J Neurosci 10:222–232.

Knudsen EI, Konishi M (1978) A neural map of auditory space in the owl. Science 200:795–797.

Knudsen EI, Konishi M (1979) Mechanisms of sound localization in the barn owl (*Tyto alba*). J Comp Physiol 133:13–21.

Knudsen EI, Esterly SD, Knudsen PF (1984a) Monaural occlusion alters sound localization during a sensitive period in the barn owl. J Neurosci 4:1001–1011.

Knudsen EI, Knudsen PF, Esterly SD (1984b) A critical period for the recovery of sound localization accuracy following monaural occlusion in the barn owl. J Neurosci 4:1012–1020.

Knudsen EI, Esterly SD, du Lac S (1991) Stretched and upside-down maps of auditory space in the optic tectum of blind-reared owls; acoustic basis and behavioral correlates. J Neurosci 11:1727–1747.

Knudsen EI, Esterly SD, Olsen JF (1994) Adaptive plasticity of the auditory space map in the optic tectum of adult and baby barn owls in response to external ear modification. J Neurophysiol 71:79–94.

Koerber C, Pfeiffer RR, Warr WB, Kiang NYS (1966) Spontaneous spike discharges from units in the cochlear nucleus after destruction of the cochlea. Exp Neurol 16: 119–130.

Konishi M (1993) Listening with two ears. Sci Am 268:66–73.

Korte M, Rauschecker JP (1993) Auditory spatial tuning of cortical neurons is sharpened in cats with early blindness. J Neurophysiol 70:1717–1721.

Kotak VC, Sanes DH (1996) Developmental influence of glycinergic transmission: regulation of NMDA receptor-mediated EPSPs. J Neurosci 16:1836–1843.

Kotak VC, Sanes DH (1997) Deafferentation weakens excitatory synapses in the developing central auditory system. Eur J Neurosci 9:2340–2347.

Kujala T, Alho K, Näätänen R (2000) Cross-modal reorganization of human cortical functions. Trends Neurosci 23:115–120.

Land PW, Rose LL, Harvey AR, Liverman SA (1984) Neonatal auditory cortex lesions result in aberrant crossed corticotectal and corticothalamic projections in rats. Brain Res 314:126–130.

Laska M, Walder M, Schneider I, von Wedel H (1992) Maturation of binaural interaction components in auditory brain stem responses of young guinea pigs with monaural or binaural conductive hearing loss. Eur Arch Otorhinolaryngol 249:325–328.

Lessard N, Paré M, Lepore F, Lassonde M (1998) Early-blind human subjects localize sound sources better than sighted subjects. Nature 395:278–280.

Levi-Montalcini R (1949) Development of the acoustico-vestibular centers in the chick embryo in the absence of the afferent root fibers and of descending fiber tracts. J Comp Neurol 91:209–242.

Lewald J (2002) Vertical sound localization in blind humans. Neuropsychologia 40: 1868–1872.

Linkenhoker BA, Knudsen EI (2002) Incremental training increases the plasticity of the auditory space map in adult barn owls. Nature 419:293–296.

Lippe WR, Steward O, Rubel EW (1980) The effect of unilateral basilar papilla removal upon nuclei laminaris and magnocellularis of the chick examined with [^{3}H]2-deoxy-D-glucose autoradiography. Brain Res 196:43–58.

Lippe WR, Fuhrmann DS, Yang W, Rubel EW (1992) Aberrant projection induced by otocyst removal maintains normal tonotopic organization in the chick cochlear nucleus. J Neurosci 12:962–969.

Lu T, Liang L, Wang X (2001) Temporal and rate representations of time-varying signals in the auditory cortex of awake primates. Nat Neurosci 4:1131–1138.

Matsushima JI, Shepherd RK, Seldon HL, Xu SA, Clark GM (1991) Electrical stimulation of the auditory nerve in deaf kittens: effects on cochlear nucleus morphology. Hear Res 56:133–142.

McAlpine D, Martin RL, Mossop JE, Moore DR (1997) Response properties of neurons in the inferior colliculus of the monaurally-deafened ferret to acoustic stimulation of the intact ear. J Neurophysiol 78:767–779.

McMullen NT, Glaser EM (1988) Auditory cortical responses to neonatal deafening: pyramidal neuron spine loss without changes in growth or orientation. Exp Brain Res 72:195–200.

McMullen NT, Goldberger B, Suter CM, Glaser EM (1988) Neonatal deafening alters nonpyramidal dendrite orientation in auditory cortex: a computer microscope study in the rabbit. J Comp Neurol 267:92–106.

McPartland JL, Culling JF, Moore DR (1997) Changes in lateralization and loudness judgements during one week of unilateral ear plugging. Hear Res 113:165–173.

Merzenich MM, Kaas JH, Wall J, Nelson RJ, Sur M, Felleman D (1983) Topographic reorganization of somatosensory cortical areas 3b and 1 in adult monkeys following restricted deafferentation. Neurosci 8:33–55.

Middlebrooks JC (1999) Virtual localization improved by scaling nonindividualized external-ear transfer functions in frequency. J Acoust Soc Am 106:1493–1510.

Middlebrooks JC, Green DM (1991) Sound localization by human listeners. Annu Rev Psychol 42:135–159.

Milbrandt JC, Holder TM, Wilson MC, Salvi RJ, Caspary DM (2000) GAD levels and muscimol binding in rat inferior colliculus following acoustic trauma. Hear Res 147: 251–260.

Miller GL, Knudsen EI (1999) Early visual experience shapes the representation of auditory space in the forebrain gaze fields of the barn owl. J Neurosci 19:2326–2336.

Mody M, Schwartz RG, Gravel JS, Ruben RJ (1999) Speech perception and verbal memory in children with and without histories of otitis media. J Speech Lang Hear Res 42:1069–1079.

Mogdans J, Knudsen EI (1992) Adaptive adjustment of unit tuning to sound localization cues in response to monaural occlusion in developing owl optic tectum. J Neurosci 12:3473–3484.

Mogdans J, Knudsen EI (1993) Early monaural occlusion alters the neural map of interaural level differences in the inferior colliculus of the barn owl. Brain Res 619:29–38.

Mogdans J, Knudsen EI (1994) Representation of interaural level difference in the VLVp, the first site of binaural comparison in the barn owl's auditory system. Hear Res 74: 148–164.

Moiseff A, Konishi M (1981) Neuronal and behavioral sensitivity to binaural time difference in the owl. J Neurosci 1:40–48.

Moore BCJ (1997) An Introduction to the Psychology of Hearing. 4th ed. San Diego, CA: Academic Press, 1997.

Moore DR (1988) Auditory brain stem of the ferret: sources of projections to the inferior colliculus. J Comp Neurol 269:342–354.

Moore DR (1990a) Auditory brain stem of the ferret: bilateral cochlear lesions in infancy do not affect the number of neurons projecting from the cochlear nucleus to the inferior colliculus. Dev Brain Res 54:125–130.

Moore DR (1990b) Auditory brain stem of the ferret: early cessation of developmental sensitivity of neurons in the cochlear nucleus to removal of the cochlea. J Comp Neurol 302:810–823.

Moore DR (1992) Trophic influences of excitatory and inhibitory synapses on neurones in the auditory brain stem. NeuroReport 3:269–272.

Moore DR (1993) Auditory brain stem responses in ferrets following unilateral cochlear removal. Hear Res 68:28–34.

Moore DR (1994) Auditory brain stem of the ferret: long survival following cochlear removal progressively changes projections from the cochlear nucleus to the inferior colliculus. J Comp Neurol 339:301–310.

Moore DR, Aitkin LM (1975) Rearing in an acoustically unusual environment—effects on neural auditory responses. Neurosci Lett, 1:29–34.

Moore DR, Irvine DRF (1981) Plasticity of binaural interaction in the cat inferior colliculus. Brain Res 208:198–202.

Moore DR, Kitzes LM (1985) Projections from the cochlear nucleus to the inferior colliculus in normal and neonatally cochlea-ablated gerbils. J Comp Neurol 240:180–195.

Moore DR, Kitzes LM (1986). Cochlear nucleus lesions in the adult gerbil: Effects on neurone responses in the contralateral inferior colliculus. Brain Res 373:268–274.

Moore DR, Kowalchuk NE (1988) Auditory brain stem of the ferret: effects of unilateral cochlear lesions on cochlear nucleus volume and projections to the inferior colliculus. J Comp Neurol 272:503–515.

Moore DR, Hutchings ME, King AJ, Kowalchuk NE (1989) Auditory brain stem of the ferret: some effects of rearing with a unilateral ear plug on the cochlea, cochlear nucleus, and projections to the inferior colliculus. J Neurosci 9:1213–1222.

Moore DR, Hutchings ME, Meyer SE (1991) Binaural masking level differences in children with a history of otitis media. Audiology 30:91–101.

Moore DR, King AJ, McAlpine D, Martin RL, Hutchings ME (1993) Functional consequences of neonatal unilateral cochlear removal. Prog Brain Res 97:127–133.

Moore DR, Russell FA, Cathcart NC (1995) Lateral superior olive projections to the inferior colliculus in normal and unilaterally deafened ferrets. J Comp Neurol 357:204–216.

Moore DR, France SJ, McAlpine D, Mossop JE, Versnel H (1997) Plasticity of inferior colliculus and auditory cortex following unilateral deafening in adult ferrets. In: Syka J (ed), Acoustical Signal Processing in the Central Auditory System. New York: Plenum, pp. 489–499.

Moore DR, Hine JE, Jiang ZD, Matsuda H, Parsons CH, King AJ (1999) Conductive hearing loss produces a reversible binaural hearing impairment. J Neurosci 19:8704–8711.

Moore DR, Hogan SC, Kacelnik O, Parsons CH, Rose MM, King AJ (2001) Auditory learning as a cause and treatment of central dysfunction. Audiol Neurootol 6:216–220.

Morest DK, Bohne BA (1983) Noise-induced degeneration in the brain and representation of inner and outer hair cells. Hear Res 9:145–151.

Morest DK, Kim J, Bohne BA (1997) Neuronal and transneuronal degeneration of auditory axons in the brain stem after cochlear lesions in the chinchilla: cochleotopic and non-cochleotopic patterns. Hear Res 103:151–168.

Mossop JE, Wilson MJ, Caspary DM, Moore DR (2000) Down-regulation of inhibition following unilateral deafening. Hear Res 147:183–187.

Mostafapour SP, Cochran SL, Del Puerto NM, Rubel EW (2000) Patterns of cell death in mouse anteroventral cochlear nucleus neurons after unilateral cochlea removal. J Comp Neurol 426:561–571.

Movshon JA, Van Sluyters RC (1981) Visual neural development. Annu Rev Psychol 32:477–522.

Mrsic-Flogel TD, King AJ, Jenison RL, Schnupp JWH (2001) Listening through different ears alters spatial response fields in ferret primary auditory cortex. J Neurophysiol 86:1043–1046.

Nordeen KW, Killackey HP, Kitzes LM (1983) Ascending projections to the inferior colliculus following unilateral cochlear ablation in the neonatal gerbil, *Meriones unguiculatus*. J Comp Neurol 214:144–153.

Nordlund B (1964) Directional audiometry. Acta Otolaryngol (Stockh) 57:1–18.

Oliver DL, Huerta MF (1992) Inferior and superior colliculi. In: Webster DB, Popper AN, Fay RR (eds), The Mammalian Auditory Pathway: Neuroanatomy. New York: Springer-Verlag, pp. 168–221.

Oliver DL, Shneiderman A (1991) The anatomy of the inferior colliculus: a cellular basis for integration of monaural and binaural information. In: Altschuler RA, Bobbin RP, Clopton BM, Hoffman DW (eds), Neurobiology of Hearing: The Central Auditory System. New York: Raven Press, pp. 195–222.

Pallas SL, Littman T, Moore DR (1999) Cross-modal respecification of callosal connectivity without altering thalamocortical input. Proc Natl Acad Sci USA 96:8751–8756.

Palmer AR, King AJ (1982) The representation of auditory space in the mammalian superior colliculus. Nature 299:248–249.

Palmer AR, King AJ (1985) A monaural space map in the guinea-pig superior colliculus. Hear Res 17:267–280.

Parks TN (1981) Changes in the length and organization of nucleus laminaris dendrites after unilateral otocyst ablation in chick embryos. J Comp Neurol 202:47–57.

Parsons CH, Lanyon RG, Schnupp JW, King AJ (1999) Effects of altering spectral cues in infancy on horizontal and vertical sound localization by adult ferrets. J Neurophysiol 82:2294–2309.

Pasic TR, Moore DR, Rubel EW (1994) Effect of altered neuronal activity on cell size in the medial nucleus of the trapezoid body and ventral cochlear nucleus of the gerbil. J Comp Neurol 348:111–120.

Paterson JA, Hosea EW (1993) Auditory behaviour and brain stem histochemistry in adult rats with characterized ear damage after neonatal ossicle ablation or cochlear disruption. Behav Brain Res 53:73–89.

Pillsbury HC, Grose JH, Hall JW 3rd (1991) Otitis media with effusion in children. Binaural hearing before and after corrective surgery. Arch Otolaryngol Head Neck Surg 117:718–723.

Popelár J, Erre JP, Aran JM, Cazals Y (1994) Plastic changes in ipsi-contralateral differences of auditory cortex and inferior colliculus evoked potentials after injury to one ear in the adult guinea pig. Hear Res 72:125–134.

Powell TPS, Erulkar SD (1962) Transneuronal cell degeneration in the auditory relay nuclei of the cat. J Anat (Lond) 96:249–268.

Purves D, Lichtman JW (1985) Principles of Neural Development. Sunderland, MA: Sinauer.

Rajan R (1998) Receptor organ damage causes loss of cortical surround inhibition without topographic map plasticity. Nat Neurosci 1:138–143.

Rajan R (2001) Plasticity of excitation and inhibition in the receptive field of primary auditory cortical neurons after limited receptor organ damage. Cereb Cortex 11:171–182.

Rajan R, Irvine DR, Wise LZ, Heil P (1993) Effect of unilateral partial cochlear lesions in adult cats on the representation of lesioned and unlesioned cochleas in primary auditory cortex. J Comp Neurol 338:17–49.

Rauschecker JP (1995) Compensatory plasticity and sensory substitution in the cerebral cortex. Trends Neurosci 18:36–43.

Rauschecker JP, Kniepert U (1994) Auditory localization behaviour in visually deprived cats. Eur J Neurosci 6:149–160.

Reale RA, Brugge JF, Chan JCK (1987) Maps of auditory cortex in cats reared after unilateral cochlear ablation in the neonatal period. Dev Brain Res 34:281–290.

Recanzone GH, Schreiner CE, Merzenich MM (1993) Plasticity in the frequency representation of primary auditory cortex following discrimination training in adult owl monkeys. J Neurosci 13:87–103.

Recanzone GH, Makhamra SD, Guard DC (1998) Comparison of relative and absolute sound localization ability in humans. J Acoust Soc Am 103:1085–1097.

Roberts JE, Burchinal MR, Davis BP, Collier AM, Henderson FW (1991) Otitis media in early childhood and later language. J Speech Hear Res 34:1158–1168.

Robertson D, Irvine DRF (1989) Plasticity of frequency organization in auditory cortex of guinea pigs with partial unilateral deafness. J Comp Neurol 282:456–471.

Robinson K, Gatehouse S (1994) Changes in intensity discrimination following monaural long-term use of a hearing aid. J Acoust Soc Am 97:1183–1190.

Röder B, Teder-Sälejärvi W, Sterr A, Rösler F, Hillyard SA, Neville HJ (1999) Improved auditory spatial tuning in blind humans. Nature 400:162–166.

Romand R (1992) Development of Auditory and Vestibular Systems 2. Amsterdam: Elsevier.

Rose S (1973) The Conscious Brain. London: Weidenfeld and Nicolson.

Rubel EW, Smith ZD, Steward O (1981) Sprouting in the avian brain stem auditory pathway: dependence on dendritic integrity. J Comp Neurol 202:397–414.

Rubel EW, Hyson RL, Durham D (1990) Afferent regulation of neurons in the brain stem auditory system. J Neurobiol 21:169–196.

Rubel EW, Popper AN, Fay RR, eds (1998) Development of the Auditory System. New York: Springer-Verlag.

Ruggero MA (1992) Physiology and coding of sound in the auditory nerve. In: Popper AN, Fay RR (eds), The Mammalian Auditory Pathway: Neurophysiology. New York: Springer-Verlag, pp. 34–93.

Russell FA, Moore DR (1995) Afferent reorganisation within the superior olivary complex of the gerbil: development and induction by neonatal, unilateral cochlear removal. J Comp Neurol 352:607–625.

Russell FA, Moore DR (1999) Effects of unilateral cochlear removal on dendrites in the gerbil medial superior olivary nucleus. Eur J Neurosci 11:1379–1390.

Russell FA, Moore DR (2002) Ultrastructural effects of unilateral deafening on afferents to the gerbil medial superior olivary nucleus. Hear Res 173:43–61.

Ryugo DK (1992) The auditory nerve: peripheral innervation, cell body morphology, and central projections. In: Webster DB, Popper AN, Fay RR (eds), The Mammalian Auditory Pathway: Neuroanatomy. New York: Springer-Verlag, pp. 23–65.

Ryugo DK, Ryugo R, Globus A, Killackey HP (1975) Increased spine density in auditory cortex following visual or somatic deafferentation. Brain Res 90:143–146.

Salvi RJ, Saunders SS, Gratton MA, Arehole S, Powers N (1990) Enhanced evoked response amplitudes in the inferior colliculus of the chinchilla following acoustic trauma. Hear Res 50:245–257.

Salvi RJ, Wang J, Ding D (2000) Auditory plasticity and hyperactivity following cochlear damage. Hear Res 147:261–274.

Sanes DH, Constantine-Paton M (1985) The sharpening of frequency tuning curves requires patterned activity during development in the mouse, *Mus musculus*. J Neurosci 5:1152–1166.

Sanes DH, Chokshi P (1992) Glycinergic transmission influences the development of dendrite shape. NeuroReport 3:323–326.

Sanes DH, Takacs C (1993) Activity-dependent refinement of inhibitory connections. Eur J Neurosci 5:570–574.

Sanes DH, Markowitz S, Bernstein J, Wardlow J (1992) The influence of inhibitory afferents on the development of postsynaptic dendritic arbors. J Comp Neurol 321: 637–644.

Sanes DH, Malone BJ, Semple MN (1998) Role of synaptic inhibition in processing of dynamic binaural level stimuli. J Neurosci 18:794–803.

Sasaki CT, Kauer JS, Babitz L (1980) Differential [^{14}C]2-deoxyglucose uptake after deafferentation of the mammalian auditory pathway—a model for examining tinnitus. Brain Res 194:511–516.

Scheffler K, Bilecen D, Schmid N, Tschopp K, Seelig J (1998) Auditory cortical responses in hearing subjects and unilateral deaf patients as detected by functional magnetic resonance imaging. Cereb Cortex 8:156–163.

Schnupp JWH, King AJ, Smith AL, Thompson ID (1995) NMDA-receptor antagonists disrupt the formation of the auditory space map in the mammalian superior colliculus. J Neurosci 15:1516–1531.

Schnupp JWH, King AJ, Carlile S (1998) Altered spectral localization cues disrupt the development of the auditory space map in the superior colliculus of the ferret. J Neurophysiol 79:1053–1069.

Schnupp JWH, Booth J, King AJ (2003) Modeling individual differences in ferret external ear transfer functions. J Acoust Soc Am 113:2021–2030.

Schwartz IR (1992) The superior olivary complex and lateral lemniscal nuclei. In: Webster DB, Popper AN, Fay RR (eds), The Mammalian Auditory Pathway: Neuroanatomy. New York: Springer-Verlag, pp. 117–167.

Schwartz IR, Higa JF (1982) Correlated studies of the ear and brain stem in the deaf white cat: changes in the spiral ganglion and the medial superior olivary nucleus. Acta Otolaryngol (Stockh) 93:9–18.

Sebkova J, Bamford JM (1981) Some effects of training and experience for children using one and two hearing aids. Br J Audiol 15:133–141.

Shatz CJ (1990) Impulse activity and the patterning of connections during CNS development. Neuron 5:745–756.

Sherman SM, Spear PD (1982) Organization of visual pathways in normal and visually deprived cats. Physiol Rev 62:738–855.

Shi J, Aamodt SM, Constantine-Paton M (1997) Temporal correlations between functional and molecular changes in NMDA receptors and GABA neurotransmission in the superior colliculus. J Neurosci 17:6264–6276.

Shinn-Cunningham B (2001) Models of plasticity in spatial auditory processing. Audiol Neurootol 6:187–191.

Shinn-Cunningham BG, Durlach NI, Held RM (1998) Adapting to supernormal auditory localization cues. II. Constraints on adaptation of mean response. J Acoust Soc Am 103:3667–3676.

Silman S (1993) Late-onset auditory deprivation. J Am Acad Audiol 4(5):xiii–xiv (editorial).

Silman S, Gelfand SA, Silverman CA (1984) Late-onset auditory deprivation: effects of monaural versus binaural hearing aids. J Acoust Soc Am 76:1357–1362.

Silverman MS, Clopton BM (1977) Plasticity of binaural interaction. I. Effect of early auditory deprivation. J Neurophysiol 40:1266–1274.

Silverman CA, Emmer MB (1993) Auditory deprivation and recovery in adults with asymmetric sensorineural hearing impairment. J Am Acad Audiol 4:338–346.

Slattery WH 3rd, Middlebrooks JC (1994) Monaural sound localization: acute versus chronic unilateral impairment. Hear Res 75:38–46.

Smith ZD, Gray L, Rubel EW (1983) Afferent influences on brain stem auditory nuclei of the chicken: n. laminaris dendritic length following monaural conductive hearing loss. J Comp Neurol 220:199–205.

Smith AL, Parsons CH, Lanyon, RG, Bizley JK, Akerman CJ, Baker GE, Dempster AC, Thompson ID, King AJ (2004) An investigation of the role of auditory cortex in sound localization using muscimol-releasing Elvax. Eur J Neurosci 19, in press.

Snyder RL, Rebscher SJ, Cao KL, Leake PA, Kelly K (1990) Chronic intracochlear electrical stimulation in the neonatally deafened cat. I: Expansion of central representation. Hear Res 50:7–33.

Snyder RL, Rebscher SJ, Leake PA, Kelly K, Cao K (1991) Chronic intracochlear electrical stimulation in the neonatally deafened cat. II. Temporal properties of neurons in the inferior colliculus. Hear Res 56:246–264.

Spigelman MN (1976) A comparative study of the effects of early blindness on the development of auditory-spatial learning. In: Jastrzembska ZS (ed), The Effects of Blindness and Other Impairments on Early Development. New York: The American Foundation for the Blind, pp. 29–45.

Spitzer MW, Semple MN (1991) Interaural phase coding in auditory midbrain: influence of dynamic stimulus features. Science 254:721–724.

Stein BE, Meredith MA (1993) The Merging of the Senses. MIT Press, Cambridge MA.

Stephenson H, Higson JM, Haggard M (1995) Binaural hearing in adults with histories of otitis media in childhood. Audiology 34:113–123.

Sterritt GM, Robertson DG (1964) Pathology resulting from chronic paraffin ear plugs: methodological problem in auditory sensory deprivation research. Percept Motor Skills 19:662.

Steward O (1989) Principles of Cellular, Molecular, and Developmental Neuroscience. New York: Springer-Verlag.

Stretavan DW, Shatz CJ, Stryker MP (1988) Modification of retinal ganglion cell axon morphology by prenatal infusion of tetrodotoxin. Nature 336:468–471.

Stuermer IW, Scheich H (2000) Early unilateral auditory deprivation increases 2-deoxyglucose uptake in contralateral auditory cortex of juvenile Mongolian gerbils. Hear Res 146:185–199.

Takahashi TT, Keller CH (1992) Simulated motion enhances neuronal selectivity for a sound localization cue in background noise. J Neurosci 12:4381–4390.

Thompson I (2000) Cortical development: binocular plasticity turned outside-in. Curr Biol 10:R348–350.

Tierney TS, Russell FA, Moore DR (1997) Susceptibility of developing cochlear nucleus neurons to deafferentation-induced death abruptly ends just before the onset of hearing. J Comp Neurol 378:295–306.

Tierney TS, Doubell TP, Xia G, Moore DR (2001) Development of brain derived neurotrophic factor and neurotrophin-3 immunoreactivity in the lower auditory brain stem of the postnatal gerbil. Eur J Neurosci 14:785–793.

Tonndorf J (1972) Bone conduction. In: Tobias JV (ed) Foundations of Modern Auditory Theory, Vol 2. New York: Academic Press, pp. 195–237.

Trune DR (1982) Influence of neonatal cochlear removal on the development of mouse cochlear nucleus: I. Number, size, and density of its neurons. J Comp Neurol 209:409–424.

Tucci DL, Rubel EW (1985) Afferent influences on brain stem auditory nuclei of the chicken: effects of conductive and sensorineural hearing loss on n. magnocellularis. J Comp Neurol 238:371–381.

Tucci DL, Born DE, Rubel EW (1987) Changes in spontaneous activity and CNS morphology associated with conductive and sensorineural hearing loss in chickens. Ann Otol Rhinol Laryngol 96:343–350.

Tucci DL, Cant NB, Durham D (2001) Effects of conductive hearing loss on gerbil central auditory system activity in silence. Hear Res 155:124–132.

Vale C, Sanes DH (2000) Afferent regulation of inhibitory synaptic transmission in the developing auditory midbrain. J Neurosci 20:1912–1921.

Wagner H (1990) Receptive fields of neurons in the owl's auditory brain stem change dynamically. Eur J Neurosci 2:949–959.

Wagner H (1993) Sound-localization deficits induced by lesions in the barn owl's auditory space map. J Neurosci 13:371–386.

Wall PD, Egger MD (1971) Formation of new connexions in adult rat brains after partial deafferentation. Nature 232:542–545.

Wallace MT, Stein BE (2000) The role of experience in the development of multisensory integration. Soc Neurosci Abstr 26:1220.

Wanet MC, Veraart C (1985) Processing of auditory information by the blind in spatial localization tasks. Perception Psychophys 38:91–96.

Webster DB (1983a) Auditory neuronal sizes after a unilateral conductive hearing loss. Exp Neurol 79:130–140.

Webster DB (1983b) Late onset of auditory deprivation does not affect brain stem auditory neuron soma size. Hear Res 12:145–147.

Webster DB, Popper AN, Fay RR, eds (1992) The Mammalian Auditory Pathway: Neuroanatomy. New York: Springer-Verlag.

Wenzel EM, Arruda M, Kistler DJ, Wightman FL (1993) Localization using nonindividualized head-related transfer functions. J Acoust Soc Am, 94:111–123.

Wiesel TN, Hubel DH (1963) Single-cell responses in striate cortex of kittens deprived of vision in one eye. J Neurophysiol 26:1003–1017.

Wightman FL, Kistler DJ (1997a) Sound localization. In: Yost WA, Popper AN, Fay RR (eds), Human Psychophysics. New York, Springer-Verlag, pp. 155–192.

Wightman FL, Kistler DJ (1997b) Monaural sound localization revisited. J Acoust Soc Am 101:1050–1063.

Wilmington D, Gray L, Jahrsdoerfer R (1994) Binaural processing after corrected congenital unilateral conductive hearing loss. Hear Res 74:99–114.

Wise LZ, Irvine DRF (1983) Auditory response properties of neurons in deep layers of cat superior colliculus. J Neurophysiol 49:674–685.

Withington DJ (1992) The effect of binocular lid suture on auditory responses in the guinea-pig superior colliculus. Neurosci Lett 136:153–156.

Withington-Wray DJ, Binns KE, Keating MJ (1990a) The maturation of the superior collicular map of auditory space in the guinea pig is disrupted by developmental visual deprivation. Eur J Neurosci 2:682–692.

Withington-Wray DJ, Binns KE, Dhanjal SS, Brickley SG, Keating MJ (1990b) The maturation of the superior collicular map of auditory space in the guinea pig is disrupted by developmental auditory deprivation. Eur J Neurosci 2:693–703.

Wong RO, Meister M, Shatz CJ (1993) Transient period of correlated bursting activity during development of the mammalian retina. Neuron 11:923–938.

Wong WT, Myhr KL, Miller ED, Wong RO (2000) Developmental changes in the neurotransmitter regulation of correlated spontaneous retinal activity. J Neurosci 20:351–360.

Wright BA, Fitzgerald MB (2001) Different patterns of human discrimination learning for two interaural cues to sound-source location. Proc Natl Acad Sci USA 98:12307–12312.

Wu GY, Malinow R, Cline HT (1996) Maturation of a central glutamatergic synapse. Science 274:972–976.

Yang L, Pollak G (1994) Binaural inhibition in the dorsal nucleus of the lateral lemniscus of the mustache bat affects responses for multiple sounds. Audit Neurosci 1:1–17.

Yin TCT, Chan JCK (1990) Interaural time sensitivity in medial superior olive of cat. J Neurophysiol 64:465–488.

Zheng W, Knudsen EI (1999) Functional selection of adaptive auditory space map by GABA$_A$-mediated inhibition. Science 284:962–965.

Zwiers MP, Van Opstal AJ, Cruysberg JR (2001) Two-dimensional sound-localization behavior of early-blind humans. Exp Brain Res 140:206–222.

5

Experience-Dependent Response Plasticity in the Auditory Cortex: Issues, Characteristics, Mechanisms, and Functions

Norman M. Weinberger

1. Introduction

The goal of this chapter is to provide a guide for understanding experience-dependent neuronal plasticity in the auditory cortex and its relation to behavior. (Unless otherwise noted, "auditory cortex" refers to the tonotopic primary auditory field, AI). It focuses on research that began in the mid-1980s concerning the question of how *learning may alter the processing and representation of acoustic information in the primary auditory cortex.* As used here, the term "plasticity" refers to systematic long-term (minutes to months) changes in the responses of neurons to sound as a result of experience. Plasticity at the level of altered neural responses is the result of various subcellular and molecular processes. Selected aspects of these substrates are included, particularly those relating to the cholinergic modulation of auditory cortical plasticity. Owing to lack of space, the subcortical auditory system cannot be reviewed, except as it directly pertains to mechanisms of cortical plasticity (see Birt et al. 1979; Cruickshank et al. 1992; Edeline and Weinberger 1992; Gonzalez-Lima and Scheich 1992; Hennevin et al. 1993; McKernan and Shinnick-Gallagher 1997).

This chapter is not a comprehensive review of the literature but rather a distillation of those topics that seem to be both important and misunderstood. The latter probably reflects the diverse backgrounds of investigators interested in experiential effects, some approaching from the field of auditory neurophysiology, others from behavioral neuroscience. The decision to be selective in choice of literature while being more comprehensive in the coverage of issues and approaches is based on two factors. First, most aspects of plasticity in the auditory system have been reviewed recently (Scheich 1991; Weinberger 1995, 1998; Irvine and Rajan 1996; Scheich et al. 1997; Kraus et al. 1998; Palmer et al. 1998; Cohen and Knudsen 1999; Edeline 1999; Rauschecker 1999; Moore et al. 2001). Moreover, research on the roles of experience in auditory processing will continue to expand. Therefore, a large-scale review focused on providing a comprehensive account of findings may become rapidly dated. Continual updates on recent findings are therefore better dealt with, for example, in periodic "Trends" type reviews. Second, there is a need to address major issues in a manner that will assist readers both in the evaluation of published findings and in contemplating their own research. The material presented here is intended to be helpful over a period of some years, by providing a framework within which past and future reports of auditory system plasticity can be evaluated. In particular, the presentation and analysis of experimental designs, which has not appeared in the literature, is equally applicable to studies of all parameters of acoustic stimulation, to all structures within the auditory system, and to all investigations of the sensory bases of learning and memory.

In concluding this Introduction, it is important to be mindful that various authors inevitably cover the same subject matter differently. Views and assumptions, whether explicit or implicit, are affected by background knowledge, scientific interests, values, and personal experiences. This tendency is intensified when the author has been deeply engaged in research on the topic under review, as is the present case. As a statement relating to "truth in writing," I believe that it is both timely (actually long overdue) and essential to achieve a high level of synthesis between the sensory neurosciences and the behavioral neurosciences, particularly for cortical function and for the neurobiology of learning and memory. Such a synthesis honors the evolutionary outcome of brain–behavior relationships, in contrast to traditional disciplinary boundaries that are rooted in the "accidental" history of the 19th century and earlier. Of note, auditory research leads all other sensory systems in this regard, having initiated sensory/learning investigation in the 1950s and sustained it, with minor interruptions, to the present (reviewed in Weinberger and Diamond 1987; Weinberger 1995).

It is difficult, if not impossible, to represent the views, rationales, and experimental approaches of all other workers in this field at a level and depth equal to that of one's own predilections. Thus, there are bound to be many disagreements. The remedies are to attempt to be explicit and even-handed, goals that I have sought but cannot claim to have achieved. The larger solution is for all concerned to put forth their views on the same issues and to encourage readers to make their own comparative conclusions. In that spirit, we can move forward.

2. "Experience-Dependence" and Learning/Memory

2.1 Background

Consideration of experience-dependent effects on the auditory system (or any system for that matter) necessarily entails the field of the neurobiology of learning and memory. The rationale for an essential role for learning and memory begins with an explication of the normal meaning of "experience." "Experience" refers to the interaction of a waking, unanesthetized organism with its environment. The interactions may be very limited, such as the receipt of a single brief acoustic stimulus, or practically unlimited, such as behaviorally dependent alterations in the reception of spectrally and temporally complex patterns of sound. An example of the former would be a sudden noise received by a stationary, awake (or even sleeping) animal. An example of the latter would be the fluctuations of simultaneously received species-specific and environmental sounds as an animal moves through space encountering various auditory scenes.

The fields of sensory neurophysiology and learning/memory have a unique relationship. They are both concerned with the processing of sensory stimuli in the brain, but from complementary points of view. Sensory neurophysiology has traditionally been interested in how the brain responds to variations in stimulus parameters, but not in how the brain changes when the behavioral significance of a stimulus changes. Conversely, learning/memory focuses on the latter, but not the former (Table 5.1). The two fields together provide for a compre-

TABLE 5.1. Complementary nature of the basic experimental paradigms of two disciplines within Neuroscience, Auditory Physiology *versus* Learning/Memory.

Neuroscience disciplines	Stimulus parameters	
	Physical	Psychological
Auditory physiology	Vary	Constant
Learning and memory	Constant	Vary

"Auditory Physiology" applies equally to other sensory systems. Both disciplines study information processing in the brain. Sensory stimuli have two basic parameters, physical and psychological. "Physical" refers to the parameters that are manipulated by changing the physical source of stimulation, for example, frequency (kHz), level (db SPL), and so forth. "Psychological" parameters refer to the acquired behavioral significance of stimuli, such as sounds. These parameters are not specified in physical units but rather by behavioral analysis, for example, the strength of a conditioned response to a given acoustic frequency. All sounds can be described by physical parameters and may also change (increase or decrease) in behavioral importance as a result of learning. Auditory Physiology varies physical parameters and determines the characteristics of neural response. To avoid incidental learning (e.g., habituation, etc.), the psychological parameters are kept constant, for example, by studying animals that are under general anesthesia. In a complementary manner, Learning/Memory experiments vary the psychological parameters by changing the relationships between stimuli (e.g., tone and food). These studies keep the physical parameters of stimuli constant, to permit interpretation of changed neural response as due to learning rather than to change in the stimuli themselves.

hensive analysis of the processing of sensory stimuli. However, they have developed separately, with little intersection until approximately the last 15 years (see Section 4). "Learning and memory" refer in the broadest sense to the acquisition and storage of information, respectively. "Acquisition" is the initial process of intake which has no fixed duration but is reasonably treated as the period of actual experience. Depending on the duration of an acoustic experience, the period of acquisition is generally seconds to minutes. "Storage" refers to the maintenance of information, from minutes to potentially a lifetime.

But why should experience-dependent auditory cortical plasticity entail learning and memory? The reason is that awake (even sleeping) organisms can acquire and store their experiences. Having received an acoustic stimulus, the resultant neuronal responses may result in nontransient changes in the brain; they may leave their mark. It matters not whether an investigator is overtly studying the acquisition and storage of information or is interested only in the immediate responses of neurons to a given set of acoustic stimuli. If the subject is not in a state of general anesthesia, it can potentially acquire and store everything about an experience. "Everything" includes at least (1) the detailed characteristics (parameters) of the sounds (e.g., spectrotemporal pattern, stimuli levels, locus in space), (2) the detailed nature of other more or less contemporary sensory (nonauditory) events (e.g, visual, somatosensory, olfactory, vestibular), (3) the organism's current physiological state (e.g., hungry, satiated, stressed); (4) the behavioral meaning, significance, or importance of the experience; and (5) integration of the new experience with prior, stored experience.

In other words, during the course of an experiment, when presumably the investigator seeks to acquire and store information about the subject's brain, the unanesthetized subject is likely to be learning and remembering as well. The critical question is how can investigators determine what the subject has acquired and stored (e.g., see Section 6.3.2). Before turning to this issue, we need to consider experimental approaches that are often thought to avoid the effects of learning and memory.

2.2 Possible Avoidance of Experience-Dependent Effects by General Anesthesia

For the neuroscientist interested in studying function without involving learning and memory processes, the occurrence of experience-dependent effects may constitute an unwelcome, and possibly unavoidable, variable. The use of general anesthesia may avoid experience-dependent effects. In particular, deep general anesthesia appears to be a state devoid of the chance for new learning and memory, unless particular hormonal events occur, such as a marked increase in levels of epinephrine (e.g., Weinberger et al. 1984a).

To avoid misunderstanding, it is important to undercut a false dichotomy, which is that data from anesthetized subjects are either superior or inferior to those obtained from waking animals. Some investigators consider that general anesthesia provides the only stable state in which acoustic processing can be precisely studied; for many workers, the nonanesthetized state simply adds noise, increasing the difficulty of understanding the process in question. A related point in favor of general anesthesia is that it affords better stimulus control. Complete stimulus control in awake animals is more difficult to achieve than in anesthetized animals because a closed stimulating system, permitting calibration at the tympanic membrane, would produce undue discomfort or pain. In addition, anesthesia permits multiple invasive procedures not possible in waking animals. However, an alternative viewpoint is that although "anesthetized studies" form the bedrock of auditory neurophysiology, it is inescapable that the anesthetized brain is not in a normal state. After all, anesthetics are drugs. Thus, it has been argued that only the study of animals in a normal behavioral state can lead to a full understanding of auditory processing.

The stance taken here is that both approaches have their advantages and limitations, and that it is important to maintain a balanced view. The extent to which auditory processing is the same under anesthesia and the unanesthetized state can be determined only empirically. Given the subject matter of this chapter—experience-dependent plasticity—it is hardly surprising that most of the findings reviewed here arise from studies of unanesthetized subjects. Anesthesia should be appreciated as the important tool that it is. But while it can tell us much, perhaps most, about the processing of acoustic stimuli, it cannot tell us everything, particularly about the auditory cortex which is more severely affected by anesthesia than, for example, the eighth nerve. As reviewed in the following section, the fact of marked learning effects on the auditory cortex is itself adequate indication that a full understanding of the auditory cortex (and many subcortical components of the auditory system) cannot be achieved only from the study of anesthetized animals.

However, it would be wrong to assume that the anesthetized state is irrelevant to the study of experience-dependent plasticity. General anesthesia does not necessarily eliminate the effects of prior experience, but only the effects of experience during the state of anesthesia. Rather, as will be seen, the effects of learning that are established in the waking state can be expressed in the state of general anesthesia (e.g., Lennartz and Weinberger, 1992a; Recanzone et al. 1993). Moreover, the anesthetized state provides an excellent control for unwanted potential effects of changes in the level of arousal or excitability during the determination of the properties of neurons whose tuning and other response parameters have been altered by learning in the waking state. Thus, it may be particularly advantageous for the study of the effects of prior experience.

3. Detecting Learning and Memory

3.1 The Importance of Behavioral Assessment of Learning

Learning and memory ultimately must be validated behaviorally because of the essential distinction between neural plasticity and learning/memory. Although it has become common for many neuroscientists to regard physiological plastic-ity as memory [e.g., long-term potentiation (LTP)], mechanisms of memory may involve one, or more likely, many types of neural plasticity. More importantly, learning and memory are behavioral constructs, valid at the level of the organ-ism, not at the level of a brain structure or circuit. Learning and memory cannot be observed directly, but rather can be inferred only from the behavior of or-ganisms under whatever experimental circumstances are set up to "interrogate" the subject. Learning and memory may be behaviorally silent unless the appro-priate question is posed by the investigator, in the form of sensitive behavioral measures within the context of a relevant experimental design. Conflating neural plasticity with memory is conceptually fuzzy and experimentally confusing. One solution to this problem is to refer to brain processes that are candidates for the storage of some information as indexing "physiological memory," whereas genuine memory can be called "behavioral memory" (e.g., Weinberger 1998). Alternatively, appropriate neurophysiological changes may be referred to as "learning-induced plasticity." The important point is to avoid confusing neural plasticity with genuine memory.

The detection of learning and memory require the use of appropriately sen-sitive behavioral measures. We will focus on examples of classical (Pavlovian) conditioning and habituation, for two reasons. First, most studies on experience-dependent plasticity in the auditory cortex employ them. Second, experimental designs to study habituation and conditioning can be accomplished rapidly and in a wide variety of preparations, from humans to analogs of conditioning for in vitro preparations.

The context for this discussion is a "model" experiment, in which the inves-tigator asks whether the development of plasticity in the auditory cortex accom-panies learning. The critical issue for understanding experience-dependent effects on stimulus processing in the auditory cortex is not merely whether changes occur, but rather the extent to which learning has *specific vs. general effects*.

3.2 Habituation

Habituation, that is, learning that a repeated (usually unexpected or novel) stim-ulus is behaviorally unimportant, is perhaps the simplest example of the acqui-sition of information. Visual observation of an animal may reveal little or no change in overt behavior. However, the recording of autonomic responses, such as heart rate or blood pressure, would reveal a systematic change in response with stimulus repetition. Decrement of behavioral and neural responses does not distinguish habituation from fatigue or refractory-like processes, but dem-onstration of specificity of decrement to the repeated stimulus solves this prob-lem. However, determination of specificity cannot be accomplished during stimulus repetition itself, but rather must be assayed after training. This is a crucial point that is elaborated in later discussion of experimental designs.

3.3 Classical Conditioning

Acquiring an association between two or more stimuli is a ubiquitous form of learning. Any regularly occurring sequences of two or more events allows for the learning of their relationship. Such associations are a basis for the common inference of cause–effect relationships (Rescorla 1988).

An accepted distinction is made between stimulus–stimulus associations and stimulus–response associations. The latter involve learning to make a particular behavioral response to a specified sensory stimulus. Obviously, stimulus–response learning should be easily detectable, because the second component is ordinarily an overt behavior. For example, pairing a tone with shock to the eyelid will come to produce an eyeblink to the tone itself. The detection of stimulus–stimulus (S–S) relationships may be more difficult but is of equal or greater importance because stimulus–stimulus associations appear to be involved in stimulus–response (S–R) associations. In the preceding example, animals learn that the tone signals the shock. Again, sensitive behavioral measures, such as heart rate, respiration, blood pressure, and pupillary dilation clearly validate the learning of an association between stimuli (e.g., the tone and shock). More-over, this S–S learning develops more rapidly, in a few trials, whereas the eye-blink conditioned response may not be evident for 50 or more trials (reviewed in Lennartz and Weinberger 1992b). If one measured only eyeblink, one would conclude that learning had not taken place until the eyeblink conditioned re-sponse appeared. The initial S–S learning would go undetected. Moreover, as we will see later (Section 5.3.2.2), plasticity in the auditory cortex during conditioning develops in 5–30 trials, well in advance of S–R learning, such as the conditioned eyeblink response.

Given these basic aspects of learning and memory, it will be helpful to obtain a brief historical perspective on studies of learning in the auditory system. This will provide a context within which contemporary approaches and findings may best be considered, and perhaps appreciated.

4. A Brief History of Neural Plasticity in the Auditory Cortex

Neural plasticity in the auditory cortex is interesting not only in itself but also as a case study in the intersection of two scientific fields that had developed quite separately, those of sensory physiology and the neurobiology of learning and memory. Furthermore, this topic provides a clear example of how assumptions constrained thought and experiment for most of the 20th century. This is not merely of historical interest because differential approaches of auditory neurophysiology vs. learning/memory continue to influence, and in many instances plague, contemporary research. Finally, because the majority of studies on learning and sensory systems have been carried out in the auditory cortex, the development of new ideas and findings in the auditory cortex has direct implications for studies of neural plasticity in other sensory systems.

The evolution of research on neuronal plasticity in the auditory cortex involves four stages: (1) sensory–motor framework excluding auditory cortex from learning and memory; (2) delineation of cortical responses to sounds in anesthetized animals; (3) documentation of plasticity during learning; and (4) discovery that learning systematically changes the basic acoustic parameters, such as frequency tuning, of neurons in the auditory cortex.

Attempts to understand the auditory cortex (as well as many other brain systems and structures) have their origin in the 19th century, within the framework of a sensory–motor conception of the nervous system. Following the early discoveries of Magendie and Bell of sensory and motor roots of the spinal cord (Cranefield 1974), much of the research program for the rest of the century concerned the extent to which the entire neuraxis was organized on sensory–motor principles (Young 1970). By the beginning of the 20th century, the sensory–motor principle had been extended from the spinal cord to the cerebral cortex. However, many regions of the cortex appeared to be neither sensory nor motor. Several of these were labeled "association cortex." The implication of this formulation was that the substrates of learning were to be found in the association cortex. Implicit in this schema was the assumption that sensory (and motor) regions of the cortex were not sites of information storage in learning and memory.

The availability of sensitive electronic amplifiers in the 1930s enabled scientists to record brain potentials that were elicited by controlled sensory stimuli, thereby initiating the modern field of sensory neurophysiology. In the auditory system, recordings in the 1940s by Woolsey and Walzl (1942) in the cat and by Tunturi (1944) in the dog delineated auditory cortical fields, in particular "tonotopic maps." Subsequent studies showed that single eighth nerve fibers exhibited specific tuning functions, responding best to one frequency [the "characteristic frequency" (CF) at threshold]. In the period immediately following World War II, auditory neurophysiology began a period of increasingly precise and sophisticated analysis of relationships between the full range of acoustic stimulus parameters and the responses of the auditory neuraxis, from the cochlea to the several fields of the auditory cortex.

The third stage, that learning induced auditory plasticity, may be dated from 1955. Robert Galambos and his colleagues performed a seminal experiment in which cats were classically conditioned by pairing an auditory conditioned stimulus (CS) with a puff of air [unconditioned stimulus (US)] to the face. Learning was validated by behavioral measurements. Evoked potentials elicited by the CS became larger during conditioning. This study also addressed the critical issue of stimulus control. To show that inadvertent changes in CS intensity (level) were not responsible, the authors also tested subjects under neuromuscular blockade, maintaining stimulus constancy at the periphery while eliminating putative contractions of the middle ear muscles (Galambos et al. 1955).

However, these findings did not unequivocally demonstrate that the plasticity was due to associative learning, because it could have reflected a general increase in excitability due to the presentation of the air puff ("sensitization"). Gluck and Rowland (1959) used a control that showed that plasticity developed only when a sound (CS) and a mild shock (US) were paired, not when the shock was present but not predicted by the sound. Another type of control to show associativity is the use of a two-tone discrimination protocol, in which one tone is followed by the US ("CS+") another tone ("CS−") is not followed by the US or any other stimulus (and the tone trials are presented in random order). Typically, both auditory plasticity and behavioral conditioned responses develop to the CS+ but not to the CS−.

Over the next decades, numerous additional studies in various animals and training situations demonstrated that the responses of the auditory cortex to sounds were affected not only by the physical sounds themselves, but also by the learned psychological or behavioral importance of acoustic stimuli (see Weinberger and Diamond 1987 for review). These findings clearly showed the error of the traditional belief that sensory cortices had purely sensory functions and were not regions directly involved in learning and memory.

However, the documentation of neural plasticity had little effect on the field of sensory neurophysiology, probably for several reasons. First, learning studies used only one or two tones, which were not interesting to sensory workers who used many stimuli. Second, there appears to have been conceptual confusion in which sensation was equated with perception so that the constructive aspect of the latter was not appreciated; hence plasticity of sensory responses was mistakenly seen as incompatible with perceptual accuracy. Neither was sensory cortical plasticity initially influential within the field of learning and memory itself. This relative neglect appeared to reflect the traditional belief in association cortex as sites of learning and memory on the one hand and emphasis on the hippocampus and other nonsensory structures on the other hand.

The fourth stage began in the mid-1980s, when studies of learning began to focus on obtained neurophysiological data that were commonly obtained in auditory neurophysiology. As this brings us to what may be described as the "Contemporary Era," we now proceed to a more detailed account of conceptual issues and empirical inquiries.

5. Experimental Designs and Findings

5.1 Introduction

An accepted canon of science is that experimental designs constrain possible results. Designs that do not include observations of certain variables cannot produce direct findings about these variables. Designs that do not adequately control for confounding variables are limited in their ability to determine the factors that produce the results. Within the present context, neural changes in the auditory system may be closely related to *what* the subjects learn, rather than to the *mere fact* of learning itself. For example, if experimenters pair a tone with reinforcement and find increased responses to that tone in the auditory cortex, they may conclude that this plasticity reflects learning about the frequency of that tone. However, the subject may have learned only that a sound is followed by reinforcement. At the very least, one would have to employ tests with many frequency values across the spectrum to determine if the cortical plasticity is specific to the training frequency.

Several experimental designs have been developed to address the issue of learning-induced specificity of plasticity in the auditory system, primarily in the auditory cortex. These are illustrated in Figure 5.1 and are discussed in turn. But first, it is necessary to consider the standard design that accounts for perhaps more than 95% of all research on the auditory system and learning over the past 50 years, the "During Trial" design.

5.2 The "During" Training Trials Design (DUR)

This design consists of recording from the auditory system during the training trials employed in habituation, conditioning, or other types of training. It has provided the vast bulk of findings and firmly established that processing in the auditory system is subject to experience.

5.2.1 Disadvantages of Relying on Recording During Training Trials

Although this design is supported by common sense, it is far more limited than generally appreciated. The overall degree of specificity of learning effects in the auditory (or any other) system cannot be determined *during* the experience itself. There are two reasons. The first is almost trivially technical rather than conceptual. The second is of critical importance, woefully ignored or misunderstood, yet easy to grasp given minimal knowledge of basic principles of learning and memory.

5.2.1.1 Comparison Stimuli

The technical limitation is that the number of comparison stimuli is too small, in the limit only a single repeated stimulus or conditioned stimulus, in habituation or classical conditioning, respectively. Conditioning paradigms can be extended to become discrete discrimination paradigms, that is, two tones are presented. One (CS+) is followed by reinforcement (e.g., food or an aversive stimulus) and the other (CS−) is unreinforced. Development of a behavioral response to the CS+ but not the CS− is evidence of discrimination. This level of specificity also demonstrates that a genuine association had been established between the CS+ and the reinforcement. (In single-stimulus conditioning, a demonstration of association requires use of a control group in which the CS is randomly related to the reinforcer, with a resultant failure to establish a conditioned response to the stimulus.)

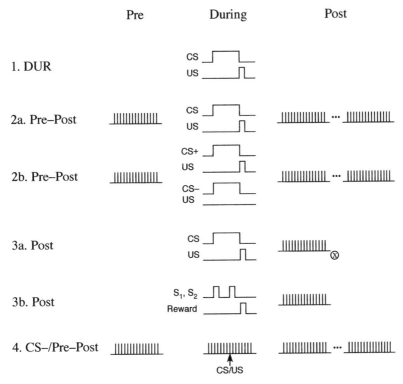

FIGURE 5.1. Schematic summary of experimental designs employed in the neuro-physiological study of learning and the auditory cortex. Depicted are four basic designs (1–4) and their treatments during three experimental periods, "Pre" (before training), "During" (during training) and "Post" (after training). Pre–Post designs 2a and 2b illustrate the fact that any training paradigm can be used. Design 2a shows single-tone conditioning and 2b illustrates two-tone discrimination conditioning. Designs 3a and 3b also illustrate the fact that any training paradigms can be used with a Post design. Design 3a illustrates the case of single-tone conditioning whereas 3b shows an example of two-tone instrumental training, in which reward is contingent on the correct response, that is, one response if the two tones (S1 and S2) are the same and another response if they are different (responses not shown). The *repeated vertical lines* represent presentation of tone bursts. The *dotted lines* in the Post period for designs 2 and 4 indicate that additional Post periods can be used to determine long-term retention, and so forth. In 3a, the "x" in the Post period signifies sacrifice of the animal for 2-DG analysis following repeated presentation of a conditioned stimulus tone. In design 4, the "CS/US" denotes that one of the frequencies in a series of tone bursts is designated as the conditioned stimulus and is paired with shock; the serial order of tones is random from one sequence to another. Illustrations are not to scale.

5.2.1.2 Confounds by Performance Factors

The second, and more critical reason, is that the *state of a subject fluctuates during training trials.* All of the nonassociative variables that occur during training trials are referred to collectively as "performance factors," because they ultimately affect the performance, that is, "read out," of associatively based behavioral and neural responses, but do not change the underlying associations (e.g., Mackintosh 1974). Performance factors during training trials include changes in general arousal level, attention, and motivational factors induced by the training milieu, such as by the expected and actual presence of an aversive or appetitive reinforcer (i.e., punishment or reward). Within a training trial, changes in arousal can greatly affect both behavior and neural processing of stimuli, both for lowered arousal (e.g., due to reduced processing, inattention, etc.) and heightened arousal (e.g., due to distraction, stress, etc.). For the auditory cortex, there are numerous reports that arousal level and motivational state alter evoked potentials and unit discharges (e.g., Murata and Kameda 1963; Teas and Kiang 1964; Wickelgren 1968a; Molnar et al. 1988; but see Oatman 1971). Some of the confounds involve changes in activity of the middle ear muscles (e.g., Baust et al. 1964; Irvine and Webster 1972). Moreover, the degree of influence of nonassociative processes can vary during a learning situation as a subject itself acquires information about the learning context, the contingencies among stimuli, and so forth. The situation is complicated by the fact that investigators may not be able to control or even detect performance factors. For example, the release of stress hormones during learning may affect processing in the auditory system, as elsewhere in the brain.

These considerations and findings indicate that neural responses to a training stimulus during training trials are likely to reflect both direct associative processes and nonassociative performance factors. Thus, the sign and magnitude of change of neural response to a training stimulus should not be interpreted to exclusively reflect learning effects on the processing and representation of acoustic information. Moreover, it is clear that the assessment of learning by measurement of behavioral change during training trials also involves the same risks. A solution to this problem is to assess the effects of a training regimen after training has been completed. Specifically, groups that have been trained differently must be assessed under identical circumstances during post-training periods (Rescorla 1988) (see also Sections 5.3 and 5.4).

The use of a nonassociative control group (e.g., CS and US presented randomly) does permit workers to determine that neural effects obtained in the standard conditioning (paired) group are attributable to the pairing process itself. However, this does not eliminate the effects of performance factors in the paired group. Similarly, the use of a discrimination paradigm within the paired group (CS+ tone paired with reinforcer, CS− tone not followed by reinforcer) does allow one to conclude that neural changes to the CS+ are associative (require pairing). However, it does not eliminate the effects of performance factors on neural response to the CS+ or the CS− for that matter. Thus, even if the number of different CS− tone frequencies were increased during training trials, in an attempt to determine specificity of plasticity (Section 5.5), performance factors could affect responses to any or all of these tones, as they do for a single CS+ and CS− tone.

In short, while one can detect changes in the response of the auditory cortex to, for example, the CS frequency during training trials, one can neither determine the *specificity* of the changes across the for example, frequency spectrum, nor be confident that the observed changes reflect *only the associative effects* of learning.

5.2.2 What Can We Deduce from Data Obtained During Training Trials?

The foregoing indicates that neural data obtained during training trials cannot be assumed to reflect only the influences of learning per se. This conclusion does not render such data without value but it does indicate that extreme caution should be exercised in their interpretation. If neuronal plasticity develops during training trials in one group compared to a nonassociative control group, then the fact of such plasticity, but not necessarily its magnitude or form, may be attributed to associative processes. Similar considerations apply to behavioral measures of learning. One can determine whether or nor behavioral signs of learning develop during training trials, but conclusions about form and magnitude need to be qualified. [Exceptions may apply to noncortical sites if they are little influenced by performance factors, for example, dentate/interpositus nuclei in direct stimulus–response eyelid conditioning, Thompson and Tracy (1995).] On the other hand, the absence of neural plasticity or a behavioral response might be due either to masking due to interference by performance factors, or to a failure to learn. Thus, as in other areas of science, interpretation of negative data presents problems. Nonetheless, while some authors assume that pairing a tone with a reinforcer produces associative learning, the actual development of an appropriate behavioral index is a sine qua non for the inference of learning and memory. This is particularly important because the training parameters sometimes used are known to be inappropriate to induce associative, for example, presenting trials at too rapid a rate.

5.2.3 Findings

The early and later use of the DUR design was discussed in the previous section on history, so the present discussion is brief. A detailed review has been provided previously (Weinberger and Diamond 1987). After the seminal study of Galambos et al. (1955), that inaugurated Western neurophysiological studies of learning in the auditory system, additional experiments were performed both for classical conditioning and also for habituation and instrumental conditioning in animals. Habituation studies reported systematic response decrements in AI in evoked potentials (Wickelgren 1968b) and unit discharges (Weinberger et al. 1975). The large majority of conditioning studies replicated and extended the findings of Galambos and associates. Over the next 30 years or so, evoked potential and multiple-unit studies reported that CS-evoked responses increased to the CS during training. While studies differed in the degree of control both for nonassociative effects and for acoustic stimulus constancy, the findings were similar across the several species studied. Single-unit studies were less consistent, reporting increased discharges but also decreased discharges in many cells (reviewed in Weinberger and Diamond 1987).

Somewhat of a pall was cast on the study of learning in the auditory cortex by a study in the rat (Hall and Mark 1967). They reported that evoked potentials to a background click increased during presentation of a visual conditioned stimulus that signaled footshock. The authors argued that prior learning effects in the auditory system simply reflected a state of fear (no doubt accompanied by heightened arousal) and therefore facilitation of acoustic CS responses did not indicate that behaviorally important sounds produce larger neural responses. In short, they argued that performance factors were responsible for the apparent effects of learning. This seems to be a case of being simultaneously right and wrong. As noted in Section 5.2.1.2, it is independently known that changes in arousal level and motivational state can alter field potentials and unit responses in AI. In fact, arousal and motivational findings are one basis for exercising extreme caution in the interpretation of findings from DUR designs. However, the negative conclusion of Hall and Mark about learning reveals an elementary error in logic. They assumed that if one variable ("fear") produces an increased response, then no other variable (e.g., association) can do so. Hence, they excluded associative learning processes as a cause of response facilitation. A later study using background stimuli and a DUR design (Kitzes et al. 1978) also used similar logic to interpret similar findings for unit discharges. As discussed in the next section, response facilitation to a CS frequency is induced by learning processes independent of performance factors.

5.3 The "Pre–Post" Training Trials Design (Pre–Post)

5.3.1 Design Considerations

The Pre–Post design has been used extensively to avoid problems of the DUR design. It has been the design of choice in my laboratory, hence bias in the form of emphasis may be unavoidable. These factors conspire to make this section longer and more detailed than Section 5.4 which explains the Post design. However, the rationale and experimental strategy of the Pre–Post design apply equally well to the Post design. Therefore, the length of the present section does not imply that the Pre–Post design is paramount. Given that all designs have some advantages and limitations, readers can determine which approaches are best suited to answer their own questions of interest. Also, of course, new experimental designs should and will be devised as well.

The Pre–Post design involves a minimum of three stages: (1) pretraining recording, (2) actual training or other designated controlled experience, and (3) post-training recording. The Pre– and Post–periods may consist of presenting many acoustic stimuli of interest (e.g., different frequencies of tone bursts), in a standard pattern used in auditory neurophysiology (e.g., 100-msec tone bursts, presented at 2/sec). This contrasts with the standard presentation of a tone during a discrete training trial (e.g., 2 sec with intertrial intervals of ~1 min). The effects of the experiential treatment, whether habituation, classical conditioning, instrumental discrimination learning, or any other task, are determined by comparing the Post data with the Pre data. If the Pre and Post data are not statistically different, then one can conclude that the experience had no effect on the processing or representation of the acoustic information under study. Conversely, significant differences between the two periods can be attributed to the intervening experience (Fig. 5.1, no. 2a,b).

It is essential that the test stimulus presentations be identical during the Pre and Post periods. Moreover, these data should be obtained in identical experimental settings, to eliminate any confounds due to changes in local environment. No reinforcement is present during these Pre and Post tests, as that would defeat the goal of avoiding performance factors. It is also helpful to perform the training in a separate experimental setting, for example, a different room with salient differences in visual and other nonauditory cues (see Section 5.3.1.1). Of course, it is important that the subjects be in the same state during the Pre and Post tests. This can be accomplished by amply habituating them to the testing (Pre and Post) situation, and this can be objectively assessed by recording heart rate or other sensitive physiological measures. Typically, heart rate is high when an animal is placed in a novel situation but such tachycardia habituates fairly rapidly over a few days of adaptation.

5.3.1.1 Advantages of the Pre–Post Design

This design permits assessment of specificity while avoiding confounding performance factors that are present during training trials. Another advantage is that post-training tests can be presented at desired intervals (hours to months) permitting detection of neural consolidation (i.e., increase in effect over time without additional training), long-term retention, or forgetting (i.e., loss of effect over time). Periods as long as 8 weeks post-training have been used (Weinberger et al. 1993). This design also is very flexible as any desired training task can be used, for example, habituation, one-tone classical conditioning, two-tone classical discrimination, one-tone instrumental training, and two-tone instrumental training, yet the assessment of receptive field plasticity can be identical. This permits a fairly direct comparison of the effects of learning and memory on auditory processing for as many types of experience as may be of interest.

At least two issues are raised with regard to this design. First, how does it eliminate or control for performance factors? Second, how does it avoid experimental extinction? (See also Diamond and Weinberger 1989; Weinberger 1998).

Performance factors are reduced, if not eliminated, by minimizing similarities between the training context and the testing circumstances. The purpose of this maneuver is to reduce or eliminate any generalization from the training environment to the testing environment. For example, if an animal receives food or shock during training, it may also associate the location ("context") of the training with these reinforcers. If tested in the same place, its arousal level and expectations could be affected.

Perhaps the most salient difference is the absence of a reinforcer, which also reduces and can even eliminate changes in state (see later). The Pre–Post design also permits acoustic stimuli to be presented with several parameters that differ from the training session. Thus, training can consist of standard, discrete conditioning trials, for example, an individual CS tone, with a duration of 2 sec, a long intertrial interval of 1–2 min (on average), and a stimulus level that is well above threshold (e.g., 60–70 db SPL). In contrast, determination of receptive fields can be accomplished necessarily using many tones to cover the relevant part of the frequency spectrum (e.g., 20 tones at quarter-octave intervals), each tone presented briefly (e.g., 100 msec duration), with brief intertone intervals (e.g., 400 msec), at stimulus levels that range widely (e.g., 0–80 db, SPL). Also, test tones can be repeated many times in any desired sequence (e.g., ascending, random), to obtain sufficient statistically reliable data. In short, the acoustic context of receptive field (RF) determination can be extremely different from that during conditioning trials.

This difference in context has proven to be generally sufficient to eliminate any behavioral or arousal response to the CS frequency when it is embedded as a brief tone in a series of test tones. Objective measures indicate that subjects do not regard the CS frequency as a conditioned stimulus during determination of receptive fields (Diamond and Weinberger 1989). In addition to the lack of performance confounds, elimination of potential effects of arousal on the CS frequency or any other tone can be accomplished by training subjects while they are awake (of course) but obtaining RFs while they are under general anesthesia (Lennartz and Weinberger 1992a; Weinberger et al. 1993).

One can also record responses to the training frequency during discrete trial presentations, as long as the likely confound of performance factors is kept in mind. Diamond and Weinberger (1989) compared changes to a CS tone during trials with changes in frequency RFs, for AII and the ventral ectosylvian field in the cat. They found that there was little correspondence between changes to the CS during training with responses to that same frequency when it was presented as one of a series of rapidly presented frequencies in the Post period. In many cases, the sign of change was opposite, for example, a decrement in response to the CS tone but a specific increase in response to that frequency during RF determination, when tuning might shift toward or to the frequency of the conditioned stimulus.

The second issue of experimental extinction is also eliminated by the Pre–Post design. As subjects do not regard the presentation of the frequency used as a CS as an actual conditioned stimulus, they do not respond to it, and therefore they do not extinguish acquired conditioned responses during post-training determination of RFs.

5.3.2 Findings

5.3.2.1 Habituation

The Pre–Post design was first used for the study of habituation in AI of the cat to rule out performance factors during stimulus repetition and to determine a degree of specificity (Westenberg and Weinberger, 1976). Two frequencies (A and B) were presented as alternating brief tone bursts ("prehabituation"). In the next stage, one tone (A) was presented repeatedly. Finally, the post-test was performed; the tones were again presented in a pattern identical to the prehabituation phase. Average evoked potentials for each tone were determined separately for the pre- and posthabituation periods and compared. Because the Pre and Post tones alternated, the average responses over time were obtained for both A and B when the subjects were in the same state; hence any differences between responses to the tones could not be attributed to differences in state from the Pre to the Post periods. The Post responses to the repeated tone (A) were significantly smaller than the Pre responses to this tone, but there was no difference for the nonrepeated (B) tone. Subsequently, B was used during habituation and the results were the inverse of the first "A" habituation experiment. These findings demonstrate that repeated acoustic stimulation produces frequency specific habituation.

The Pre–Post design has been expanded to determine the entire frequency RF ("tuning curves") of auditory cortical neurons (Condon and Weinberger 1991). Habituation produced a decreased response that was specific to the frequency that had been repeatedly presented; frequencies 0.125 octaves from the habituated frequency exhibited no response decrement. In short, habituation does produce frequency-specific decrement of response in the auditory cortex.

5.3.2.2 Conditioning

The first use of the Pre–Post design with receptive field analysis was in a single unit study of AII and ventral ectosylvian fields in the cat (Diamond and Weinberger 1986). CS-specific plasticity was obtained in a group that received a single brief (20–45 trials) session of tone-shock pairing but not for sensitization controls (tone-shock unpaired), showing that the effects were associative. Analysis of pupillary dilation revealed that behavioral associative learning had developed. Some cells developed a CS-specific increase while others developed a CS-specific decrease in the RF. Extinction (additional CS presentation without the shock US) produced loss of the RF plasticity. The findings received little notice, probably because these "secondary" auditory fields were not well understood, particularly compared to AI.

The first study of RF plasticity performed in AI revealed that CS-specific increases responses (not decreases) developed during classical conditioning (tone-shock pairing) in the guinea pig. Moreover, responses to the pretraining best frequency (BF) and other frequencies tended to decrease. When these opposing changes were sufficiently large, tuning shifted toward the frequency of the CS and in some cases shifted *to* the CS, which became the new BF (Bakin and Weinberger 1990). An example is presented in Figure 5.2. If the CS frequency selected were distant from the pretraining BF (e.g., 2 octaves) and thus has a relatively weak initial response, then tuning shifts were less strong. RF plasticity is *associative,* as it requires stimulus pairing; sensitization training (no pairing) produces only a general increase in response to all frequencies across the RF (Bakin and Weinberger 1990; Bakin et al. 1992).

Several other characteristics of RF plasticity make it an attractive candidate for a process that operates in normal concert with sensory coding processes to subserve the storage of behaviorally relevant auditory information. First, RF plasticity is *highly specific* to the CS frequency; responses to frequencies a small fraction of an octave away are attenuated. Second, it exhibits *generality* across different types of training, as it develops in instrumental avoidance conditioning (Bakin et al. 1996) as well as classical conditioning. Third, RF plasticity develops during two-tone classical *discrimination* training, that is, increased responses to the CS+ frequency but decreased responses to the CS−, BF, and other frequencies (Edeline et al. 1990a; Edeline and Weinberger 1993). It also develops in discriminative instrumental avoidance conditioning (Bakin et al. 1996). Fourth, RF plasticity *develops very rapidly*, after only five training trials, as rapidly as the first behavioral (e.g., cardiac) signs of association (Edeline et al. 1993). Fifth, RF plasticity exhibits *long-term retention*, enduring for the longest periods tested, up to 8 weeks after a single 30 trial conditioning session (Weinberger et al. 1993). Sixth, RF plasticity *consolidates*, that is, continues to develop increased responses to the frequency of the CS *vs.* decreased responses to other frequencies in the absence of further training over hours (Weinberger et al. 1993; Edeline and Weinberger 1993) and days (Galván and Weinberger 2002).

Kisley and Gerstein (1999) questioned whether the shifts in tuning were due to learning-induced plasticity vs. spontaneous variation in tuning, because they observed changes in tuning over days. Spontaneous variation could explain neither why shifts in tuning during conditioning are toward or to the CS frequency but not away from the CS, nor why they develop only in animals receiving paired CS and US, not in those exposed to unpaired stimuli. Nonetheless, the issue of spontaneous tuning changes over days is important. Therefore, Kisley and Gerstein (2001) subsequently studied the effects of time over days per se and the effects of conditioning, using the Pre–Post design. In so doing, they extended inquiry to another type of motivation, appetitive reinforcement, and another species, the rat.

On the fourth day of seven days of recording, rats underwent a single brief (30-trial) session of classical conditioning. The US was independently assessed positive reinforcement consisting of stimulation of the medial forebrain bundle within the lateral hypothalamic/ventral tegmental area. Subjects were studied under light ketamine anesthesia throughout, rather than in an undrugged waking state. Behavioral evidence of conditioning to the tone (freezing) was nonetheless obtained. The authors did find that the entire tuning curves became less correlated over days in the absence of conditioning, but they did not track changes

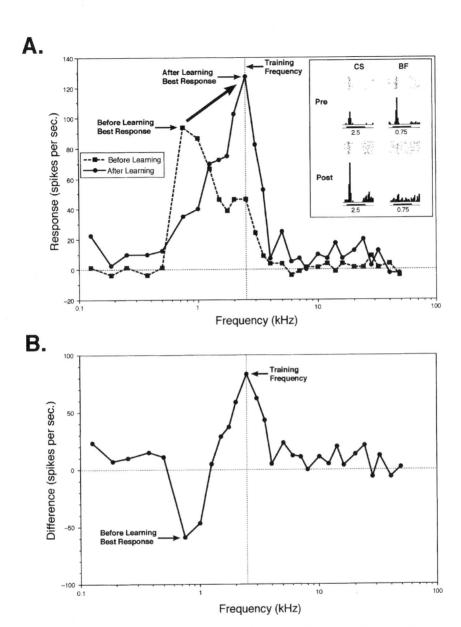

FIGURE 5.2. Learning shifts receptive field tuning to the frequency of the conditioned stimulus in tone-shock conditioning (30 trials). The best frequency before learning was 0.75 kHz. The subject was trained with a CS of 2.5 kHz. **(A)** Pre and Post training receptive fields. Note that tuning shifted so that 2.5 kHz became the best frequency. The *inset* shows the opposite changes of response in poststimulus time histograms for the pretraining BF and the CS frequency during the Pre and Post periods. **(B)** Difference in receptive fields due to conditioning (Post minus Pretraining RFs). Note that the maximum increase of response was at the CS frequency and the maximum decrease in response was at the pretraining best frequency.

in BF per se. It is possible that the reduced correlations reflect, to a greater or lesser extent, the fact that responses to all frequencies within a tuning curve received equal weighting in the statistical analysis. Thus, decreased correlations might reflect spontaneous changes in weak responses to frequencies distant from the BF, that is, at the lower and upper limits of the tuning curves. Galván et al. (2001) tracked BFs for 14 days in the guinea pig and found no directional drift of tuning but rather a random variation in BF of approximately 0.25 octaves.

Most important for the present topic, Kisely and Gerstein also found that conditioning does produce CS-specific plasticity, including shifts of tuning toward or to the frequency of the conditioned stimulus (Fig. 5.3). Also in agreement with prior studies, this RF plasticity was associative because it required CS–US pairing. The learning effects were above and beyond the spontaneous changes. [See Galván and Weinberger (2002) for a detailed consideration of the issue of spontaneous tuning changes related to effects of learning.]

As RF plasticity is not an artifact of spontaneous changes in tuning, neither is it an artifact of state. As previously noted, although animals exhibit arousal and related responses to sustained (e.g., 2–5 sec) CS frequencies during training trials, they do not exhibit any behavioral (e.g., cardiac) responses to the frequency of the CS when it is presented as one of a number of rapidly presented, brief (e.g., 200 msec) sequential tone pips during RF determination (Diamond and Weinberger 1989). Moreover, animals trained in the waking state exhibit RF plasticity when tested under deep general anesthesia (Lennartz and Weinberger 1992a; Weinberger et al. 1993).

Learning-induced tuning plasticity is not limited to animals. The same paradigm of classical conditioning (tone paired with a mildly noxious stimulus) produces concordant CS-specific associative changes in the primary auditory cortex of humans (Molchan et al. 1994; Schreurs et al. 1997; Morris et al. 1998).

In summary, RF plasticity has major characteristics of associative memory. It is not only associative, but also is highly specific, discriminative, rapidly acquired, retained at least for many weeks, develops consolidation over hours and days, and exhibits generality across training tasks, types of motivation, and species. Thus, RF plasticity in the auditory cortex reflects the learned importance of experiences.

5.3.3 A Note on the Seductiveness of the "During Design" and the Failure to Appreciate Performance Confounds

The preceding section on the DUR design emphasized the problem of confounds by performance factors during training trials. Also noted was that this issue is greatly underappreciated and often ignored. Because of its critical importance, it is well worthwhile to underscore the problem so that it will be explicitly considered in future research. I have selected two examples; readers are invited to note other instances in the literature.

The first concerns the inordinate common sense appeal, and thus the staying

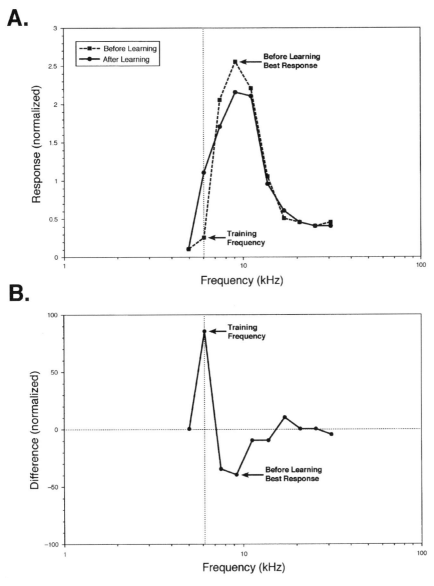

FIGURE 5.3. CS-specific learning-induced plasticity of receptive field tuning for positive motivational conditioning, in which the unconditioned stimulus was rewarding stimulation of the lateral hypothalamic/ventral tegmental area. **(A)** Receptive fields before and after conditioning (30 trials) in which a tone (6.13 kHz, *vertical line*) was paired with stimulation; the pretraining BF was 9.19 kHz). Note that response to the CS frequency increased and the response to the BF decreased after conditioning. **(B)** Difference (Post minus Pretraining RFs). The largest increase was at the CS frequency and the greatest decrease was at the pretraining BF. Tuning did not shift in this example, probably because pretraining responses to the CS frequency were weak. [A, redrawn from Kisely and Gerstein (2001); B was estimated from the published Pre and Post graphs.]

power, of the DUR design. A striking example within the present context concerns our own failure to adequately recognize the confounding effects of performance factors for many years. Although we used a Pre–Post design to avoid these factors to study habituation in the mid-1970s (Westenberg and Weinberger 1976), we subsequently continued to record single-unit activity in classical conditioning exclusively during training trials both in primary auditory cortex (Weinberger et al. 1984b) and in secondary cortical fields (Diamond and Weinberger 1984). It was only after trying and failing to understand adequately the complicated results of these studies that we finally appreciated the performance confounds that limit the DUR design and started to use the Pre–Post design.

The second concerns the failure to distinguish between neuronal effects observed during training trials vs. those obtained in receptive field analysis during Post training sessions. For example, Armony et al. (1998) attempted to test the hypothesis that the basolateral amygdala was involved in linking plasticity in the magnocellular medial geniculate nucleus to the auditory cortex via the cholinergic nucleus basalis (Weinberger et al. 1990a). Using a DUR design, they found little effect of amygdala lesions on short latency auditory cortical plasticity and erroneously concluded that the amygdala was not involved in receptive field plasticity. However, they never obtained receptive fields, but rather assumed that changes observed during training trials would be the same as those observed in post-training RF determination. As discussed previously, this assumption had been shown to be wrong (Diamond and Weinberger 1989). The role of the amygdala in receptive field plasticity remains an open question.

5.4 The "Post" Design (Post)

The Post design consists of recording only after the treatment experience. Because of the absence of Pre data, a minimum of two groups are needed for this design, one that receives the learning experience and a control group that either does not undergo formal learning or undergoes a different learning protocol. The Post design is fairly common in sensory and behavioral neuroscience, having been used to determine the effects of peripheral sensory denervation on cortical organization for several years. It has been applied to the study of cortical plasticity of tonotopic organization following selective cochlear lesions (e.g., Robertson and Irvine 1989). In the study of learning and the auditory cortex, the Post design is required when data can be obtained only once from a subject, for example, in metabolic studies or for detailed physiological mapping of the auditory cortex (Fig. 5.1, no. 3a,b).

Gonzalez-Lima and Scheich (1984a,b) appear to have been the first to use the Post design for the study of auditory system plasticity in conditioning. In the first study of the auditory cortex (Gonzalez-Lima and Scheich 1986), gerbils received tone paired with strong aversive electrical stimulation of the mesencephalic reticular formation. A large number of elegantly designed control groups were also studied. Following training, all groups received continual presentation of the CS stimulus during an injection of 2-deoxyglucose (2-DG) in a single Post session. Analysis of patterns of 2-DG uptake revealed a CS-specific increase in metabolic activity for the cortical area that represented the CS frequency. Subsequent studies amplified these findings (reviewed in Gonzalez-Lima 1992).

Another important example of findings yielded by the Post design was provided in the study of long-term instrumental frequency discrimination training in the monkey. Recanzone et al. (1993) trained subjects using 400–750 trials per day for some months (e.g., 60–80 sessions, minimum of ~27,000–36,000 trials), with gradually more difficult frequency discriminations until asymptotic performance was attained. Prior study of RF shifts of tuning toward/to the CS frequency in the guinea pig (Bakin and Weinberger 1990) had led to the prediction that the representation of behaviorally important frequencies would be expanded, because the tonotopic organization of the auditory cortex is the distribution of best frequencies (more correctly, characteristic frequencies, i.e., BF at threshold) (Weinberger et al. 1990a). In the Recanzone et al. study, there was no single CS frequency but rather different frequency bands for reinforced and nonreinforced frequencies at the conclusion of discrimination training. These authors did find that the representation of the training frequency band was increased, that the effects were specific to the different frequency ranges used for different subjects, and that the effects were not found in naïve animals. Moreover, bandwidths of tuning function for relevant frequencies were reduced, that is, tuning became sharper.

As summarized in the preceding, RF studies using only 30 trials (as few as 5 trials) for single-tone training and only 60 trials for discrimination training (30 trials each CS+, CS−) produced CS-specific tuning shifts. In contrast to the findings of Recanzone et al., increased sharpness of tuning was not prominent. Therefore, together, the results suggest a two-stage process. First tuning shifts develop, producing an expanded representation of behaviorally important frequencies. Second, tuning becomes sharper for cells within the expanded representation. (See Section 7 for further discussion.) Time-sampling studies are needed to test this two-stage model.

5.5 Continuous Pre–Post Design: Multiple CS-Stimuli (CS−/Pre–Post)

5.5.1 Introduction

Ohl and Scheich (1996, 1997) have argued that it is wrong to use the standard Pre–Post design (as described above) because the acoustic context of training (the "During" period) is different from the acoustic context of testing (Pre and Post periods), for example, when receptive field tuning data are obtained. "Our special aim was to keep the expected probability of occurrence for each of the stimuli [i.e., tones] during all phases of the experiment constant and equal to each other." This methodological issue is an important one particularly because the novel "CS−/Pre–Post" design devised by these workers yields results different than the Pre–Post and Post designs.

5.5.2 Design

The CS−/Pre–Post design consists of three stages, Pre, DUR, and Post, during which tone bursts of many frequencies are presented at short intertone intervals (250 msec to 3 sec) to obtain receptive fields. [The number of different frequencies is estimated to be 9 or 10 in one study, based on published figures (Ohl and Scheich 1996; their Figs. 6 and 7) and stated to be 4–30 in a second study (Ohl and Scheich 1997); the number of different frequencies was constant within a subject.) These three stages are continuous. They differ from the Pre–Post design in which the Pre and Post periods are identical in stimulus presentation, but the DUR (training period) consists of some accepted behavioral training paradigm, for example, classical or instrumental conditioning with one-tone, or two-tone discrimination conditioning (Fig. 5.1, no. 4). In the CS−/Pre–Post design, the DUR stage differs from the Pre and Post stages only by pairing a particular frequency (CS+) with shock. From the subject's standpoint, an experiment consists of many tones presented rapidly over a prolonged period of time, and during some of that time a shock is received following a particular frequency, and at some later time, the shock is no longer received. Presentation of all frequencies was semirandom. Sets of approximately 10 tones were repeated (at least 10 times) but no two series had the same tone order. Thus, the CS frequency might appear in any serial position (1–10 for 10 tones). In short, the CS−/Pre–Post design constitutes a discrimination experiment with one CS+ frequency and perhaps up to 30 CS− frequencies. As with the Pre–Post design, the effects of training were assessed by comparing the Post with the Pre receptive fields. Heart rate was recorded from the subjects (gerbils).

5.5.3 Findings

The authors did *not* find selective increased responses to the CS+ frequency in RFs, nor shifts of tuning toward/to this frequency. Conversely, they reported CS+ decreases, and also increased responses at adjacent lower and higher frequencies. They argue that learning does not facilitate the processing of important sounds, nor increase their area of representation in the auditory cortex. Rather, they propose that the CS+ frequency is coded by neurons in adjacent (lower and higher) regions of the tonotopic map, while responses to the CS+ frequency are relatively unchanged or decreased.

5.5.4 Issues and Assumptions

Such opposite findings and interpretations are exciting because they provide an opportunity to sharpen, perhaps rethink and broaden, conceptions about learning and the auditory cortex. The following discussion is necessarily a bit detailed, but hopefully worthwhile because it concerns not merely particular conflicting findings but more generally important issues in the design of experiments aimed at elucidating experience-related functions of the auditory cortex.

The most striking difference between the current and previous studies, that reported increased response and representation to the CS frequency, is in their experimental designs. The prior studies all used either a Pre–Post or a Post design. In all of these cases, the training context was different from the context of determining tuning in the auditory cortex. Moreover, the learning of frequency discrimination was behaviorally validated in prior discrimination studies (Edeline and Weinberger 1993; Recanzone et al. 1993). In contrast, the current study used a single CS+ frequency but many (e.g., 10) CS− frequencies. Each tone pip in this continuous presentation design constitutes a training trial, with tail shock presented only after the CS+ frequency. While this protocol is a form of discrimination training, no prior use of this design could be found in the literature, so there seems to be no validation that subjects can learn to discriminate one CS+ from many CS− frequencies in a continuous presentation design.

Heart rate was recorded, but no quantified data were provided in this paper and the authors noted that behavioral discrimination was not determined. A second experiment using the CS−/Pre–Post design did present examples of cardiac behavior for one animal (Ohl and Scheich 1997, their Fig. 1, p. 687). These showed that heart rate was slower overall during the conditioning phase (DUR) and during a Post conditioning phase than during Pre or silent periods. However, heart rate was pooled across the presentation of all frequencies. Therefore, even if these data are accepted as statistically valid for the group of subjects, the findings could indicate only that training altered heart rate. They fail to speak to the issues of (1) whether or not cardiac deceleration was elicited by the CS+ frequency and (2) whether or not animals learned the discrimination. To resolve these issues it is necessary to (1) obtain cardiac responses to the CS+ and (2) compare them to heart rate responses to the CS− frequencies. Unfortunately, this is not possible using the multiple CS−/Pre–Post design because training consists of the rapid continual presentation of tones. As noted previously, intertone intervals (each tone constituting a training trial) ranged from 250 msec to 3 sec. With such brief intervals, there is not enough time to determine the change in heart rate to one tone before the next tone is presented.

The findings that responses to the CS+ decrease while responses to adjacent lower and higher frequencies increase may reflect a failure of discrimination learning. Subjects exposed to the CS−/Pre–Post paradigm undoubtedly learned something. Such learning and its correlated plasticity might not be so paradoxical. The problem is that the nature of what was learned seems unknowable within the constraints of the current CS−/Pre–Post design. The discrimination aspect of what was learned could be assessed, in the standard and accepted manner of determining behavioral generalization gradients after training, by presentation of the CS+ and each of the CS− stimuli with sufficiently long intertone intervals (e.g., 30–60 sec) so that heart rate responses to each stimulus could be assessed (e.g., McLin et al. 2002b). However, then the testing would be in a context different from training and thus violate the goal of keeping the acoustic context the same during training and post-training assessment.

In summary it seems that the presumptive advantage of the CS−/Pre–Post design is balanced by the disadvantage of being unable to determine if behavioral discrimination learning has occurred. One possible resolution of this problem would be to lengthen the intertone intervals, both to increase the opportunity for learning the discriminations and enable appropriate analysis of behavior. In any event, although the present use of the CS−/Pre–Post design has yielded findings that are not interpretable within the intended framework of behavioral learning and memory, nonetheless, they broaden the discourse.

5.5.5 Context Reexamined

The particular findings aside, let's return to a basic design issue, the rationale for the CS−/Pre–Post design. As explained in the preceding, it was to maintain the same context throughout the stages of the experiment, by keeping constant the probability of each tone frequency. If the CS−/Pre–Post design maintains the same context, then we should expect to observe the same neural changes during training as during post training. This outcome would support the authors' assumption that performance factors (e.g., changes in state, motivation) did not affect cortical responses during the introduction of the tail shock. However, neural responses during training were not analyzed, so the assumption remains unsupported. On the contrary, behavioral data suggest the converse. Heart rate was slower during both training and the Post period compared to the Pre period. This shows that the behavioral state of the subjects was *different* in the Post vs. Pre periods. As the training effects were determined by subtracting the Pre receptive fields from the Post receptive fields, the RF changes could reflect differences in state as well as effects due to the (presumptive) discrimination learning (Table 5.2).

But why attempt to render the context of learning identical to the context of the Post period, when the effects of learning are assessed? Because changes in response to the CS+ were found to be different during training trials vs. during determination of receptive fields after training (Diamond and Weinberger 1989). (Our previous discussion of the DUR design attributes this to unavoidable performance factors during training.) But why is that a problem? Ohl and Scheich (1996) note that the size of an area in a neural map can increase with increased stimulus use (e.g., Merzenich and Sameshima 1993) and decrease with decreased stimulus use (e.g., Robertson and Irvine 1989): "Therefore it appears important to design an experiment in which the mere probability structure of the stimulus input can be ruled out as participating in the associative learning process and as contributing to the experimentally determined effects" (Ohl and Scheich 1996, p. 1002). In essence, their view is that presenting many tones during receptive field determination, but only one or two tones during simple conditioning or discrimination training, can produce results that are erroneously judged to reflect associative processes.

TABLE 5.2. Comparison of the factors of State and Learning in two experimental designs during three periods: Pre Training, During Training (e.g., discriminative conditioning), and Post Training.

CS−/Pre–Post Design

Factor	Pre Period	During	Post Period	Post–Pre
State	0	+	+	+ − 0 = +
Learning	0	+	+	+ − 0 = +

Pre–Post Design

Factor	Pre Period	During	Post Period	Post–Pre
State	0	+	0	0 − 0 = 0
Learning	0	+	+	+ − 0 = +

The last column denotes the assessment of the effects of training by subtracting receptive fields in the Pre period from receptive fields obtained in the Post period. "0" denotes whatever ground state or status of learning is present during a period. "+" indicates a change from the ground states. Both designs begin with baselines of no learning and similar quiet levels of arousal state. During training (DUR period) both designs entail changes for both state and the establishment of learned associations. The designs *differ* in that the CS−/Pre–Post design yields the same state in the Post as in the DUR period (indexed by continued conditioned behavioral responses in both periods). Therefore, subtracting the Pre from the Post data confounds state and associative processes (Ohl and Scheich, 1997), In contrast, the Pre–Post design can avoid common states, due to a marked change in context between DUR and Post periods (e.g., Bakin and Weinberger, 1990). However, associative neural effects are retained despite the planned absence of behavioral expression in the Post phase. Thus, subtracting the Pre from the Post data provides estimates of the effects of learning per se, independent of state effects.

This concern is misplaced. Standard controls for nonassociative effects during conditioning routinely employ a two-group design, in which one group receives paired CS and US and the other group receives the same density of stimulation, except the CS and US are not paired. This tactic does maintain the same probability of stimulation between groups. It is this well-established, standard control design that permits the conclusion that the effects of an experience depend upon association of the CS and the US. Embedded with a Pre–Post design, nonassociative controls have revealed CS-specific associative increases in the area of CS representation (e.g., Gonzalez-Lima and Scheich 1986) and CS-specific shifts of tuning (Bakin and Weinberger 1990). Moreover, two-tone discrimination training maintains an equal probability of presentation of the CS+ and the CS−, while manipulating only the associative relationship with the unconditioned stimulus (reinforcement). This paradigm also yields increase responses to the CS+ frequency plus decreased responses to the CS− frequency (e.g., Edeline and Weinberger 1993). Therefore, it would be erroneous to conclude that the CS−/Pre–Post design is necessary to reveal genuine associative effects on receptive field and representational map plasticity.

But, is there an actual advantage to induce maintaining identical learning-related acoustic contexts in the training vs. Pre and Post periods? There are certainly clear disadvantages in doing so. From a practical standpoint, it would be impossible to induce learning-related plasticity in the anesthetized state because learning is not adequate under general anesthesia. Yet assessment of effects with anesthetized subjects offers considerable experimental benefits. As noted previously, a constant arousal level can be maintained and extensive mapping of the organization of the auditory cortex is possible. From a broader point of view, memory ordinarily transcends the context of the original learning. Otherwise recall would be impaired unless we returned to the original training context. (This is not to deny that the learning context can enter into memories, but rather that context is not paramount.) The benefits of the capacity to learn and remember would be extremely limited were they operative only within the context of the original experience. Therefore, that receptive field plasticity established in one context can be "read out" consistently over weeks in another context (Weinberger et al. 1993) comports well with normal circumstances of memory retention and recall.

In concluding this section, neuroscience has only begun to explore the effects of context on the representation and storage of information in the auditory cortex (as elsewhere in the brain). This is one issue of practically unlimited opportunities for investigation. Whether a given change in acoustic (or other) context is important in a given situation is a matter that needs extensive study. Additional studies are needed that include recording during both training and Pre/Post periods. The question should yield to systematic use of appropriate behavioral indicators of learning and memory. But unless workers avoid adding unsupported assumptions and maintain a sufficiently broad knowledge base of both learning and auditory physiology, future exploration is unlikely to shed more light on the subject, while inadvertently reducing the candlepower of inquiry.

6. Mechanisms

6.1 Introduction

Investigation of mechanisms of experience-dependent tuning plasticity in the auditory cortex has been multilevel and focused on three interlocking issues: (1) the locus of receptive field change; (2) synaptic changes involving "Hebbian-like" mechanisms, and (3) the role of neuromodulators. A preliminary model (Weinberger et al. 1990a) attempted to capture all three aspects of mechanism in the case of tone-shock pairing, and therefore may serve as a point of departure.

The model involves three loci of convergence of tone excitation and the direct or indirect effects of shock afferentation. Each convergence was thought to involve "Hebbian type" mechanisms, that is, the coactivation of pre- and post-synaptic elements. The interacting hypotheses are as follows. Tone information ascends the lemniscal auditory pathway and reaches the auditory cortex relatively unchanged, from the periphery through the ventral medial geniculate body (MGv). Tone information also reaches the nonlemniscal medial (magnocellular) division of the medial geniculate body (MGm), where it (1) converges (is associated) with nociceptive information from shock that ascends the spinothalamic pathway. The result is facilitation of MGm responses to the tone on subsequent trials. The MGm projects mainly to apical dendrites of pyramidal cells in Layer I of the auditory cortex, where its facilitated discharges (2) converge with (lingering) excitatory effects of the immediately preceding tone on pyramidal cells. This "Hebbian" convergence is sufficient to promote short-term plasticity of tuning (i.e., for short-term memory). However, the convergence of MGv and MGm input to pyramidal cells is too weak to induce strong plasticity because the excitatory postsynaptic potentials from the MGm to Layer I decrement along the apical dendrites which has a large length constant. Resultant MGm EPSPs would be small at the convergence zone with the MGv input below Layer I. But, the facilitated MGm response also is projected to the cholinergic nucleus basalis, via the basolateral and then central nucleus of the amygdala, where its effect produces a release of acetylcholine (ACh) in the auditory cortex. The release of ACh, acting at cortical muscarinic receptors (3) converges with cortical excitation from the effects of the tone (via the direct MGv and indirect MGm paths), thus producing long-term plasticity via its established long-term increases in cellular excitability. Responses to the CS tone are thereby strengthened, and increased responses to this frequency successfully compete with inputs from other frequencies, often producing a shift in tuning (for further details, see Weinberger et al. 1990a,b).

Some aspects of the model have received experimental support while others have been either proven wrong or have been challenged on a variety of grounds (for a detailed evaluation of this model, see Weinberger 1998). It is not the purpose of this section to defend the model but rather to indicate the issues involved in formulating and testing such models. (Some evidence both pro and con will be included in the following sections for illustrative purposes.) It will be helpful to begin with an alternative model that is a modification of the original model.

Suga and colleagues have extended the domain of inquiry to the corticofugal system, specifically to the projections of the auditory cortex to the central nucleus of the inferior colliculus, and also to the somatosensory cortex, in the bat. They have reported CS-specific tuning shifts both in the auditory cortex and in the inferior colliculus. Moreover, they report that collicular tuning shifts develop before auditory cortical shifts, although they disappear within a few hours whereas cortical shifts last at least 24 hours. Further, muscimol applied to the somatosensory cortex prevents cortical and collicular shifts. [But see Edeline et al. 2000 for large diffusion of muscimol.] The findings do not include quantified data validating behavioral learning but leg flexion and body movement were reported after 40 trials of a 60-trial training session (Gao and Suga 2000). Whether associative learning occurred is unknown.

Gao and Suga did not explicate their model in great detail but delineated its main features. First, the auditory and somatosensory cortices receive tone and shock (nociceptive) information, respectively. Both cortices project to the amygdala, indirectly via association cortex; the amydgala is thought to be the site of convergence, that is, learning. The amygdala then effects the release of ACh from the nucleus basalis, as previously hypothesized (Weinberger et al. 1990a). The resultant auditory cortical plasticity then produces shifts in the colliculus and these are fed back to the cortex to strengthen what would otherwise be weak plasticity. [For a slightly modified version of this model, see Suga and Ma 2003; see Weinberger 2004 for a detailed critique.]

6.2 Subcortical Sites of Plasticity: Commonalities and Differences

We can now consider the issue of sites of plasticity. The CS-specific plasticity in AI might be a passive reflection of plasticity projected to it from the auditory thalamus, the medial geniculate. The ventral and magnocellular subdivisions both project to AI. It is possible that both contribute to the *initiation* of RF plasticity in the cortex because both develop CS-specific RF plasticity. The MGm develops associative increased responses to the CS tone during training and expresses CS-specific RF plasticity immediately after learning and for at least 1 hour (the longest period tested). However, RF plasticity in AI cannot merely be a reflection of projected plasticity from this nucleus because MGm RFs are much more complex, multipeaked, and broadly tuned than those of auditory cortical cells (Edeline et al. 1990b; Edeline and Weinberger 1992; Lennartz and Weinberger 1992a). The MGv develops narrowly tuned RF plasticity but this dissipates within one hour (Edeline and Weinberger 1991). [This finding shows that the Weinberger et al. model is wrong in postulating no plasticity in the MGv.] Therefore, whatever may be the roles of the MGm and MGv in the initiation of AI plasticity, they cannot account for the selective CS-specific long-term maintenance of plasticity (tracked to 8 weeks, Weinberger et al. 1993). Therefore, it is reasonable to conclude that active plastic processes within AI contribute to the tuning plasticity therein observed.

The Gao and Suga model does not address the auditory thalamus and the Weinberger et al. model does not address the inferior colliculus. But the models may be in agreement given the transient RF plasticity in the MGv. This is what might be expected if transient plasticity in the inferior colliculus is projected to the auditory cortex, presumptively via the MGv. However, the Gao and Suga model is based on slowly developing RF plasticity in AI, perhaps requiring one hour. Incompatible with the Gao and Suga hypothesis of subcortical initiation of slowly developing cortical plasticity is the fact that RF plasticity in AI of the guinea pig develops within only five training trials, that is, about as rapidly as possible (Edeline et al. 1993).

The two models postulate very different roles for the amygdala. Gao and Suga hold it to be the site of CS–US convergence and learning, whereas the Weinberger et al. model treats the amygdala as part of the associative machinery but not the prime site. Evaluation of prior literature on aversive conditioning reveals that it would be premature to assign a primary function for learning to the amygdala (Cahill et al. 2001), particularly in light of the finding that destruction of the basolateral amygdala does not prevent fear conditioning (Cahill et al. 2000) but does impair unconditioned freezing, which is the behavioral assay on which the amygdalar hypothesis is largely based (Vazdarjarnova 2000). The Gao and Suga model ignores several facts. First, acoustic and nociceptive information converge directly in the MGm (Lund and Webster 1967; Wepsic 1966). Second, plasticity (increased responses to the CS) develops in the MGm during training (Gabriel et al. 1975; Ryugo and Weinberger 1978; Edeline et al. 1990a,b; McEchron et al. 1995; O'Connor et al. 1997), is long-lasting (Edeline et al. 1990b), and is evident as CS-specific RF plasticity after conditioning (Edeline and Weinberger 1992; Lennartz and Weinberger 1992a). Third, analogs of learning-induced plasticity reveal that the MGm develops long-term potentiation (Gerren and Weinberger 1983) and that tone paired with stimulation of the MGm induces heterosynaptic long-term potentiation in AI (Weinberger et al. 1995). Fourth, the MGm is necessary for the amygdala to receive acoustic input during conditioning (Iwata et al. 1986). Moreover, the MGm develops synaptic plasticity during conditioning (McEchron et al. 1996) and does so with a shorter latency than does the amygdala (McEchron et al. 1996). Nonetheless, some workers assign CS–US convergence plasticity only to the amygdala (Quirk et al. 1997) on the apparently physiologically impossible assumption that the MGm provides pure acoustic input to the amygdala without somehow also transferring the effects of its plasticity, as indexed by its increased discharges to the CS.

However, the roles of the MGm and the BLA are by no means resolved. For example, recent findings report that plasticity in the MGm is dependent on the amygdala (Maren et al. 2001; Poremba and Gabriel, 2001). On the other hand, during appetitive conditioning (tone/food), the MGm develops strong plasticity in waking and continues to express plasticity during paradoxical sleep, whereas the BLA exhibits weaker plasticity and no expression of plasticity during paradoxical sleep (Maho and Hennevin 2002). The authors conclude that the amygdala is more involved in strong emotional states, such as in aversive conditioning whereas the MGm signals the importance of the CS for both aversive and appetitive conditioning, that is, regardless of emotional sign. They also suggest that plasticity first develops in the MGm and then the results of this plasticity are sent to the lateral amygdala, which adds its own plasticity that concerns the strength of motivation and/or sign of emotion. This conception is fully compatible with the view that the plasticity which develops in the MGm affects AI via its monosynaptic projections to upper lamina. Given that RF and map plasticity develop in appetitive (Recanzone et al. 1993; Kisley and Gerstein 2001) as well as aversive learning, the MGm seems more closely tied to AI plasticity than does the amygdala.

At this stage of inquiry, the convergent and particularly divergent findings provide an opportunity to formulate and test hypotheses that will ultimately provide a complete circuit level account of at least the roles of the subcortical auditory system and the amygdala in learning induced auditory cortical tuning plasticity. They also point to the finding that cortical tuning plasticity, however initiated, continues to develop after the cessation of any putative subcortical auditory influences. Perhaps neuromodulatory systems acting directly on the auditory cortex are involved.

6.3 Neuromodulatory Systems

6.3.1 Introduction

The nucleus basalis (NB) cholinergic system has received a great deal of attention as a substrate for CS-specific receptive field plasticity in AI. Several lines of evidence implicate the nucleus basalis, and its cholinergic projections to the cortex, in learning-induced RF plasticity in the ACx. (For original studies of the somatosensory cortex, see Dykes 1997.) However, one should not assume that ACh is the only neuromodulator that can affect processing in the auditory cortex. For example, tuning can be altered by norepinephrine (Manuta and Edeline 1997). Pairing a tone with activation of the dopaminergic ventral tegmental reward area can increase the representation of that frequency (Bao et al. 2001). Serotonin can regulate level-dependent response functions (Juckel et al. 1999) and exhibits increased levels in AI during initial stages of avoidance training (Stark and Scheich 1997). Therefore, ACh is not viewed as an exclusive modulator of the auditory cortex. Nonetheless, systematic studies to date do provide an emerging picture and progress with this system can serve as a template for the study of other modulatory systems.

6.3.2 The Nucleus Basalis Cholinergic System

The NB is the major source of cortical acetylcholine (Johnston et al. 1979; Lehmann et al. 1980; Mesulam et al. 1983). [We use the term "nucleus basalis" rather than "basal forebrain cholinergic cells" for brevity and to refer more specifically to cells designated as the "Ch4" group in primates, which innervates the neocortex, vs. other forebrain groups that largely innervate subcortical structures.] Cortical projections do have some topography. For example, the auditory cortex in the rat is innervated by cholinergic cells within the ventral caudal globus pallidus and the caudal substantia innominata (Bigl et al. 1982; Saper 1984; Moriizumi and Hattori 1992).

The established links between the cholinergic system and memory provide additional impetus for investigating its role in auditory cortical plasticity and behavioral memory. For example, pharmacological blockade of the cholinergic system impairs many forms of memory (e.g., Deutsch 1971; Blozovski and Hennocq 1982; Rudy 1996; Potter et al. 2000). Cholinergic agonists and cholinesterase antagonists can facilitate memory (e.g., Stratton and Petrinovich 1963; Flood et al. 1981; Gower 1987; Introini-Collison and McGaugh 1988), promote recovery of memory from brain damage (Russell et al. 1994), and achieve rescue from memory deficits in transgenic mice (Fisher et al. 1998). Also, several noncholinergic treatments that facilitate memory, such as adrenergic agents and stress hormones, affect memory via actions on the cholinergic system (Salinas et al. 1997).

Lesions of the NB using excitatory amino acid agonists disrupt learning in many tasks and the effects may be specific to certain memory processes, for example, impairment of acquisition but not retrieval of conditioned taste aversion (Miranda and Bermudez-Rattoni 1999). However, excitotoxic lesions are not selective to cholinergic neurons, also affecting γ-aminobutyric-ergic (GABA-ergic) and other noncholinergic cells. In contrast, selective immunotoxic cholinergic lesions apparently impair attention rather than learning or memory (reviewed in Everitt and Robbins 1997; Wenk 1997). Yet, recent findings indicate that immunotoxic lesions leave a small amount of ACh that is sufficient to support learning (Gutierrez et al. 1999). Thus, the role of NB cholinergic neurons in learning and memory is still controversial.

Stimulation of the NB releases ACh in the cortex (Casamenti et al. 1986; Kurosawa et al. 1989; Rasmusson et al. 1992) and produces widespread EEG activation, which is the waking cortical state optimal for learning. Conversely, lesions of the NB reduce cortical levels of ACh and impair cortical activation (Celesia and Jasper 1966; LoConte et al. 1982; Riekkinen et al. 1992; Jimenez-Capdeville and Dykes 1996). The discharge rate of identified cholinergic projection cells increases during cortical activation and decreases during cortical slow waves (reviewed in Duque et al. 2000).

ACh produces long-lasting modification of receptive fields in primary sensory cortices (e.g., Sillito and Kemp 1983; Metherate et al. 1988). In AI, iontophoretic application of cholinergic agents produces long-lasting modification of frequency tuning via muscarinic receptors (McKenna et al. 1989; Ashe et al. 1989). Stimulation of the NB produces atropine-sensitive, persistent modification of evoked responses in the auditory cortex, including facilitation of field potentials, cellular discharges, and EPSPs elicited by medial geniculate stimulation (Metherate and Ashe 1991, 1993). NB stimulation also facilitates neuronal discharges to tones in AI (Hars et al. 1993; Edeline et al. 1994a,b). Pairing a tone with iontophoretic application of muscarinic agonists to AI produces pairing-specific, atropine-sensitive, modification of RFs that include shifts of tuning toward the frequency of the paired tone; however, this protocol is more likely to produce frequency-specific decreases (Metherate and Weinberger 1990). This outcome indicates that NB stimulation and iontophoretic administration of muscarinic agonists cannot be assumed to substitute directly for each other. As emphasized by Rasmusson (2000), while electrical stimulation of the nucleus basalis may be viewed as relatively crude, it has the advantage that spatial and temporal characteristics of ACh release are more consistent with the normal anatomy and physiology of this system.

ACh is preferentially released into relevant sensory cortical areas at the time of learning (Orsetti et al. 1996; Butt et al. 1997). ACh applied to AI augments RF shifts during classical conditioning while cortical atropine prevents such RF shifts (Ji et al. 2001). Studies of neuronal responses in the NB are consistent with an important role for the NB/ACh in cortical plasticity. NB cells exhibit frequency tuning for pure tones, indicating that specific acoustic information can reach this structure (Chernychev and Weinberger, 1998). Stimuli that signal appetitive or aversive reinforcement elicit increased responses of NB cells during learning (Travis and Sparks 1968; Richardson and Delong 1986; Wilson and Rolls 1990; Whalen et al. 1994). Of particular relevance, cells in the NB develop increased discharges to the CS+ during tone-shock conditioning before the development of neuronal plasticity in AI and thus could be causal to the cortical RF plasticity (Maho et al. 1995). Furthermore, NB neurons projecting to AI selectively increase transcription of the gene for ACh's synthetic enzyme, choline acetyltransferase, during tone-shock conditioning, indicating that acoustic learning engages specific cholinergic subcellular mechanisms (Oh et al. 1996).

If learning-induced plasticity in AI develops via engagement of the NB, then NB stimulation should be able to substitute for a standard reinforcer, such as food or shock, although no motivational reinforcement would be involved. The NB cholinergic system is capable of inducing RF plasticity that has the same characteristics as learning-induced RF plasticity. Pairing a tone with NB stimulation for only 30 trials induces RF plasticity. This plasticity is associative, as is the case for standard behavioral learning, because it requires stimulus pairing (Bakin and Weinberger 1996). Moreover, as in the case of learning, two-tone discriminative RF plasticity develops when one tone is paired with NB stimulation while another is presented alone (30 trials each of the CS+ and CS−, Dimyan and Weinberger 1999). Moreover, like behavioral learning, a single session of training that produces NB-induced RF plasticity also produces consolidation (i.e., increased strength of effect without additional training) when tested over 24 hours (the longest period tested) (Bjordahl et al. 1998). NB-induced RF plasticity is blocked by atropine directly applied to AI, showing that the engagement of muscarinic receptors in AI is necessary for this change in frequency tuning (Miasnikov et al. 2001). As in the case of learning-induced plasticity, RF plasticity would be expected to produce an increased representation of the paired tone frequency because the tonotopic map is the distribution (at threshold) of BFs, including those that have shifted toward or to the CS frequency. This has been obtained (Kilgard and Merzenich 1998; Kilgard et al. 2001).

All of these findings support the hypothesis that the NB can induce behavioral memory. However, neither these findings, nor any others in the literature, have directly linked NB activation to behavioral memory. McLin et al. (2000a,b,

2003) asked whether NB mechanisms are sufficient to produce a predicted specific behavioral memory. Rats were trained by pairing a 6.0-kHz tone with NB stimulation; a control group received unpaired stimulation ("Post" design). After training, they were tested in another context (different room and illumination conditions) in the absence of any NB stimulation. The specificity of behavioral effects was assessed by recording heart rate and respiration, using the well-established metric of the stimulus generalization gradient which is obtained when an animal trained with one stimulus is subsequently tested with many stimuli. If paired NB stimulation induced associative memory for the training tone, then this tone (6.0 kHz) should later elicit the largest behavioral responses of all tones tested, that is, occupy the peak of the frequency generalization function.

The pairing did produce CS-specific behavioral memory as demonstrated by the fact that cardiac and respiratory behavior exhibited frequency generalization gradients with the CS tone of 6.0 kHz at the peak (Fig. 5.4). These findings show that NB stimulation produces behavioral associative effects that are highly specific to the CS frequency. The subjects behaved as though they had learned that 6.0 kHz had acquired increased behavioral significance through a learning experience. The findings do meet the dual criteria of associativity and specificity that are the long-accepted standard as sufficient to infer memory from behavioral change. Viewing the data alone, one could not determine whether a standard reinforcer had been used versus NB stimulation. However, NB stimulation itself is apparently not rewarding or punishing as it is not part of any known motivational system (e.g., Pennartz 1995). It seems to act as an effective but neutral "cortical activation mechanism" (see Olds and Peretz 1960; Wester 1971). These findings indicate that paired tone NB stimulation not only can induce cortical plasticity but also is sufficient for the formation of specific auditory associative memory.

6.4 "Hebbian-Like" Processes

Two lines of inquiry have asked whether or not learning-related plasticity in AI involves neuronal interactions related to coactivation strengthening of functional connections, or negative coactivation weakening of synaptic strengths. The term "Hebbian-like" is used as a neutral phrase for such processes. Detailed critical reviews of Hebbian-like processes, outside the scope of this chapter, should be consulted (Cruikshank and Weinberger 1996; Edeline 1999).

Cruikshank and Weinberger (1996) manipulated covariance for single neurons in AI of the anesthetized guinea pig by the use of juxtacellular current that either increased or decreased responses to tones (CS+ and CS−, respectively). They found evidence supporting the covariance hypothesis, that is, after treatment, responses to tones with increased covariance were significantly strengthened relative to tones associated with decreased covariance. The probability of covariance plasticity was greater when the EEG was nonsynchronized (not dominated by slow waves). The findings indicate that Hebbian-like processes are

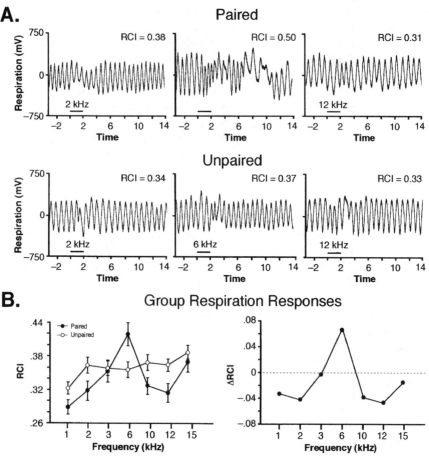

FIGURE 5.4. CS-specific induction of behavioral memory (indexed by respiration changes) following pairing tone (6 kHz) with NB stimulation in another context. (A) Examples of individual respiration records (with value of respiration change index, RCI) to three frequencies (2, 6, and 12 kHz) for one animal each from the paired and unpaired groups. The largest response was at 6 kHz for the paired animal (RCI = 0.50). The *horizontal bar* indicates tone duration. (B) Frequency generalization functions. *Left panel*, group mean ($\pm$ SE) change in respiration to all tones for both groups. The maximal response was at the CS frequency of 6 kHz for the paired group, but not for the unpaired group. *Right panel*, the group difference function (Paired minus Unpaired) shows a high degree of specificity of respiratory responses to 6 kHz. [From McLin et al. (2002a).]

effective in changing synaptic strengths and that such processes may be gated by factors regulating cortical state.

They then tested the hypothesis that covariance plasticity would be facilitated by adding stimulation of the nucleus basalis, to release ACh and activate the EEG (Cruikshank and Weinberger 2001). Cells were morphologically identified (biocytin) as being mainly pyramidal, distributed from Layer III–VI. They found no net increase in plasticity, but actually evidence that adding NB stimulation decreased plasticity. An intriguing finding was that facilitated plasticity was confined to a pyramidal cell that had by far the greatest development of apical dendrites that extended into Layers I and II, in contrast to "nonplastic" cells that lacked upper lamina apical dendrites. These observations cannot be evaluated statistically, but do emphasize the importance of morphological data to complement physiological findings.

A complementary study, using direct application of ACh and norepinephrine (NE) in the anesthetized guinea pig, also failed to find facilitation of functional connectivity (cross-correlations of cell pairs) during manipulation of covariance (Ahissar et al. 1996). In both approaches, the failure to increase covariance plasticity by cholinergic intervention, either by NB stimulation or direct application of ACh to the cortex, might be attributed to many factors, including timing of cholinergic treatment, and thus additional studies are needed. However, it is possible that ACh itself does not strengthen functional connections because iontophoretic application of ACh in AI of the anesthetized guinea pig increases excitability without altering cross-correlation functions between pairs of cells (Shulz et al. 1997).

Ahissar et al. (1992, 1998) have focused on functional connectivity between pairs of neurons within the context of behavior in waking animals. Two neurons in AI of a waking monkey were induced to fire together by presenting a tone that excited one unit contingent on the spontaneous firing of a second cell; this facilitated the functional connection, observed after the treatment. Weakening of functional connectivity was caused by preventing simultaneous firing. Importantly, the effects were significantly stronger when the tone was behaviorally relevant, that is, when the monkey could receive reward on detecting a change between tone and noise or between two tones. Therefore, the temporal contingency between cells may be necessary for cortical plasticity but is not sufficient. The authors conclude that neuronal plasticity in AI obeys a Hebbian-like associative rule. The generality of the findings may be mitigated by the use of a single trained animal but the findings underscore the potential importance of studying AI within a behavioral context.

Support for the view that behavioral context appears to be a critical factor in enabling cortical plasticity is provided from an entirely different experimental direction. In a distinctively novel approach, Talwar and Gerstein (2001) broached the issue of the functional significance of an expanded representation of a given frequency (e.g., Recanzone et al. 1993). Rather than inferring that such plasticity necessarily confers processing advantages to the frequency in question, they directly induced an expanded representation, by the use of intracortical microstimulation (Dinse et al. 1993; Maldonado and Gerstein 1996) and determined if this treatment affected behavioral frequency discrimination. However, they found no effect either on signal detectability or response bias. They conclude that cortical reorganizations are behaviorally effective if they develop within an appropriate behavioral context.

6.5 Summary

It is premature to draw firm conclusions about mechanisms of experience-dependent plasticity in the auditory cortex, particularly as the dimensions of inquiry are still being delimited and investigation beyond the cholinergic system has not yet achieved great depth. The "however" is that the outlines of a reductionistic account are beginning to emerge.

First, a minimal starting point, often assumed, is that auditory cortex is indeed an active site of plastic processes in learning. (This conclusion transcends the directly documented ability of experimenters to impose plasticity on the cortex. While all invasive techniques are helpful, the central issue is what happens in normal life, when we learn that a sound has behavioral significance.) The evidence is that the nature and time course of thalamic and midbrain auditory plasticity in learning cannot explain all cortical effects. The question is whether they can fully explain any of the plasticity in AI. From a conservative stance, this can be considered an open question. Certainly, the roles of the magnocellular medial geniculate and the amygdala need to be clarified. Inclusion of corticofugal mechanisms into the dialogue is a step forward, as is identification of all putative influences on the auditory cortex.

The NB cholinergic system is currently foremost among subcortical systems that may enable the auditory cortex to store specific information, particularly in the case of frequency processing and representation. The case for a cholinergic enabling mechanism, which itself is not carrying precise acoustic information, is becoming increasingly strong. Anatomical, physiological, pharmacological, and behavioral findings all implicate the NB/ACh system. RF plasticity induced in a variety of tasks, types of motivation, and species also can be induced by substituting NB stimulation for appetitive or aversive reinforcers and this plasticity does require the engagement of muscarinic receptors in AI. Beyond that, it has now been shown that this treatment is sufficient to induce actual specific behavioral memory (see above, McLin et al. 2002a, 2003). Much more needs to be done but it is extremely unlikely that the cholinergic system will prove to be irrelevant to the ultimate explanation.

In contrast, the study of cellular relationships is at a very early stage. Yet, it has already been shown that an analytic understanding of plasticity in the auditory cortex will have to involve waking subjects with conditions of behavioral control. One can foresee a broad range of experiments that address issues to achieve synthesis of behavioral, cellular, and modulatory approaches. For example, Does the facilitatory effect of behavioral context on covariance plasticity (Ahissar et al. 1992, 1998) depend on the nucleus basalis cholinergic system? There is also a prior question: Does the formation of functional neuronal ensembles underlie the acquisition and storage of information in the auditory cortex?

Future inquiry might then take the broad form of the study of learning in the auditory system within the framework of controlled, quantified, appropriate behaviors that can be linked causally to organized neural plasticity in auditory cortex, and to other levels of the auditory system. Lesions and inactivation of tissue will continue to play a role, but in the absence of behavioral processes conclusions are likely to be limited to highly sophisticated conjectures. We can expect to see an increasing use of molecular and genetic techniques, including markers for synapses whose strengths have recently undergone change, and the use of inducible, targeted translation and transcription factors. However, and finally, all mechanistic approaches need to be solidly anchored in behavior.

7. Functions of Learning-Induced Plasticity in the Auditory Cortex

7.1 Introduction

One can discern a progression of shifting questions, the queries overlapping across time, even decades, with respect to learning and the auditory cortex. The first question was, Is the auditory cortex involved in learning and memory? Although the answer was affirmative, the functional significance was unclear and in effect the answer was simply "added" to the list of brain regions "involved" in the acquisition and storage of information. The second question, initiated with studies of learning and plasticity of receptive fields and frequency representation in the 1980s, increasing greatly in the 1990s and beyond, has been, Are the effects of learning highly specific? In distinction to the first question, the second continues to capture the interest of an increasing number of auditory, and other, neuroscientists. It has been suggested (Section 4) that this marked attention developed because the stimulus parameters and types of data (e.g., tuning curves) obtained are standard within auditory physiology, so that the question was fundamentally, Does learning systematically modify the processing and representation of acoustic information? The answer to this question is also affirmative. To date, experiments to answer this question have been directed almost exclusively to the issue of frequency processing. It seems evident that the domain of study should and will be expanded to encompass all stimulus parameters, but there has been a very good start. While experiments directed to the second question need to be performed for the indefinite future, two somewhat divergent "third" questions have arisen. Question 3A is, What are the mechanisms of learning-induced cortical plasticity? The status of this developing line of study has been discussed in the preceding; an adequate answer will require many years of multilevel investigation. Question 3B is, What is the functional/behavioral significance of tuning plasticity? We turn to consideration of this issue.

7.2 Perceptual Learning and Learning/Memory ("Rapid Associative Learning")

There are two major hypothesized functions, which originate from two disciplines, "perceptual learning" and "learning/memory." However, this terminology is confusing because perceptual learning must be a subclass of learning/memory. The awkward terminology reflects the historically separate developments of the sensory sciences and the field of learning/memory. The term "learning/memory" has been used previously in this chapter without distinguishing types of learning, because emphasis was placed on finding common ground. However, the distinction can no longer be avoided. Therefore, I will use the term "rapid associative learning" to distinguish the initial type of acquisition and storage of information that develops, for example, during classical conditioning.

7.2.1 Perceptual Learning

One proposed function of cortical plasticity emanates from the field of perception in general and auditory perception in particular. Given its focus on the nature and capacity of acoustic processing, plasticity is viewed as providing for greater perceptual acuity. Reductionistic analysis of behavioral perceptual learning seeks underlying neuronal mechanisms. For example, Recanzone et al. (1993) found that improvement in frequency discrimination in monkeys was accompanied by increased representation of relevant frequency bands. Of note, a fundamental characteristic of perceptual learning, and of the physiological studies of Recanzone et al. is that the effects often involve thousands of trials over days or weeks. A recent study of human discrimination perceptual learning, also revealing frequency-specific effects, involved 4,000–5,000 trials (Irvine et al. 2000). Auditory perceptual learning obtained for a variety of acoustic parameters, may be somewhat specific to the training parameters (e.g., Cansino and Williamson 1997; Tremblay et al. 1998; Irvine et al. 2000).

Perceptual learning experiments often reveal two stages of learning. "Fast" learning is evident within the first training session, requiring only minutes; "slow" learning is triggered by practice, requires a latent period of hours to become manifest, and is incremental over days of training and large numbers of trials. The two types of learning may reflect different processes: "fast" indexes learning the routine needed to solve the particular task while "slow" indexes increased acuity due to remodeling of the processing system itself. The prolonged amount of training ("slow" learning) necessary to attain high levels of improvement in perception is characteristic of *skill learning* (Karni and Bertini 1997).

7.2.2 Rapid Associative Learning

A second hypothesized function of AI plasticity originates in the field of the neurobiology of learning and memory. Given the focus on mnemonic processes, plasticity is seen as evidence that at least some acoustic experience is stored within the auditory cortex. Supporting data include the findings that receptive field plasticity during classical and instrumental one-tone and two-tone discrimination training possesses all of the major characteristics of associative memory (reviewed in Section 5.3). In contrast to cortical changes in perceptual learning, RF plasticity in classical conditioning develops rapidly, in as few as five trials (Edeline et al. 1993). CS-specific plasticity in the human also develops rapidly, within a single session of two-tone discrimination training (Morris et al. 1998).

Of course, not all of the aspects of an acoustic experience are stored in AI. For example, the physical context or place of the experience as well as the detailed nature of reinforcers (e.g., food) undoubtedly involve their relevant sensory systems as well as systems dealing with motivational and other state factors. The "total memory" of an acoustic experience (as with any experience) most likely involves a highly distributed network involving both multiple cortical and multiple subcortical circuits and neural elements and transmitter.

Within the auditory cortex, tuning shifts toward/to the CS frequency and the resultant increased representation are hypothesized to provide a *Memory Code* for the acquired behavioral significance of sound, viz., the greater the importance of a frequency, the larger the area that becomes tuned to that frequency. That is, under at least some circumstances, the brain stores the acquired significance of sensory experiences by increasing the number of neurons that become tuned to the relevant stimulus. Such a memory code could explain aspects of selective attention, for example, why one is more likely to hear one's own name in a noisy room than a random name. For example, while background noise may be controlling many neurons, having "spare" neurons tuned to our names increased the probability that the name will engage some of the "spares." (Of course, this example is for illustrative purposes only, there are no "name" cells, but the principle holds for networks or cell ensembles that may be "tuned" to our names.) Also, it could explain why the loss of memory in aging and brain degeneration is less severe for the most important memories of a lifetime; as the memory is represented by more cells, important information has a "safety factor" in numbers (see Weinberger 2001a for details). Alleviation of deafness by electrical stimulation of the inner ear (cochlear implants) involves new learning and thus the Memory Code hypothesis predicts increased cortical representation of important learned sounds in such individuals. Initial animal studies support this hypothesis; the degree of increased representation of a frequency is directly proportional to the level of its behavioral importance (Rutkowski et al. 2002).

7.2.3 The Necessary Distinction Between Perceptual Learning and Rapid Associative Learning

It would appear odd to emphasize distinctions between perceptual learning and rapid associative learning because they both involve plasticity in AI and this plasticity is compatible. Thus, both involve facilitated processing of behaviorally important frequencies.

However, this distinction is important both on practical and basic neurobio-logical grounds. Regarding the former, most neuroscientists who study learning and memory regard all learning effects within sensory cortices as *perceptual learning*. This "rush to judgment" avoids the apparent embarrassment of in-volving sensory cortex in "genuine memory," thus "saving higher cortical areas" for "genuine memory." The situation is exacerbated by the findings that AI is involved in both perceptual learning and rapid associative learning. Therefore, we need to address the differences.

The neurobiological distinction can be approached by asking, After a bout of perceptual learning, what is changed in auditory cortex? Perhaps after percep-tual learning, the "machinery" of AI has been altered. The AI "machine" now analyzes the same physical stimulus differently, for example, at a finer grain. More neurons become tuned to behaviorally important frequencies and these neurons develop narrower bandwidths, that is, become more sharply tuned (Re-canzone et al. 1993).

This certainly constitutes a type of learning by any definition. But, interest-ingly, the term "perceptual memory" does not accompany "perceptual learning." This makes sense if improvement in sensory analysis alters the gateway to mem-ory but increased acuity by itself is not necessary for memory as the term is normally understood, that is, as the "contents" of experience. That is, the "anal-ysis machinery" need not be improved for information to be acquired and stored at whatever level of auditory acuity that currently characterizes A1. For ex-ample, RF plasticity is induced in the range of 5–30 trials compared to the thousands of trials usually needed for perceptual learning. Clearly, RF plasticity, which emphasizes the frequency of the CS, occurs long before actual improve-ments in acuity develop for that particular frequency domain. However, the extant level of acuity, which has been shaped by genetics and prior perceptual learning, will determine the *precision* with which the information is analyzed, encoded and can be stored. In short, whereas rapid associative learning produces mnemonic content, literally "rapidly-acquired associative *memory*," perceptual learning per se is not a mechanism for storing content memory. But it does limit what can be stored. If the machinery of the auditory cortex cannot distin-guish between two frequencies, then from the standpoint of the cortex, they are the same although not physically identical. (Of course, the eighth nerve is the first gateway to what may be discriminated, learned and stored, but itself is not the locus of plasticity that could encode auditory memories as far as is known.)

"Rapid associative learning" is appropriate for stimulus–stimulus associative memory that, as its name states, develops in a relative small number of trials. After exposure to stimulus–stimulus relationships, the auditory cortex has both acquired the association (that the CS sound signals another event, e.g., food) and also stored the fact of its importance as well. In short, these rapidly de-veloping associations provide some of the new *content* that comprises the "stuff" of memory. On the other hand, perceptual learning is thought to involve mod-ification of the machinery by which the auditory cortex can distinguish sounds and thereby learn about them and remember them.

7.2.4 Relationship Between Rapid Associative Learning and Perceptual Learning

The foregoing suggests that rapid associative learning and perceptual learning may reflect two stages in the modification of information processing in the auditory cortex. Let's consider the main aspects of an experiment in which animals are trained to discriminate between two tones. If the subject responds (e.g., depresses a bar) to the CS+, it receives food; if it responds to the CS−, it does not and receives a "time out," thereby increasing the time until the next CS+ is likely to appear. Aside from learning the general rules of the situation ("this is a place where food is available," "food sometimes comes after the bar is pressed," etc.), what kinds of auditory learning would be taking place?

An initial result would be learning that the CS+ is associated with food, that is, the CS+ signals food contingent upon a bar press. This rapid associative learning would be accompanied by the development of CS+ specific RF plasticity, including tuning shifts toward and to the frequency of the CS+. Perhaps nothing further would develop, that is, there would be no perceptual learning, unless the subject were challenged to make finer discriminations between the tones. If the experimenters institute this requirement, there might be further shifts of tuning, so that even more of AI preferentially becomes tuned to the CS+. Regardless of the detailed time course of tuning shifts (we known they can begin within five trials), perceptual learning would develop only in the course of a challenging discrimination training situation. Unchallenged, the auditory cortex would be unlikely to increase its frequency acuity.

7.2.5 Some Misconceptions Concerning Functions of the Primary Auditory Cortex

Some auditory neuroscientists wonder if the auditory cortex has any auditory function because lesions of AI may seem to have no effect on measures chosen for assessment, such as frequency discrimination. However, Talwar et al. (2001) found that reversible cortical inactivation yields not merely perceptual deficits, but even deafness. They point out that chronic lesions may not reveal the normal functions of the area that has been destroyed because of recovery processes.

Some learning/memory neuroscientists wonder if the auditory cortex has any learning/memory function because it appears to be "non-essential for classical conditioning," for example, the establishment of tone–eyeblink conditioned responses. However, this is based on an erroneously narrow understanding of classical conditioning, which first involves the learning of relationships between the CS and the US and never is simply the learning of a particular stimulus–response link (see Section 3.3). Learning that a particular sound has gained increased behavioral importance permits adaptive behavior that transcends any particular set of muscle contractions. Unless organisms are "interrogated" appropriately, AI functions will not be revealed.

In the cases of both perceptual learning and rapid associative learning involving AI, the cortical plasticity should be viewed broadly as subserving far more than an immediate behavioral response to an acoustic stimulus. Both the finer grain analysis provided by the former and the storage of content by the latter should provide substrates for adaptive behavior for the indefinite future, in whatever situations organisms may find themselves.

7.2.6 Summary

In summary, the findings to date indicate that two types of learning are involved in the functional reorganization of the primary auditory cortex. In the course of perceptual learning, subjects are thought to first employ rapid associative learning in which the content of experience is encoded. Rapid associative learning should be more common, as it is more easily established and should develop in the course of training that produces perceptual learning. Happily, the two types of learning are compatible at the neural level, although they have somewhat different functions for the organism. In fact, the plasticity involved is more than merely compatible because perceptual learning builds on the foundation of the receptive field plasticity induced in rapid associative learning. In so doing, perceptual learning literally provides for "fine tuning" of the auditory cortex. This hypothetical schema is open to experimental test.

8. General Summary

The primary auditory cortex had been known to develop changes in response to sounds that become behaviorally important during learning. However, the relationship of such plasticity to the processing and representation of information was unknown until experiments initiated in the mid 1980s investigated the specificity of learning-induced neural plasticity. An analysis of the advantages and disadvantages of various experimental designs indicates that experience-dependent plasticity is best understood by comparing receptive fields after a learning experience with receptive fields before the experience. In contrast, recordings obtained during training trials, that is, the learning experience itself, are subject to confounding "performance (state) factors" and hence do not merely reveal the actual nature of learning-induced plasticity. Animal studies of metabolic activity, and particularly of receptive field and map plasticity, using "Pre–Post" training designs (or a related variant, the "Post" design) revealed that learning induces highly specific changes in the functional organization of AI. Responses to, for example, conditioned stimuli are increased whereas responses to other frequencies often decrease. The opposing changes are often sufficient to shift tuning toward or even to the frequency of the conditioned stimulus, thereby increasing the area of representation of such signal stimuli in AI. Receptive field plasticity has all of the major characteristics of a substrate of memory, that is, associativity, specificity, rapid induction, consolidation, and very long term retention. This plasticity develops in various animals and in humans, in several different tasks and with both appetitive and aversive motivational (reinforcement) situations.

Studies of mechanisms have focused on determination of the involvement of subcortical auditory nuclei, "Hebbian" (cellular covariance) processes, and the NB cholinergic system. While the medial geniculate and inferior colliculus may be involved in the initiation of AI plasticity, they are not responsible for its maintenance, which has been tracked for up to 8 weeks. While there is some positive evidence that plasticity involves cellular covariance processes, there is insufficient information to draw conclusions. The evidence for involvement of the NB cholinergic projections to AI is consistent and strong, including its ability to induce receptive field and map plasticity when NB stimulation is paired with a preceding tone and its ability to induce specific behavioral memory under the same circumstances.

Current findings support the view that experience-dependent plasticity in AI has two functions: perceptual learning and rapid associative learning. The latter is thought to rapidly (5–30 trials) shift tuning, increase representational area and encode the "contents" of experience. One function may be to comprise a *Memory Code*" for the learned importance of sound, the greater the importance, the greater the number of neurons that become tuned to the stimulus. Perceptual learning appears to increase the acuity of AI after prolonged training, possibly by further increasing the area representing important frequencies but more critically by reducing the bandwidth (sharpening tuning) within the increased area of representation.

9. A Perspective in Closing

A theme that permeates research on experience-dependent plasticity, as it does this chapter, is that of dissolving disciplinary walls and boundaries. The fields of auditory physiology and the neurobiology of learning/memory traditionally have trod parallel, nonintersecting paths. Their separate routes were set in the 19th century, and continued undiverted through perhaps the first 80 or more years of the 20th century. They produced the very dominant, if implicit, assumption and belief within auditory physiology that the auditory cortex is a "sound analyzer" but not a "learning machine." They produced the same conclusion within the neurobiology of learning and memory, which added only that the hippocampus and nonsensory areas of the cortex are the main "learning machines." But these disciplinary boundaries are distinctions that the brain does not honor. We should accept this long-standing "decision" of Nature and proceed to draw upon the best of both fields as neuroscientific problem solvers.

There are two ways of viewing the involvement of learning in the functions of the auditory cortex. The first is to regard learning as another parameter, along with stimulus parameters, for example, frequency, sound level. Within this framework, there may be an implicit assumption that auditory function can be satisfactorily understood without regard to behavior, so that learning is regarded merely as modulator of "normal" function but not endemic to auditory processing. The second is that as the waking, behaving brain is the brain that evolved, an adequate account of the auditory system requires investigation within the context of waking, behaving organisms. Unfortunately, this would greatly complicate matters, rendering much experimentation very difficult and given current techniques, many studies would be impossible. But when has Nature organized herself for the benefit of neuroscientists?

Between these positions, we can note both an intermediate stance and a marked irony. The stance is that we should use all the techniques at our disposal (and invent some better ones) while determining the extent to which auditory function as understood in the anesthetized state or reduced preparation is valid for the behaving organism. A scan of the literature indicates that this stance receives far more lip service than serious application. The degree of importance of behavioral state, or lack thereof, for the auditory system is undoubtedly different at different levels of the neuraxis, but it certainly must be taken into account for the auditory cortex.

The irony is that without the foundational studies of auditory system function in the anesthetized state, neurophysiological studies of learning and memory, whether perceptual or rapid associative in nature, could not have been initiated with any hope of achievement. If nothing else, that is why we should continue to actively pursue a broadly integrative approach. When our field produces the first generation, or at least a substantial and sustaining cohort, of neuroscientists adept at conceptual and experimental research in both auditory and behavioral neuroscience, we will be in a far better position to understand the auditory system than we might possibly imagine.

And in retrospect, where might auditory information be acquired and where should auditory memories dwell, if not within the only brain system that has the organization and ability to precisely analyze, distinguish, and identify sound?

Acknowledgments. This research was supported by grants DC-02346 and DC-02938 from NIDCD and MH-57235 to N.M.W.) We thank Jacquie Weinberger for preparation of the manuscript and Gabriel Hui for preparation of the figures. I am grateful to Dr. Jean-Marc Edeline for his critical comments but all errors are completely the responsibility of the author.

References

Ahissar E, Vaadia E, Ahissar M, Bergman H, et al. (1992) Dependence of cortical plasticity on correlated activity of single neurons and on behavioral context. Science 257: 1412–1415.

Ahissar E, Haidarliu S, Shulz DE (1996) Possible involvement of neuromodulatory systems in cortical Hebbian-like plasticity. J Physiol 90:353–360.

Ahissar E, Abeles M, Ahissar M, Haidarliu S, et al. (1998) Hebbian-like functional plasticity in the auditory cortex of the behaving monkey. Neuropharmacology 37:633–655.

Armony JL, Quirk GJ, LeDoux JE (1998) Differential effects of amygdala lesions on early and late plastic components of auditory cortex spike trains during fear conditioning. J Neurosci 18:2592–2601.

Ashe JH, McKenna TM, Weinberger NM (1989) Cholinergic modulation of frequency receptive fields in auditory cortex: II. Frequency-specific effects of anticholinesterases provide evidence for a modulatory action of endogenous ACh. Syn 4:45–54.

Bakin JS, Weinberger NM (1990) Classical conditioning induces CS-specific receptive field plasticity in the auditory cortex of the guinea pig. Brain Res 536:271–286.

Bakin JS, Weinberger NM (1996) Induction of a physiological memory in the cerebral cortex by stimulation of the nucleus basalis. Proc Natl Acad Sci USA 93:11219–11224.

Bakin JS, Lepan B, Weinberger NM (1992) Sensitization induced receptive field plasticity in the auditory cortex is independent of CS-modality. Brain Res 577:226–235.

Bakin JS, South DA, Weinberger NM (1996) Induction of receptive field plasticity in the auditory cortex of the guinea pig during instrumental avoidance conditioning. Behav Neurosci 110:905–913.

Bao S, Chan VT, Merzenich MM (2001) Cortical remodelling induced by activity of ventral tegmental dopamine neurons. Nature 412:79–83.

Baust W, Berlucchi G, Moruzzi G (1964) Changes in the auditory input during arousal in cats with tenotomized middle ear muscles. Arch Ital Biol 102:675–685.

Bigl V, Woolf NJ, Butcher LL (1982) Cholinergic projections from the basal forebrain to frontal, parietal, temporal, occipital, and cingulate cortices: a combined fluorescent tracer and acetylcholinesterase analysis. Brain Res Bull 8:727–749.

Birt D, Nienhuis R, Olds ME (1979) Separation of associative from non-associative short latency changes in medial geniculate and inferior colliculus during differential conditioning and reversal in rats. Brain Res 167:129–138.

Bjordahl TS, Dimyan MA, Weinberger NM (1998) Induction of long term receptive field plasticity in the auditory cortex of the waking guinea pig by stimulation of the nucleus basalis. Behav Neurosci 112:467–479.

Blozovski D, Hennocq N (1982) Effects of antimuscarinic cholinergic drugs injected systemically or into the hippocampo-entorhinal area upon passive avoidance learning in young rats. Psychopharmacology 76:351–358.

Butt AE, Testylier G, Dykes RW (1997) Acetylcholine release in rat frontal and somatosensory cortex is enhanced during tactile discrimination learning. Psychobiol 25:18–33.

Cahill L, Vazdarjanova A, Setlow B (2000) The basolateral amygdala complex is involved with, but is not necessary for, rapid acquisition of Pavlovian "fear conditioning." Eur J Neurosci 12:3044–3050.

Cahill L, McGaugh JL, Weinberger NM (2001) The neurobiology of learning and memory: some reminders to remember. Trends Neurosci 24:578–581.

Cansino S, Williamson SJ (1997) Neuromagnetic fields reveal cortical plasticity when learning an auditory discrimination task. Brain Res. 764:53–66.

Casamenti F, Deffenu G, Abbamondi A, Pepeu G (1986) Changes in cortical acetylcholine output induced by modulation of the nucleus basalis. Brain Res Bull 16:689–695.

Celesia GG, Jasper HH (1966) Acetylcholine released from cerebral cortex in relation to state of activation. Neurology 16:1053–1063.

Chernychev BV, Weinberger NM (1998) Acoustic frequency tuning of neurons in the basal forebrain of the waking guinea pig. Brain Res 793:79–94.

Cohen YE, Knudsen EI (1999) Maps versus clusters: different representations of auditory space in the midbrain and forebrain. Trends Neurosci 22:128–135.

Condon CD, Weinberger NM (1991) Habituation produces frequency-specific plasticity of receptive fields in the auditory cortex. Behav Neurosci 105:416–430.

Cranefield PF (1974) The Way In and The Way Out. Francois Magendie, Charles Bell and the Roots of the Spinal Nerves with a Facsimile of Charles Bell's Annotated Copy of His Idea of a New Anatomy of the Brain. Mount Kisco, NY: Futura.

Cruikshank SJ, Weinberger, NM (1996) Receptive field plasticity in adult auditory cortex induced by Hebbian covariance. J Neurosci 16:861–875.

Cruikshank SJ, Weinberger, NM (2001) In vivo hebbian and basal forebrain stimulation treatment responses in morphologically identified auditory cortical cells. Brain Res 891:78–93.

Cruikshank SJ, Edeline JM, Weinberger NM (1992) Stimulation at a site of auditory-somatosensory convergence in the medial geniculate nucleus is an effective unconditioned stimulus for fear conditioning. Behav Neurosci 106:471–483.

Deutsch JA (1971) The cholinergic synapse and the site of memory. Science 174:788–794.

Diamond DM, Weinberger NM (1984) Physiological plasticity of single neurons in auditory cortex of cat during acquisition of the pupillary conditioned response. II. Secondary field (AII) Behav Neurosci 98:189–210.

Diamond DM, Weinberger NM (1986) Classical conditioning rapidly induces specific changes in frequency receptive fields of single neurons in secondary and ventral ectosylvian auditory cortical fields. Brain Res 372:357–360.

Diamond DM, Weinberger NM (1989) The role of context in the expression of learning-induced plasticity of single neurons in auditory cortex. Behav Neurosci 103:471–494.

Dimyan MA, Weinberger NM (1999) Basal forebrain stimulation induces discriminative receptive field plasticity in auditory cortex. Behav Neurosci 113:691–702.

Dinse HR, Recanzone GH, Merzenich MM (1993) Alterations in correlated activity parallel ICMS-induced representational plasticity. NeuroReport 5:173–176.

Duque A, Balatoni B, Detari L, Zaborszky L (2000) EEG correlation of the discharge properties of identified neurons in the basal forebrain. J Neurophysiol 84:1627–1635.

Dykes RW (1997) Mechanisms controlling neuronal plasticity in somatosensory cortex. Can J Physiol Pharm 75:535–545.

Edeline J-M (1999) Learning-induced physiological plasticity in the thalamo-cortical sensory systems: a critical evaluation of receptive field plasticity, map changes and their potential mechanisms. Prog Neurobiol 57:165–224.

Edeline J-M, Weinberger NM (1991) Thalamic short term plasticity in the auditory system: associative retuning of receptive fields in the ventral medial geniculate body. Behav Neurosci 105:618–639.

Edeline J-M, Weinberger NM (1992) Associative retuning in the thalamic source of input to the amygdala and auditory cortex: receptive field plasticity in the medial division of the medial geniculate body. Behav Neurosci 106:81–105.

Edeline J-M, Weinberger NM (1993) Receptive field plasticity in the auditory cortex during frequency discrimination training: selective retuning independent of task difficulty. Behav Neurosci 107:82–103.

Edeline J-M, Neuenschwander-El Massioui N, Dutrieux G (1990a) Frequency-specific cellular changes in the auditory system during acquisition and reversal of discriminative conditioning. Psychobio 18:382–393.

Edeline J-M, Neuenschwander-El Massioui N, Dutrieux G (1990b) Discriminative long-term retention of rapidly induced multiunit changes in the hippocampus, medial geniculate and auditory cortex. Behav Brain Res 39:145–155.

Edeline J-M, Pham P, Weinberger NM (1993) Rapid development of learning-induced receptive field plasticity in the auditory cortex. Behav Neurosci 107:539–551.

Edeline J-M, Hars B, Maho C, Hennevin E (1994a) Transient and prolonged facilitation of tone-evoked responses induced by basal forebrain stimulation in the rat auditory cortex. Exp Brain Res 96:373–386.

Edeline J-M, Maho C, Hars B, Hennevin E (1994b) Nonawaking basal forebrain stimulation enhances auditory cortex responsiveness during slow-wave sleep. Brain Res 636:333–337.

Everitt BJ, Robbins TW (1997) Central cholinergic systems and cognition. Annu Rev Psychol 48:649–684.

Fisher A, Brandeis R, Chapman S, Pittel Z, et al. (1998) M1 muscarinic agonist treatment reverses cognitive and cholinergic impairments of apolipoprotein E-deficient mice. J Neurochem 70:1991–1997.

Flood JF, Landry DW, Jarvik ME (1981) Cholinergic receptor interactions and their effects on long-term memory processing. Brain Res 215:177–185.

Gabriel M, Saltwick SE, Miller JD (1975) Conditioning and reversal of short-latency multiple-unit responses in the rabbit medial geniculate nucleus. Science 189:1108–1109.

Galambos R, Sheatz G, Vernier VG (1955) Electrophysiological correlates of a conditioned response in cats. Science 123:376–377.

Galván VV, Weinberger NM (2002) Long-term consolidation and retention of learning: induced tuning plasticity in the auditory cortex of the guinea pig. Neurobiol Learn Mem 77:78–108.

Galván VV, Chen J, Weinberger NM (2001) Long term frequency tuning of local field potentials in the auditory cortex of the waking guinea pig. JARO 2:199–215.

Gao E, Suga N (2000) Experience-dependent plasticity in the auditory cortex and the inferior colliculus: role of the corticofugal system. Proc Natl Acad Sci USA 97:8081–8086.

Gerren R, Weinberger NM (1983) Long term potentiation in the magnocellular medial geniculate nucleus of the anesthetized cat. Brain Res 265:138–142.

Gilbert C, Ito M, Kapadia M, Westheimer G (2000) Interactions between attention, context, and learning in primary visual cortex. Vision Res 40:1217–1226.

Gluck H, Rowland V (1959) Defensive conditioning of electrographic arousal with delayed and differentiated auditory stimuli. Electroencephalogr Clin Neurophysiol 11:485–491.

Gonzalez-Lima F (1992) Brain imaging of auditory learning functions in rats: studies in fluorodeoxyglucose autoradiograph and cytochrome oxidase histochemistry. In: Gonzalez-Lima F, Finenstadt Th, Scheich H (eds), Advances in Metabolic Mapping Techniques for Brain Imaging of Behavioral and Learning Functions. NATO ASI Series D, vol. 68. Boston/London: Kluwer Academic, pp. 39–109.

Gonzalez-Lima F, Scheich H (1984a) Neural substrates for tone-conditioned bradycardia demonstrated with 2-deoxyglucose: I. Activation of auditory nuclei. Behav Brain Res 14:213–233.

Gonzalez-Lima F, Scheich H (1984b) Classical conditioning enhances auditory 2-deoxyglucose patterns in the inferior colliculus. Neurosci Lett 51:79–85.

Gonzalez-Lima F, Scheich H (1986) Neural substrates for tone-conditioned bradycardia demonstration with 2-deoxyglucose: II. Auditory cortex plasticity. Behav Brain Res 20:281–293.

Gower AJ (1987) Enhancement by secoverine and physostigmine of retention of passive avoidance response in mice. Psychopharmacology 91:326–329.

Gutierrez H, Gutierrez R, Silva-Gandarias R, Estrada J, et al. (1999) Differential effects of 192IgG-saporin and NMDA-induced lesions into the basal forebrain on cholinergic activity and taste aversion memory formation. Brain Res 834:136–141.

Hall RD, Mark G (1967) Fear and the modification of acoustically evoked potentials during conditioning. J Neurophysiol 30:893–910.

Hars B, Maho C, Edeline J-M, Hennevin E (1993) Basal forebrain stimulation facilitates tone-evoked response in the auditory cortex of awake rat. Neuroscience 56:61–74.

Hennevin E, Maho C, Hars B, Dutrieux, G (1993) Learning-induced plasticity in the medial geniculate nucleus is expressed during paradoxical sleep. Behav Neurosci 107: 1018–1030.

Introini-Collison IB, McGaugh JL (1988) Modulation of memory by post-training epinephrine: Involvement of cholinergic mechanisms. Psychopharmacology 94:379–385.

Irvine DR, Rajan R (1996) Injury and use-related plasticity in the primary sensory cortex of adult mammals: possible relationship to perceptual learning. Clin Exp Pharm Physiol 23:939–947.

Irvine DR, Webster WR (1972) Arousal effects on cochlear potentials: investigation of a two-factor hypothesis. Brain Res 39:109–119.

Irvine DR, Martin RL, Klimkeitt E, Smith R (2000) Specificity of perceptual learning in a frequency discrimination task. J Acoust Soc Am 108:2964–2968.

Iwata J, LeDoux JE, Meeley MP, Arneric S, et al. (1986) Intrinsic neurons in the amygdaloid field projected to by the medial geniculate body mediate emotional responses conditioned to acoustic stimuli. Brain Res 383:195–214.

Ji W, Gao E, Suga N (2001) Effects of acetylcholine and atropine on plasticity of central auditory neurons caused by conditioning in bats. J Neurophysiol 86:211–225.

Jimenez-Capdeville ME, Dykes RW (1996) Changes in cortical acetylcholine release in the rat during day and night: differences between motor and sensory areas. Neuroscience 71:567–579.

Johnston MV, McKinney M, Coyle JT (1979) Evidence for a cholinergic projection to neocortex from neurons in the basal forebrain. Proc Nat Acad Sci USA 76:5392–5396.

Juckel G, Hegerl U, Molnar M, Csepe V, et al. (1999) Auditory evoked potentials reflect serotonergic neuronal activity—a study in behaving cats administered drugs acting on 5-HT1A autoreceptors in the dorsal raphe nucleus. Neuropsychopharmacology 21: 710–716.

Karni A, Bertini G (1997) Learning perceptual skills: behavioral probes into adult cortical plasticity. Curr Opin Neurobiol 7:530–535.

Kilgard MP, Merzenich MM (1998) Cortical map reorganization enabled by nucleus basalis activity. Science 279:1714–1718.

Kilgard MP, Pandya PK, Vazquez J, Gehi A, et al. (2001) Sensory input directs spatial and temporal plasticity in primary auditory cortex. J Neurophysiol 86:326–338.

Kisley MA, Gerstein GL (1999) Long term variation of frequency response curves recorded from neuronal populations of auditory cortex: random variability or plasticity? Soc Neurosci Abstr 25:392.

Kisley MA, Gerstein GL (2001) Daily variation and appetitive conditioning-induced plasticity of auditory cortex receptive fields. Eur J Neurosci 13:1993–2003.

Kitzes LM, Farley GR, Starr A (1978) Modulation of auditory cortex unit activity during the performance of a conditioned response. Exp Neurol 62:678–697.

Kraus N, McGee TJ, Koch DB (1998) Speech sound representation, perception, and plasticity: a neurophysiologic perspective. Audiol Neuro-Otol 3:168–182.

Kurosawa M, Sato A, Sato Y (1989) Stimulation of the nucleus basalis of Meynert increases acetylcholine release in the cerebral cortex in rats. Neurosci Lett 98:45–50.

Lehmann J, Nagy JI, Atmadia S, Fibiger HC (1980) The nucleus basalis magnocellularis: the origin of a cholinergic projection to the neocortex of the rat. Neuroscience 5: 1161–1174.

Lennartz RC, Weinberger NM (1992a) Frequency-specific receptive field plasticity in the medial geniculate body induced by Pavlovian fear conditioning is expressed in the anesthetized brain. Behav Neurosci 106:484–497.

Lennartz RC, Weinberger NM (1992b) Analysis of response systems in Pavlovian conditioning reveal rapidly vs. slowly acquired conditioned responses: support for two-factors and implications for neurobiology. Psychobiol 20:93–119.

LoConte G, Bartolini L, Casamenti F, Marconcini-Pepeu I et al. (1982) Lesions of cholinergic forebrain nuclei: changes in avoidance behavior and scopolamine actions. Pharm Biochem Behav 17:933–937.

Lund RD, Webster KE (1967) Thalamic afferents from the spinal cord and trigeminal nuclei. An experimental anatomical study in the rat. J Comp Neurol 130:313–328.

Mackintosh NJ (1974) The Psychology of Animal Learning. London: Academic Press.

Maho C, Hennevin E (2002) Appetitive conditioning-induced plasticity is expressed during paradoxical sleep in the medial geniculate but not in the lateral amygdala. Behav Neurosci 116:807–823.

Maho C, Hars B, Edeline J-M, Hennevin E (1995) Conditioned changes in the basal forebrain: relations with learning-induced cortical plasticity. Psychobiol 23:10–25.

Maldonado PE, Gerstein GL (1996) Neuronal assembly dynamics in the rat auditory cortex during reorganization induced by intracortical microstimulation. Exp Brain Res. 112:431–441.

Manunta Y, Edeline J-M (1997) Effects of noradrenaline on frequency tuning of rat auditory cortex neurons. Eur J Neurosci 9:833–847.

Maren S, Yap SA, Goosens KA (2001) The amygdala is essential for the development of neuronal plasticity in the medial geniculate nucleus during auditory fear conditioning in rats. J Neurosci 21:1–6.

McEchron MD, McCabe PM, Green EJ, Llabre MM, et al. (1995) Simultaneous single unit recording in the medial nucleus of the medial geniculate nucleus and amygdaloid central nucleus throughout habituation, acquisition, and extinction of the rabbit's classically conditioned heart rate. Brain Res 682:157–166.

McEchron MD, Green EJ, Winters RW, Nolen TG, et al. (1996) Changes of synaptic efficacy in the medial geniculate nucleus as a result of auditory classical conditioning. J Neurosci 16:1273–1283.

McKenna TM, Ashe JH, Weinberger NM (1989) Cholinergic modulation of frequency receptive fields in auditory cortex: I. Frequency-specific effects of muscarinic agonists. Synapse 4:30–44.

McKernan MG, Shinnick-Gallagher P (1997) Fear conditioning induces a lasting potentiation of synaptic currents in vitro. Nature 390:607–611.

McLin DE III, Miasnikov AA, Weinberger NM (2002a) Induction of behavioral associative memory by stimulation of the nucleus basalis. Proc Natl Acad Sci USA 99: 4002–4007.

McLin DE III, Miasnikov AA, Weinberger NM (2002b) The effects of electrical stimulation of the nucleus basalis on the electroencephalogram, heart rate and respiration. Behav Neurosci 116:795–806.

McLin DE III, Miasnikov AA, Weinberger NM (2003) CS-specific gamma, theta, and alpha EEG activity detected in stimulus generalization following induction of behavioral memory by stimulation of the nucleus basalis. Neurobiol Learn Mem 79:152–176.

Merzenich MM, Sameshima K (1993) Cortical plasticity and memory. Curr Opin Neurobiol 3:187–196.

Mesulam MM, Mufson EJ, Wainer BH, Levey AI (1983) Central cholinergic pathways in the rat: an overview based on an alternative nomenclature. Neuroscience 114:64–76.

Metherate R, Ashe JH (1991) Basal forebrain stimulation modifies auditory cortex responsiveness by an action at muscarinic receptors. Brain Res 559:163–167.

Metherate R, Ashe JH (1993) Nucleus basalis stimulation facilitates thalamocortical synaptic transmission in the rat auditory cortex. Synapse 14:132–143.

Metherate R, Weinberger NM (1990) Cholinergic modulation of responses to single tones produces tone-specific receptive field alterations in cat auditory cortex. Synapse 6:133–145.

Metherate R, Tremblay N, Dykes RW (1988) Transient and prolonged effects of acetylcholine on responsiveness of cat somatosensory cortical neurons. J Neurophysiol 59:1231–1252.

Miasnikov A, McLin D III, Weinberger NM (2001) Muscarinic dependence of nucleus basalis induced conditioned receptive field plasticity. NeuroReport 12:1537–1542.

Miranda MI, Bermudez-Rattoni F (1999) Reversible inactivation of the nucleus basalis magnocellularis induces disruption of cortical acetylcholine release and acquisition, but not retrieval, of aversive memories. Proc Natl Acad Sci USA 96:6478–6482.

Molchan SE, Sunderland T, McIntosh AR, Herscovitch P, et al. (1994) A functional anatomical study of associative learning in humans. Proc Natl Acad Sci USA 91:8122–8126.

Molnar M, Karmos G, Csepe V, Winkler I (1988) Intracortical auditory evoked potentials during classical aversive conditioning in cats. Biol Psychol 26:339–350.

Moore DR, Schnupp JWH, King AJ (2001) Coding the temporal structure of sounds in auditory cortex. Nature Neurosci 4:1055–1056.

Moriizumi T, Hattori T (1992) Separate neuronal populations of the rat globus pallidus projecting to the subthalamic nucleus, auditory cortex and pedunculopontine tegmental area. Neuroscience 46:701–710.

Morris JS, Friston KJ, Dolan RJ (1998) Experience-dependent modulation of tonotopic neural responses in human auditory cortex. Proc R Soc Lond 265:649–657.

Murata K, Kameda K (1963) The activity of single cortical neurones of unrestrained cats during sleep and wakefulness. Arch Ital Biol 101:306–331.

Oatman LC (1971) Role of visual attention on auditory evoked potentials in unanesthetized cats. Exp Neurol 32:341–356.

O'Connor KN, Allison TL, Rosenfield ME, Moore JW (1997) Neural activity in the medial geniculate nucleus during auditory trace conditioning. Exp Brain Res 113:534–556.

Oh JD, Edwards RH, Woolf NJ (1996) Choline acetyltransferase mRNA plasticity with Pavlovian conditioning. Exp Neurol 140:95–99.

Ohl FW, Scheich H (1996) Differential frequency conditioning enhances spectral contrast sensitivity of units in auditory cortex (field Al) of the alert Mongolian gerbil. Eur J Neurosci 8:1001–1017.

Ohl FW, Scheich H (1997) Learning induced dynamic receptive field changes in primary auditory cortex of the unanaesthetized Mongolian gerbil. J Comp Physiol 181:685–696.

Olds J, Peretz B (1960) A motivational analysis of the reticular activating system. Electroencephalogr Clin Neurophysiol 12:445–454.

Orsetti M, Casamenti F, Pepeu G (1996) Enhanced acetylcholine release in the hippocampus and cortex during acquisition of an operant behavior. Brain Res 724:89–96.

Palmer CV, Nelson CT, Lindley George A IV, (1998) The functionally and physiologically plastic adult auditory system. J Acous Soc Amer 103:1705–1721.

Pennartz CM (1995) The ascending neuromodulatory systems in learning by reinforcement: comparing computational conjectures with experimental findings. Brain Res Rev 21:219–245.

Poremba A, Gabriel M (2001) Amygdalar efferents initiate auditory thalamic discriminative training-induced neuronal activity. J Neurosci 21:270–278.

Potter DD, Pickles CD, Roberts RC, Rugg MD (2000) Scopolamine impairs memory performance and reduces frontal but not parietal visual P3 amplitude. Biol Psych 52: 37–52.

Quirk GJ, Armony JL, LeDoux JE (1997) Fear conditioning enhances different temporal components of tone-evoked spike trains in auditory cortex and lateral amygdala. Neuron 19:613–624.

Rasmusson DD (2000) The role of acetylcholine in cortical synaptic plasticity. Behav Brain Res 115:205–218.

Rasmusson DD, Clow K, Szerb JC (1992) Frequency-dependent increase in cortical acetylcholine release evoked by stimulation of the nucleus basalis magnocellularis in the rat. Brain Res 594:150–154.

Rauschecker, JP (1999) Auditory cortical plasticity: a comparison with other sensory systems. Trends Neurosci 22:74–80.

Recanzone GH, Schreiner CE, Merzenich MM (1993) Plasticity in the frequency representation of primary auditory cortex following discrimination training in adult owl monkeys. J Neurosci 13:87–103.

Rescorla RA (1988) Behavioral studies of Pavlovian conditioning. Annu Rev Neurosci 11:329–352.

Richardson RT, DeLong MR (1986) Nucleus basalis of Meynert neuronal activity during a delayed response task in monkey. Brain Res 399:364–368.

Riekkinen P Jr, Riekkinen M, Sirvio J, Miettinen R, et al. (1992) Loss of cholinergic neurons in the nucleus basalis induces neocortical electroencephalographic and passive avoidance deficits. Neuroscience 47:823–831.

Robertson D, Irvine DR (1989) Plasticity of frequency organization in auditory cortex of guinea pigs with partial unilateral deafness. J Comp Neurol 282:456–471.

Rudy JW (1996) Scopolamine administered before and after training impairs both contextual and auditory-cue fear conditioning. Neurobiol Learn Mem 65:73–81.

Russell RW, Escobar ML, Booth RA, Bermudez-Rattoni F (1994) Accelerating behavioral recovery after cortical lesions: II. In vivo evidence for cholinergic involvement. Behav Neural Biol 61:81–92.

Rutkowski R, Than KH, Weinberger, NM (2002) Evidence for area of frequency representation encoding acquired stimulus importance in rat primary auditory cortex. Soc Neurosci Abstr 28:530.

Ryugo DK, Weinberger NM (1978) Differential plasticity of morphologically distinct neuron populations in the medial geniculate body of the cat during classical conditioning. Behav Biol 22:275–301.

Salinas JA, Introini-Collison IB, Dalmaz C, McGaugh JL (1997) Postraining intraamygdala infusion of oxotremorine and propranolol modulate storage of memory for reductions in reward magnitude. Neurobiol Learn Mem 68:51–59.

Saper CB (1984) Organization of cerebral cortical afferent systems in the rat: II. Magnocellular basal nucleus. J Comp Neurol 222:313–342.

Scheich H (1991) Auditory cortex: comparative aspects of maps and plasticity. Curr Opin Neurobiol 1:236–247.

Scheich H, Stark H, Zuschratter W, Ohl FW, et al. (1997) Some functions of primary auditory cortex in learning and memory. Adv Neurol 73:179–193.

Schreurs BG, McIntosh AR, Bahro M, Herscovitch P, et al. (1997) Lateralization and behavioral correlation of changes in regional cerebral blood flow with classical conditioning of the human eyeblink response. J Neurophysiol 77:2153–2163.

Shulz DE, Cohen S, Haidarliu S, Ahissar E (1997) Differential effects of acetylcholine on neuronal activity and interactions in the auditory cortex of the guinea-pig. Eur J Neurosci 9:396–409.

Sillito AM, Kemp JA (1983) Cholinergic modulation of the functional organization of the cat visual cortex. Brain Res 289:143–155.

Stark H, Scheich H (1997) Dopaminergic and serotonergic neurotransmission systems are differentially involved in auditory cortex learning: a long-term microdialysis study of metabolites. J Neurochem 68:691–697.

Stratton L, Petinovich L (1963) Post-trial injections of an anti-cholinesterase drug and maze learning in two strains of rats. Psychopharmacology 5:47–54.

Talwar SK, Gerstein GL (2001) Reorganization in awake rat auditory cortex by local microstimulation and its effect on frequency-discrimination behavior. J Neurophysiol 86:1555–1572.

Talwar SK, Musial PG, Gerstein GL (2001) Role of mammalian auditory cortex in the perception of elementary sound properties. J Neurophysiol 85:2350–2358.

Teas DC, Kiang NY (1964) Evoked responses from the auditory cortex. Exp Neurol 10: 91–119.

Thompson RF, Tracy, JA (1995) Cerebellar localization of a memory trace. In: McGaugh JL, Bermudez-Rationi F, et al. (eds), Plasticity in the Central Nervous System: Learning. Hillsdale, NJ: Lawrence Erlbaum, pp. 107–127.

Travis RP, Sparks DL (1968) Unitary responses and discrimination learning in the squirrel monkey: the globus pallidus. Physiol Behav 3:187–196.

Tremblay K, Kraus N, McGee T (1998) The time course of auditory perceptual learning: Neurophysiological changes during speech-sound training. NeuroReport 9:3557–3560.

Tunturi AR (1944) Audio frequency localization in the acoustic cortex of the dog. Am J Physiol 141:397–403.

Vanderwolf CH, Cain DP (1994) The behavioral neurobiology of learning and memory: a conceptual reorientation. Brain Res Rev 19:264–297.

Vazdarjanova A (2000) Does the basolateral amygdala store memories for emotional events? Trends Neurosci 23:345.

Weinberger NM (1995) Dynamic regulation of receptive fields and maps in the adult sensory cortex. Annu Rev Neurosci 18:129–158.

Weinberger NM (1998) Physiological memory in primary auditory cortex: characteristics and mechanisms. Neurobiol Learn Mem 70:226–251.

Weinberger NM (2001) Memory codes: a new concept for an old problem. In: Gold P, Greenough W (eds), Memory Consolidation: Essays in Honor of James L. McGaugh. Washington DC: American Psychological Association, pp. 321–342.

Weinberger NM (2004) Specific long-term memory traces in primary auditory cortex. Nature Rev Neurosci 4:279–290.

Weinberger NM, Diamond DM (1987) Physiological plasticity of single neurons in auditory cortex: rapid induction by learning. Prog Neurobiol 29:1–55.

Weinberger NM, Oleson TD, Ashe JH (1975) Sensory system neural activity during habituation of the pupillary orienting reflex. Behav Biol 15:283–301.

Weinberger NM, Gold PE, Sternberg DB (1984a) Epinephrine enables Pavlovian fear conditioning under anesthesia. Science 223:605–607.

Weinberger NM, Hopkins W, Diamond DM (1984b) Physiological plasticity of single neurons in auditory cortex of cat during acquisition of the pupillary conditioned response: I Primary field (AI). Behav Neurosci 98:171–188.

Weinberger NM, Ashe JH, Metherate R, McKenna TM, et al. (1990a) Retuning auditory cortex by learning: a preliminary model of receptive field plasticity. Concept Neurosci 1:91–132.

Weinberger NM, Ashe JH, Metherate R, McKenna TM, et al. (1990b) Neural adaptive information processing: A preliminary model of receptive field plasticity in auditory cortex during Pavlovian conditioning. In: Gabriel M, Moore J. (eds), Neurocomputation and Learning: Foundations of Adaptive Networks. Cambridge, MA: Bradford Books/The MIT Press, pp. 91–138.

Weinberger NM, Javid R, Lepan B (1993) Long-term retention of learning-induced receptive field plasticity in the auditory cortex. Proc Natl Acad Sci USA 90:2394–2398.

Weinberger NM, Javid R, Lepan B (1995) Heterosynaptic long term facilitation of sensory evoked responses in the auditory cortex by stimulation of the magnocellular medial geniculate body. Behav Neurosci 109:10–17.

Wenk GL (1997) The nucleus basalis magnocellularis cholinergic system: one hundred years of progress. Neurobiol Learn Mem 67:85–95.

Wepsic JG (1966) Multimodal sensory activation of cells in the magnocellular medial geniculate nucleus. Exp Neurol 15:299–318.

Westenberg IS, Weinberger NM (1976) Evoked potential decrements in auditory cortex. II. Critical test for habituation. Electroencephalogr Clin Neurophysiol 40:356–369.

Wester K (1971) Habituation to electrical stimulation of the thalamus in unanaesthetized cats. Electroencephalogr Clin Neurophysiol 30:52–61.

Whalen PJ, Kapp BS, Pascoe, JP (1994) Neuronal activity within the nucleus basalis and conditioned neocortical electroencephalographic activation. J Neurosci 14:1623–1633.

Wickelgren WO (1968a) Effect of state of arousal on click-evoked responses in cats. J Neurophysiol 31:757–768.

Wickelgren WO (1968b) Effect of acoustic habituation on click-evoked responses in cats. J Neurophysiol 31:777–784.

Wilson FAW, Rolls ET (1990) Learning and memory is reflected in the responses of reinforcement related in the primate basal forebrain. J Neurosci 10:1254–1267.

Woolsey CN, Walzl EM (1942) Topical projection of nerve fibers from local regions of the cochlea to the cerebral cortex of the cat. Johns Hop Hosp Bull 71:315–344.

Young, RM (1970) Mind, Brain and Adaptation in the Nineteenth Century. Oxford: Clarendon Press.

6

The Avian Song Control System: A Model for Understanding Changes in Neural Structure and Function

ELIOT A. BRENOWITZ AND SARAH M.N. WOOLLEY

1. Introduction

Songs are complexly structured vocalizations that are widely used in communication by birds. Song behavior in the oscine passerines, songbirds, shares many features with human speech (Doupe and Kuhl 1999). Both song and speech are *learned* early in life in these species. There is an intimate relationship between the auditory and song/speech control systems. Young birds and humans must hear adults in order to imitate song/speech, and they show innate predispositions to learn conspecific vocalizations. Song and speech acquisition are characterized by an early perceptual phase in which models of the communication sounds are listened to and memorized. These sensory models, or templates, subsequently guide vocal production. Auditory feedback from an individual bird or human's own vocalizations is necessary for both the development and maintenance of normal song and speech. Neural hierarchies involving projections from forebrain regions to midbrain areas and then to brain stem nuclei regulate both song and speech. For these reasons, birdsong has emerged as the leading animal model for studies of the neural basis of learned vocal communication.

Song and speech are characterized by extensive plasticity, both during development and in adulthood. In this chapter we focus on plasticity of song behavior and the neural circuits that regulate song in birds at different times of life. Comprehensive reviews of other aspects of the song control system can be found elsewhere (in particular, see the special issues of *Journal of Neurobiology* (November 5, 1997) and *Journal of Comparative Physiology (A)* (December 2002) on the neurobiology of birdsong. We begin by providing an introduction to song behavior, the neural song control circuits, and the effects that gonadal steroid hormones have on song. We discuss plasticity in juvenile birds, seen in the development of the song control circuits and the stages of song learning. Selective auditory responses to a bird's own song within the song control system are considered. We review the relationship between changes in auditory feedback and song behavior in adult birds. Plasticity of adult song circuits is discussed, with particular reference to ongoing neurogenesis, as well as seasonal and social factors. We conclude by considering directions for future research.

1.1 Avian Systematics

There are approximately 9,000 species of living birds. About 5,300 species belong to the order Passeriformes, which consists of two suborders, the oscines or songbirds (e.g., finches, sparrows, and blackbirds), and the suboscines, which includes such birds as flycatchers, antbirds, and ovenbirds. The order is considered to be a single evolutionary lineage derived from a common ancestor (i.e., monophyletic), and the oscines and suboscines are each viewed as monophyletic lineages within the Passeriformes (Raikow 1982; Sibley et al. 1988). There are about 4,000 species of songbirds, nearly all of which use vocalizations for communication. Within this large group of species there is extensive diversity in various aspects of vocal behavior including the timing of vocal learning, sex patterns of song production, the number of songs that are learned (i.e., repertoire size), and the seasonality of song behavior. This diversity presents excellent opportunities for comparative studies of the relationship between the structure and function of brain regions and song behavior. In addition to the songbirds, some parrots and hummingbirds have been shown to learn their songs (reviewed in Brenowitz and Kroodsma 1996).

1.2 Song Structure

Songs have well-defined acoustic structures that are unique to each species. The structure of a song can be visualized using a sound spectrograph, which plots the sound frequencies against time (see Fig. 6.1). Song consists of different structural components (reviewed in Brenowitz et al. 1997). The simplest individual sounds are called song "elements" or "notes" while a sequence of one or more notes that occurs consistently is called a song "syllable." A series of one or more syllables that is repeated in a predictable sequence is referred to as a song "phrase" or "motif," and a specific combination of phrases that occurs consistently as a unit is a song "type."

There is great diversity among songbird taxa in the structural complexity of song behavior. In some species, such as the white-throated sparrow (*Zonotrichia albicollis*), a male produces only one simple song type. At the other extreme, a male brown thrasher (*Toxostoma rufum*) has a repertoire of several thousand different song types. Most bird species have song repertoire sizes that fall between these extremes (Brenowitz and Kroodsma 1996).

Songs are generally distinguished from calls. Songs are long (several seconds to minutes) in duration, complexly structured, learned with reference to auditory models, and require auditory feedback. Calls, by contrast, tend to be short, simply structured, usually do not require sensory learning, and can be produced in the absence of auditory feedback.

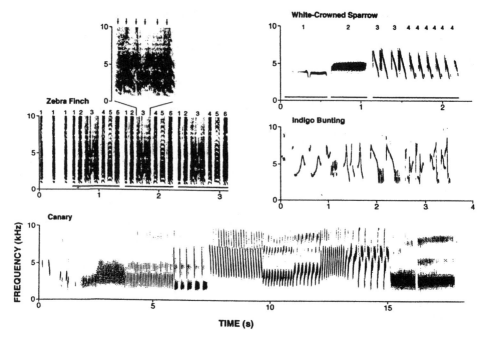

FIGURE 6.1. Examples of birdsongs from four species, presented as sound spectrographs. The zebra finch song starts with introductory notes (see text for discussion of song terminology) and is organized into a motif that can be repeated several times (*bars*). Each motif consists of a sequence of syllables, which are identified by the numbers above the spectrograph. Syllables can contain one or more notes (or elements). As an example, the syllable shown at higher resolution has six notes (denoted by *arrows*). Each note and syllable has a spectrally complex pattern, with either broadband or harmonic structure. The white-crowned sparrow and indigo bunting songs are spectrally simpler and illustrate different temporal patterns. For the white-crowned sparrow song, phrases (*bars*) and syllables (*numbers*) are marked. The song types of the former three species are relatively discrete and short. In contrast, the canary sings continuously for long periods of time. The song consists of a series of phrases, with each phrase having one or two syllables repeated several times. [Reprinted with permission from Brenowitz et al. (1997).]

1.3 The Behavioral Functions of Birdsong

Song serves two main functions in birds (reviewed in Catchpole and Slater 1995). It can play an important role in aggressive behavior, usually between members of the same sex, and such songs are most often found in defense of a territory used for breeding and/or feeding. The second main function of song occurs in the context of courtship. In most songbird species males use song to attract females to their territories. Females may select among potential mates on the basis of individual song characteristics, and the male's song may directly stimulate reproductive behavior in females. In addition to these two main functions, song may be used in other behavioral contexts, such as in mediating dominance behavior among members of a social group.

1.4 Song and the Seasons

Breeding occurs seasonally in most species of birds that live in temperate and subtropical latitudes, at times of the year when the resources necessary for successfully rearing offspring are most abundant. In such species, singing is also seasonal. Males sing at a high rate early in the breeding season when they first establish territories. Once males have mated with one or more females, the rate of song production drops considerably. Outside the breeding season, males may sing only occasionally or not at all.

2. Song Learning in Juveniles and Adults

2.1 Basic Principles of Song Development

The development of song behavior in birds shows several basic characteristics. (1) Songs must be learned by auditory exposure to adult conspecific song. (2) Young birds demonstrate an innate auditory preference for attending to and learning their own species' songs. (3) Song development occurs in a two-stage process of sensory memorization and vocal motor practice that depends critically on hearing. (4) There is an early sensitive period for song acquisition. (5) Some species, called age-limited learners, learn song only during the first year of life. Other species, called open-ended learners, learn new song patterns as adults.

2.2 Song Must Be Learned

Early experimental studies on birdsong indicated that songs must be learned by exposure to adult song. Birds taken into the laboratory in ovo or as nestlings and raised in isolation develop highly abnormal songs lacking the acoustic structure and stereotypy characteristic of normal adult song (Thorpe 1958, 1961; Nottebohm 1968; Marler 1970; Fig. 6.2). Konishi (1965) deafened juvenile white-crowned sparrows *(Zonotrichia leucophrys)* and found that they developed highly abnormal songs. These findings implicated auditory experience specifically as the sensory input required for normal song development. The "isolate" songs of birds raised without hearing a song model and even the songs of deafened birds show some species-typical acoustic features (Thorpe 1958; Konishi 1964, 1965; Price 1979; Marler and Sherman 1983). This observation suggests that innate characteristics of brain circuitry and/or the vocal apparatus predispose each species to produce certain sounds. Birds raised in the laboratory copy the songs of conspecific adults to which they are exposed as juveniles, or

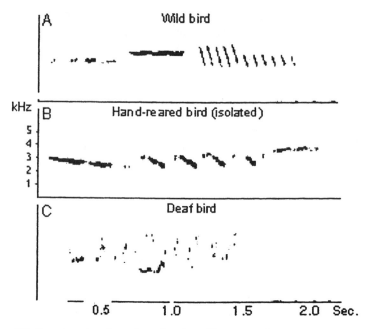

FIGURE 6.2. Song learning depends on hearing. The songs of three male white-crowned sparrows raised with different auditory experience are shown. (A) The song of a wild bird, showing the characteristic structure of white-crowned sparrow song. (B) The song of a bird raised in isolation from other birds and therefore without an adult tutor. The song shows some structural aspects typical of white-crowned sparrow song, but has abnormal phonology. (C) The song of a bird deafened as a juvenile, showing highly abnormal structure. [Modified with permission from Konishi (1985).]

in some cases, even copy songs presented on tape (Thorpe 1958, 1961; Marler 1970; Waser and Marler 1977). Some improvisation occurs such that no two birds' songs are exactly alike, but the pupil develops a song that is clearly attributable to the available song tutor or tape recording (Marler and Waser 1977; Waser and Marler 1977).

Other early evidence that birdsongs are learned came from field studies in which the acoustic features of songs from different populations of the same species were examined. Marler (1952) observed that individuals within distinct populations of chaffinches *(Fringilla ceoleb)* sang songs with similar acoustic features. Those features were not shared between populations. This observation suggested that young birds copy their songs from the previous generation. Marler and Tamura (1964) later showed that white-crowned sparrows around Berkeley, California share song elements within but not between populations, demonstrating the occurrence of song "dialects." Many studies have now shown that song learning is highly dependent on the level of exposure to adult song and the social conditions under which learning occurs (see Tchernichovski and Nottebohm 1998; Tchernichovski et al. 1999; Nordby et al. 2001).

2.3 Innate Preferences for Conspecific Vocalizations

Studies measuring heart rate, begging calls, and perch hops as assays for recognition in young birds indicate that songbirds have the ability to discriminate between their own species' vocalizations and those of other species (Dooling and Searcy 1980; ten Cate 1991; Adret 1993; Nelson and Marler 1993; Whaling et al. 1997). Changes in these measures specifically during presentation of conspecific song suggest that a bird's state of arousal increases when it hears its own species' song. This ability is not limited to songbirds, but does suggest that early auditory preferences may guide song learning. Indeed, active preferences for hearing conspecific songs have been demonstrated specifically during song learning. Juvenile males "choose" (usually by pecking a key or perch hopping near a speaker) to hear playback of conspecific song over other song stimuli during the sensitive period for song learning (ten Cate 1991; Adret 1993; Braaten and Reynolds 1999).

The innate preference for species-typical songs is not only a matter of recognition but also appears to guide a bird's selection of song material for memorization and eventual production. Juvenile songbirds preferentially learn their own species' songs over other song models that are presented in an equivalent manner and with equal frequency (Thorpe 1958; Marler 1970; Marler and Peters 1977). In fact, young birds will rarely acquire the songs of other species, even when the two species' songs are similar. Marler and Peters (1977) found, for example, that swamp sparrows (*Melospiza georgiana*) learn only the songs of their own species and apparently ignore the songs of a species that occupies the same habitat (i.e., sympatric), the song sparrow *(Melospiza melodia)*. This selective learning occurs even though the spectral sensitivities and song spectra of the different species are similar (see Dooling and Searcy 1981; Okanoya and Dooling 1987). In the case of sympatric species, innate preferences for conspecific song could be advantageous for young birds hearing the songs of many species. Such preferences may ensure that juvenile birds sharing acoustic space with multiple species acquire the "correct" song of their own species. Without such predispositions, young birds might learn heterospecific songs, which would prevent them from subsequently mating with a conspecific.

While it is true that juveniles normally acquire conspecific song, exceptions to this rule can be induced experimentally. Young birds can sometimes be convinced to learn heterospecific songs if they are deprived of an appropriate conspecific song model (Price 1979; Baptista and Morton 1981; Baptista and Petrinovich 1984, 1986; Petrinovich and Baptista 1987). Another way that this can happen is by cross-fostering birds such that young males are raised by adults of another species (Immelmann 1969; Slater et al. 1988; Clayton 1989). When birds learn heterospecific song, the song copies tell us three important things about how genetic predispositions and experience interact during song learning. First, the fact that birds can learn the songs of other species indicates that learning preferences do not simply reflect the limitations of the vocal apparatus or motor control circuitry. Instead, it suggests that sensory mechanisms that attach special meaning to certain acoustic stimuli typical of conspecific song may exist for each species. For example, young swamp sparrows will not acquire the song phrases of heterospecific birds if a conspecific song model is available (Marler and Peters 1977). Evidence that song learning may be cued by relatively simple acoustic features characteristic of conspecific song comes from a study by Soha and Marler (2000). They tutored young white-crowned sparrows with modified conspecific songs in which the whistle common to the beginning of white-crowned sparrow song had been removed. The same birds were also tutored with heterospecific songs to which the white-crowned sparrow whistle had been attached. The authors found that the heterospecific songs with white-crowned sparrow whistles attached were learned, while the conspecific and heterospecific songs without the beginning whistle were not learned. These results suggest that an acoustic signal universal to the conspecific song may cue the attention of a young songbird in the process of learning song and facilitate the learning of acoustic material that follows that conspecific signal. Second, the songs of cross-tutored birds are usually somewhat inaccurate copies of the heterospecific tutor's song. This inaccuracy arises because the copied songs contain features (usually temporal) of conspecific song that are not found in the heterospecific tutor's song. This result suggests an innate propensity to produce species-typical vocalizations. Third, the fact that cross-fostering can lead to cross-tutoring suggests that social factors play a role in what song material is acquired. Beecher and colleagues (2000a,b; Nordby et al., 2000) have recently shown that song sparrows in the field or in the laboratory preferentially copy the songs of adult tutors with which they have had the most social interaction.

2.4 Two Stages of Song Learning: Sensory Acquisition and Sensory–Motor Integration

Birds do not simply begin producing well-formed songs. Rather, this skill develops over time and depends critically on hearing. Although the timing of song development varies among species (see later), the sequence of behavioral changes and the experiential requirements for normal song development are similar among species. Song learning can be divided into two stages, which are described here (Fig. 6.3). The model of song learning that first defined these stages has been called the auditory template theory (Konishi 1964, 1965; Marler and Tamura 1964; Marler 1970).

2.4.1 Sensory Acquisition

The first stage of song learning is called sensory acquisition, or memorization. During this stage, juvenile birds listen to and memorize, or form "templates"

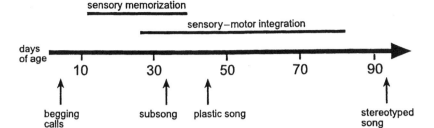

FIGURE 6.3. Song learning occurs in two major stages, sensory memorization and sensory–motor integration. The timing of these stages, and of the different phases of song development, are shown for the zebra finch, an age-limited song learner. In this species memorization overlaps with sensory–motor integration. The relative timing of the learning stages varies among species. The overall sequence of learning stages and of song development are the same across species, however. [Modified with permission from S. Bottjer (2002).]

of, adult songs. At the age most birds first hear the songs of adult conspecifics (this is thought to be around 10 days of age), they themselves produce only begging calls. The evidence for memory formation during this time comes from laboratory studies showing that exposure to a song model before the juvenile sings is sufficient for an accurate copy to be produced later, long after the model has been removed (Thorpe 1958; Marler and Tamura 1964; Waser and Marler 1977). Thus the juvenile must remember the song model until the age at which he develops his own song. In nature, the timing of sensory acquisition may vary among species and individuals such that there is not a long period of time between memorizing and producing song. The relative timing of sensory acquisition and sensory–motor integration is discussed in the following section.

2.4.2 Sensory–Motor Integration

The second stage of song learning is called sensory–motor integration because, during this time, the memorized template is gradually translated into a vocal motor program. It begins when juveniles start to produce song-like vocalizations and ends when birds have reached full song. Full song is characterized by production of a stereotyped, well-formed rendition of the memorized song model and coincides with sexual maturity. Like sensory acquisition, this stage depends critically on hearing because a bird uses auditory feedback to monitor and shape its own vocal output. Deafening birds just before or during sensory–motor integration prevents normal song development, whereas isolation does not (Konishi 1964, 1965; Marler and Waser 1977; Price 1979). The abnormal songs of birds that have been deafened before full song is reached show: (1) a lack of unit organization (notes are not grouped into syllables, syllables are not ordered into motifs); (2) atypical syllable and note structure; and (3) a high degree of variability among different renditions of the same note or syllable type. A correlation exists between the stage of sensory–motor integration at which deafening occurs and the subsequent song abnormalities; birds deafened at later stages produce more normal songs (Konishi and Nottebohm 1969).

Song behavior follows a sequence of developmental phases that is consistent across species (Fig. 6.4). First, young juveniles begin to produce quiet, highly variable, and unstructured vocalizations that are termed subsong. Later, vocal-

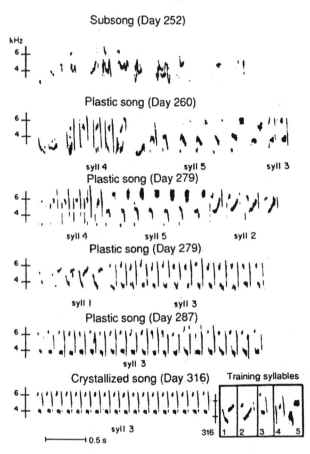

FIGURE 6.4. The different phases of the sensory–motor stage of song development. This series of sound spectrograms shows the development of song in a male swamp sparrow. The syllables with which this juvenile was tutored are shown in a *box* at the bottom right. Subsong is produced first and is characterized by low-intensity and poorly structured sounds. Plastic songs are louder and better structured, but show low stereotypy. A bird's plastic song often contains syllables that will be dropped from the adult repertoire. With continued practice, plastic song progressively matches the tutor's song more closely. Lastly, some syllables are dropped from the bird's repertoire and song stabilizes or crystallizes in its adult form. [Reprinted with permission from Marler and Peters (1982b).]

izations become louder and better structured, and they begin to resemble the species-typical song. These vocalizations are called plastic song because they are much less stereotyped than adult song and contain acoustic elements that will not be retained in the adult song. There is evidence that birds may sing more during this period, potentially to increase vocal practice before the onset of adulthood (Johnson et al. 2002). Over time, birds hone their vocalizations to a series of acoustically well-defined notes and syllables, which are produced in a predictable sequence to form a song type or motif. In species that sing multiple song types, birds may switch song types within a singing bout. At sexual maturity, when circulating levels of gonadal steroids rise, birds stabilize their song output. This is called song crystallization, and marks the completion of sensory–motor integration.

2.4.3 Timing of Sensory Acquisition and Sensory–Motor Integration

The timing of these stages varies widely among species, depending on how quickly they mature and if they are seasonal or opportunistic breeders. Seasonal breeders mate only during the spring and summer, when yearly temperatures reach their maximum, day length is long, and food is plentiful. Opportunistic breeders are prepared to mate throughout the year and do so when crucial resources such as food and water access are available. Song development in white-crowned sparrows, which are seasonal breeders, takes place over the first year of life. When exposed to tape recordings of conspecific song in the laboratory, birds memorize song material during the first roughly 20–50 days of age (Marler 1970; although see Baptista and Petrinovich 1984, 1986; Nelson 1998). These laboratory-raised birds do not begin to sing until 8–10 months later, however. Adult song crystallizes in the spring or early summer of their first year and is coincident with the onset of the first breeding season. Thus, for this species, sensory acquisition and sensory–motor integration are separated by months, under these laboratory conditions. By contrast, Zebra Finches *(Taeniopygia guttata)* mature rapidly and are opportunistic breeders. They are sexually mature by 3 months of age and can breed year around in the laboratory. In this species, sensory acquisition and sensory–motor integration overlap in time. Sensory acquisition occurs between roughly 20 and 60 days of age, and sensory–motor integration occurs between 30 and 90 days of age (Immelmann 1969; Price 1979; Eales 1985). Thus, a juvenile zebra finch can be memorizing and practicing song at the same time.

2.5 The Sensitive Period for Song Learning

In most species studied, song learning is restricted to the first year of life. In other species, song learning can also occur in adulthood. Even in these cases, however, learning is restricted to a particular time of year. The differences between these two types of learners are discussed in Section 2.7. Here, we discuss the sensitive period for initial song development.

Song learning involves both memorization of a song and producing a copy of that song. We measure a bird's ability to memorize song by what is eventually produced. The sensitive period is defined as the time period during which exposure to adult song leads to the copying of that song. A male zebra finch will eventually produce the song he heard between 20 and 60 days after hatching, for example, and not before or after that time (Immelmann 1969; Eales 1985). Sensitive periods for song learning have been described in chaffinches (Thorpe 1958), white-crowned sparrows (Marler 1970; Baptista and Petrinovich 1984, 1986; Nelson 1998), eastern marsh wrens (*Cistothorus palustris*) (Kroodsma 1978), zebra finches (Immelmann 1969; Price 1979; Eales 1985), Bengalese finches (*Lonchura striata domestica*) (Immelmann 1969; Dietrich 1980; Clayton 1987; Slater et al. 1991), and song and swamp sparrows (Marler and Peters 1987; Marler and Peters 1988a,b).

The exact timing of the sensitive period depends on the circumstances of the learning. Social interaction between tutor and pupil, including visual, auditory, and physical contact, can increase the likelihood that songs are copied. Juveniles will often copy songs from live tutors when they have ceased to copy songs from tape, or will not learn from tapes at any age (Price 1979; Baptista and Morton 1981, 1988; Baptista and Petrinovich 1984, 1986; Clayton 1987; Eales 1989). In addition, if exposure to a song model is withheld, the sensitive period can be extended. Eales (1985) tested the flexibility of the zebra finch sensitive period by raising juvenile males in isolation until adulthood. Their songs were highly abnormal and never stabilized as they should have. When Eales exposed the young isolate males to other adults, she found that some birds copied the other males' songs at an age well beyond the normal sensitive period. Another example of how the sensitive period interacts with experience comes from studying marsh wrens. This species learns most of its song patterns in the first 60 days of life, but it can also acquire some songs the following year, as the breeding season approaches. The amount of song material a young adult marsh wren will acquire in this later learning period depends on how much he learned earlier. Birds that hatch early in the season have more opportunity to learn song from adult conspecifics in their first year than do birds that hatch late in the season. If a bird hatches early and acquires much song material early, then he will learn less song material in the following year (Kroodsma and Pickert 1980). Thus, the sensitive period is not just a matter of age, but also depends on the amount and quality of exposure to a song model. Evidence for the role of social factors in song memorization and production comes from a series of studies done by Beecher and colleagues. In the field, young male song sparrows memorize song types after fledging and before selecting and defending their own territories. These young males are referred to as "floaters." A young male selectively memorizes and produces song types that are shared by territorial adult males in the area where the floater will subsequently establish his own breeding territory (Beecher et al. 2000 a,b; Nordby et al. 2000).

2.6 Action-Based Learning: A Selection Model of Song Development

Some species produce more acoustic elements in plastic song than are retained in the stable adult song. For example, swamp sparrows sing up to 12 distinct acoustic elements during plastic song, but retain only three or four of those in their crystallized songs (Marler and Peters 1981, 1982). A model of song learning called action-based learning attempts to explain why some plastic song elements are crystallized and some are dropped (Marler and Nelson 1992; Marler 1997). This model proposes a type of song learning that occurs by selecting only some of the song material that can be produced to include in the permanent song repertoire. Action-based learning is distinguished from the initial juvenile song acquisition because it may be more guided by social reinforcement, may be less dependent on sensitive periods, and does not require sequential memorization and vocal practice. There is evidence suggesting that the selection of song material as birds go from plastic song to crystallized song is related to what other birds in the immediate social environment sing. There is good evidence in territorial species that, during the plastic phase, birds retain song elements that are shared by other conspecific adults in their acoustic space. Saddlebacks (*Philesturnus carunculatus*), white-crowned sparrows, and field sparrows (*Spizella pusilla*) share song or note types with conspecifics on neighboring territories, for example (Jenkins 1978; DeWolfe et al. 1989; Nelson 1992). Thus, which plastic song elements are retained in the adult song is not simply programmed. Instead it appears to be the result of a process of selection that is guided by experience. Whether or not the plastic song elements that are absent in the adult song are retained in memory beyond the time of song crystallization remains unknown. There is evidence that white-crowned sparrows may retain syllables in memory that they do not produce in their stable songs. In laboratory-reared birds, lesions of the song control nucleus, lMAN, induce changes in adult song (Benton et al. 1998). Those changes include singing syllables that are not included in the stable song but are similar to plastic song syllables that were deleted before song crystallization. In addition, Hough et al. (2000) found that intact adult white-crowned sparrows seasonally reexpress song types that have been deleted from their crystallized songs. This finding provides further evidence that adult white-crowned sparrows retain multiple song types in memory even though their crystallized songs consist of only one song type, which does not change from year to year.

2.7 Two Types of Song Learners: Age-Limited and Open-Ended

Some songbirds do not change their stable vocalizations as adults. These birds are called age-limited learners because new song elements are not acquired after the first year or so of life. This pattern of juvenile learning and adult stability appears to be the most common among songbird species that have been studied. Other birds are able to alter their songs in adulthood (Nottebohm et al. 1986; McGregor and Krebs 1989; Eens 1992; Mountjoy and Lemon 1995). These species are called open-ended learners because they learn new song elements beyond their first year of life. Song changes in adult open-ended learners are restricted seasonally, with the greatest song change occurring in early fall, just after the breeding season. The hormonal and neural changes that are correlated with seasonal song acquisition are discussed in Section 6.

While several age-limited learners such as zebra finches and white-crowned sparrows have already been mentioned, island canaries (*Serinus canaria*) and European starlings (*Sturnus vulgaris*) are examples of open-ended learners. Canaries initially learn to sing over the first 10 months of age. Song stabilizes prior to the first spring breeding season, but then becomes unstable again in summer and early fall. Syllables are added, modified, and dropped such that song repertoires (which stabilize again before breeding) are larger and contain new elements by the second breeding season (Nottebohm et al. 1986). These modifications can go on for subsequent years, but the size of the song repertoire does not increase. European starlings also alter their song repertoires from year to year by adding, modifying, and dropping elements seasonally (Eens 1992; Mountjoy and Lemon 1995). Unlike canaries, starlings can continue to increase their repertoire size after 2 years of age. The amount of new song material learned decreases with age, however (Mountjoy and Lemon 1995). Thus even for open-ended learners, age appears to constrain song learning. The neural and/or hormonal mechanisms that underlie this age constraint on song learning are not yet known.

3. Anatomy of the Song Control Circuits and Mechanisms of Song Production

3.1 Neural Circuits

3.1.1 Two Basic Pathways

Song behavior in songbirds is regulated by a discrete network of interconnected nuclei. This network consists of two pathways that are involved in song learning and production (Fig. 6.5). The motor pathway controls the production of song. This circuit consists of projections from the thalamic nucleus uvaformis (Uva) and the nidopallial nucleus interfacialis (NIf) to the nidopallial nucleus HVC. HVC projects to the robust nucleus of the arcopallium (RA) in the telencephalon, and RA projects both to the dorsomedial part of the intercollicular nucleus (ICo) in the midbrain (not shown in Fig. 6.5) and to the tracheosyringeal part of the hypoglossal motor nucleus in the brain stem (nXIIts). Motor neurons in nXIIts send their axons to the muscles of the sound-producing organ, the syrinx. When these motorneurons are stimulated, the syringeal muscles contract and move the

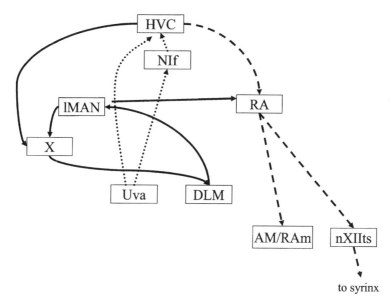

FIGURE 6.5. A schematic showing projections of the major nuclei in the song control system. The motor pathway (*dotted* and *dashed lines*) controls the production of song. The *dotted lines* indicate inputs to HVC from the thalamic nucleus Uva and the nidopallial nucleus NIf. The *dashed lines* indicate the descending projections from HVC in the nidopallium to RA in the arcopallium, and thence to the vocal nucleus nXIIts, the respiratory nucleus RAm, and the laryngeal nucleus Am in the medulla. The *solid lines* indicate the anterior forebrain pathway that is essential for song learning and perception. It indirectly connects HVC to RA, via area X (thought to be a basal ganglia homolog), DLM in the thalamus, and lMAN in the nidopallium. lMAN also projects to area X. AM, Nucleus ambiguus; DLM, medial portion of the dorsolateral nucleus of the thalamus; lMAN, lateral portion of the magnocellular nucleus of the anterior nidopallium; NIf, nucleus interface; RA, robust nucleus of the arcopallium; RAm, nucleus retroambigualis; Uva, nucleus uvaeformis; X, area X; nXIIts, tracheosyringeal part of the hypoglossal nucleus.

medial and lateral labia into the expiratory air stream; this sets the labia into vibration to produce sound. Contraction of the syringeal muscles also changes the shape and/or length of the vocal tract, which can influence the frequency composition of sounds produced by the labia (Suthers et al. 1999). The projection from RA onto the motor neurons in nXIIts is myotopically organized (Vicario 1991). Neuronal activity in the premotor nuclei HVC and RA is synchronized with the production of sound by the syrinx (Yu and Margoliash 1996; Hahnloser et al. 2002). If nuclei in the motor pathway are lesioned, a bird may adopt appropriate posture and beak movements but will not produce song (Nottebohm et al. 1976).

RA also projects to nucleus retroambigualis (RAm) and nucleus ambiguus

(Am) in the medulla (Wild 1997). RAm contains many respiratory related neurons that fire in phase with expiration. Am contains motoneurons that innervate the larynx. This pattern of descending projections from RA likely is important for coordination of syringeal, laryngeal, and respiratory muscle activity during song production; birds produce sound only during expiration.

The second, anterior forebrain, pathway (AFP) is believed to be essential for song learning and recognition. This pathway consists of projections from HVC to area X, then to the dorsolateral nucleus of the thalamus (DLM), from DLM to the lateral portion of the magnocellular nucleus of the anterior nidopallium (lMAN), and finally to RA. In addition, lMAN neurons that project to RA send collaterals to area X, thus providing the potential for feedback within this pathway. The projections within this pathway are roughly topographically organized (reviewed in Bottjer and Johnson 1997). Lesions of lMAN, DLM, or area X in juvenile male zebra finches prevent the development of normal song (reviewed in Nordeen and Nordeen 1997). Juvenile males with lesions of area X persist in producing songs that are plastic in structure, as though they are unable to crystallize on a stereotyped version of the species song. In contrast, if lMAN is lesioned in juvenile male zebra finches, they produce songs with aberrant but stable structure. Observations in adult canaries, which can develop new songs as adults, appear consistent with results from juvenile zebra finches; lesions of lMAN made in adult male canaries in mid-September, when song is seasonally plastic in structure, lead to a progressive decline in syllable diversity (Nottebohm et al. 1990).

It is not yet clear what role the AFP plays in the production of stereotyped song in adult birds. Chronic recordings from awake male zebra finches indicate that neuronal activity in lMAN and area X changes during song production (Hessler and Doupe, reviewed in White 2001). This neuronal activity in the AFP is consistent with increased transcription of the immediate early gene *ZENK* in lMAN and area X of male zebra finches during singing (reviewed in Ribeiro and Mello 2000). Studies of the behavioral effects of AFP lesions in adult birds have yielded variable results, as discussed in Section 5.

3.1.2 Auditory Input to the Song System

Song learning and maintenance depend on auditory input from other birds and auditory feedback from self-generated song. There are multiple inputs from the auditory system to the song nuclei and associated regions of the caudal telencephalon (Fig. 6.6.) (reviewed in Vates et al. 1996; Margoliash 1997). The field L complex in the nidopallium (analogous to mammalian auditory cortex) is the primary forebrain recipient of auditory input from the thalamic nucleus ovoidalis (Ov) complex. Ov consists of "shell" and "core" components. Ov core projects to the field L complex, whereas the Ov shell projects to both the field L complex and the caudomedial nidopallium (NCM). There are reciprocal connections between the different regions of the field L complex. There are also reciprocal connections between the caudolateral mesopallium (CLM) and all regions of

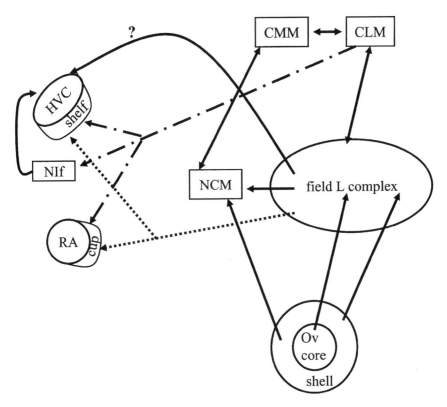

FIGURE 6.6. Inputs from the auditory system to the song nuclei and associated regions of the caudal telencephalon. The field L complex in the nidopallium is the primary forebrain recipient of auditory input from the thalamic Ov complex. Field L projects widely to regions of the caudal telencephalon (NCM, CMM, and CLM). The song system receives auditory input via NIf and possibly HVC. CLM, caudolateral mesopallium; CMM, caudomedial mesopallium; NCM, caudomedial nidopallium; NIf, nucleus interfacialis; Ov, nucleus ovoidalis complex; RA, robust nucleus of the arcopallium. [Based on information in Vates et al. (1996) and Margoliash (1997).]

field L, between CLM and the caudomedial mesopallium (CMM), and between CMM and NCM. CLM projects to a region immediately ventral to the song nucleus HVC, referred to as the "shelf," and to a region surrounding the song nucleus RA, referred to as the "cup." CLM projects as well to NIf, which in turn projects to HVC. This projection from NIf appears to be the major source of auditory input to the song control circuits. Regions of field L project to the HVC shelf and RA cup. Some investigators have reported that regions of field L also project directly onto neurons in HVC. This issue remains a source of debate, however (reviewed in Margoliash 1997).

3.1.3 A Circuit for Song Perception and Memory

As stated in the preceding, there are auditory projections to NCM and CMM, regions not regarded as part of the song learning and production circuits. These two regions may play a role in song perception and memory, however. There is strong induction of the immediate early gene *ZENK* (which is also known as *ngfi-a* in rats and *egr-1* in mice) in these regions in response to presentation of conspecific song, but only weak induction in response to heterospecific song or artificial sound stimuli (reviewed in Ribeiro and Mello 2000). Neurons in NCM show electrophysiological responses to playback of conspecific songs and other complex auditory stimuli (reviewed in Ribeiro and Mello 2000). Both the *ZENK* induction and auditory responses in NCM decrease with repeated presentation of the same conspecific song. This decrease in responsiveness persists for 1–2 days and is reminiscent of habituation. The formation of these persistent memories of song could be important in the context of recognizing the songs of territorial neighbors and mates. Inactivation studies have not yet been performed to demonstrate that NCM is necessary for song perception and memory, owing in part to its large size and diffuse distribution.

3.2 Steroid Hormones and the Song Control System

Gonadal steroid hormones and their metabolites in the brain have important effects on the development of song circuits, song learning, the activation of adult song behavior, and seasonal plasticity of the adult song system. The influence of these hormones on the song system is consistent with the fact that song is an integral part of reproduction in birds. Sensitivity of the song system to gonadal steroids may confer the adaptive benefits of synchronizing song behavior to seasonal reproductive cycles, and enabling the sexually dimorphic development of the song control circuits. We review hormone contributions to juvenile song learning and adult production, and the distribution of receptors for these hormones in the song circuits in this section. Steroid influences on development of the song system and seasonal plasticity are reviewed in later sections.

3.2.1 Gonadal Steroids and Song

Gonadal steroid hormones influence the song learning process. Testosterone, for example, is important for song crystallization, which usually occurs around the time of sexual maturation. Male swamp sparrows and song sparrows castrated as juveniles acquire a sensory model of conspecific song, and go through the early stages of motor learning. These castrated birds, however, do not achieve crystallized song unless they are treated with exogenous testosterone (Marler et al. 1988). Similarly, zebra finches castrated as juveniles develop abnormal songs that are variable in structure (reviewed in Bottjer and Johnson 1997). These results suggest that testosterone is not necessary for sensory learning or early motor learning, but is critical for the development of stereotyped motor patterns of song. Testosterone may stimulate song crystallization by inducing changes in the distribution of synaptic inputs from lMAN and HVC onto neurons in RA (reviewed in Nordeen and Nordeen 1997), and by altering the expression and kinetics of receptors for N-methyl-D-aspartate (NMDA) at synapses. These ideas are discussed below.

Steroid hormones are also important in the activation of singing in adult birds. Castration reduces or completely eliminates song production in males. Implanting these castrated males with testosterone reinstates song (reviewed in Bottjer and Johnson 1997; Schlinger and Brenowitz 2002). Seasonal changes in song production also correlate with changes in circulating levels of testosterone. In some species, adult females can be stimulated to sing when implanted with testosterone (reviewed in Bottjer and Johnson 1997; Schlinger and Brenowitz 2002).

3.2.2 Gonadal Steroid Receptors

Sex steroids appear to exert their organizational and activational effects by acting directly upon neurons in song regions of the brain. Like other vertebrates, songbirds have androgen receptors (AR) and estrogen receptors (ER) in diencephalic and mesencephalic structures, as well as in limbic regions of the telencephalon such as the hippocampus (reviewed in Schlinger and Brenowitz 2002). Songbirds are unique among vertebrates, however, in also having AR and ER in nonlimbic regions of the telencephalon. Androgen receptors are expressed at high levels in HVC, RA, lateral and medial lMAN, and NIf, as well as in the midbrain vocal region ICo, nXIIts in the brain stem, and the muscles of the syrinx. The presence of AR in these regions was first demonstrated with in vivo and in vitro steroid-binding methods, and subsequently with immunocytochemistry (ICC) (see Schlinger and Brenowitz 2002). Expression of the mRNA for the AR was shown in these nuclei using in situ hybridization methods (reviewed in Schlinger and Brenowitz 2002). The message for AR is also expressed in area X, NCM, and in brain stem respiratory nuclei that are connected to vocal nuclei. (Gahr and Wild 1997; Metzdorf et al. 1999; Perlman et al. 2003).

ER are not as widely distributed within the song system as are AR. ER are present in HVC and ICo of all species of songbirds examined, as shown by in vivo steroid-binding, ICC, and in situ hybridization methods (Gahr et al. 1987, 1993; Jacobs 1999; Schlinger and Brenowitz 2002).

3.3 Development of Song Circuits in Juveniles

The two major song control circuits develop at different times, and the timing suggests that these pathways play roles in different phases of song learning. The development of the anterior forebrain pathway coincides with the initial sensory phase of song learning. Before hatching, canaries already have most connections in this pathway: HVC neurons that project to area X, DLM neurons that project to lMAN, and lMAN neurons that project to RA (reviewed in Alvarez-Buylla and Kirn 1997). In male zebra finches, functional connections are already present between all of the nuclei in the anterior forebrain pathway (AFP) by 15 days after hatching (P15), when sensory song learning is occurring (reviewed in Bottjer and Johnson 1997).

The motor pathway develops later than the AFP, and plays an important role in the sensory–motor phase of song learning. Axonal connections between HVC and RA in this pathway increase in number over the period when young male zebra finches first begin to sing (Foster and Bottjer 1998). HVC-to-RA projection neurons continue to be born into adulthood in both zebra finches and canaries (reviewed in Alvarez-Buylla and Kirn 1997). The addition of new RA-projecting neurons to HVC occurs at a higher rate in adult canaries than in adult zebra finches, however, perhaps reflecting the difference between the open-ended and age-limited song learning in these two species. Ongoing neuronal recruitment to adult HVC is discussed below.

The volumes of the song nuclei change after hatching in parallel with the development of song behavior. In zebra finches, HVC, RA, and area X increase in size sharply between P12 and 53, with growth of HVC preceding that of its efferent nuclei (reviewed in Bottjer and Johnson 1997; Nordeen and Nordeen 1997). This time period coincides with the sensory and early sensory–motor phases of song learning. The core portion of lMAN, however, decreases in size between P25 and 53, losing over half its neurons. This decrease in the volume of lMAN may reflect a developmental change in its function. As described in the preceding, lesions of lMAN early in this period completely disrupt song development. Birds that receive lMAN lesions once song has become stereotyped (i.e., crystallized), however, continue to be able to produce well-structured song (reviewed in Bottjer and Johnson 1997).

As discussed previously, there are extensive species differences in the number of song types that birds learn to sing (i.e., repertoire size). Several studies have reported correlations between the volumes of HVC/RA and repertoire size, both within and between species (Nottebohm et al. 1981; reviewed in Brenowitz 1997). In zebra finches, HVC neuron number correlates with the number of song syllables that are learned from other birds (reviewed in Nordeen and Nordeen 1997).

One hypothesis to explain this correlation is that cues associated with song learning influence the growth of the song nuclei; the more songs a young bird learns, the larger his song nuclei become (Nottebohm et al. 1981). The available evidence suggests, however, that while experience may influence subtle features such as dendritic spine density in HVC (Airey et al. 2000), it does not determine the overall growth of song nuclei. This conclusion is supported by a study in which captive juvenile eastern marsh wrens were tutored with either a small repertoire (five song types) or a large repertoire (45 song types); wild birds sing about 50 song types (reviewed in Brenowitz 1997). As adults the first group sang only five to six song types, but the second group sang 35–46 song types. Despite the pronounced differences in adult song repertoire size between these two groups, however, there were no differences between them in the volumes of HVC and RA. Males that sang only a few song types were therefore just as likely to have large song nuclei as males that sang many songs. Further support for the conclusion that early learning experience does not determine the growth of the song nuclei comes from the observation that deafening young male zebra finches does not alter the normal developmental time course of the changes that occur in neuron number in HVC and area X (reviewed in Nordeen and Nordeen 1997). Other factors such as genetic differences, the amounts of steroid hormones present in eggs and young birds (discussed below), and/or early nutrition may influence the growth of the song nuclei. The size to which the song nuclei grow may, in turn, determine the number of song types a young bird can subsequently learn, if "brain space" does in fact constrain song learning (Nottebohm et al. 1981; Brenowitz 1997).

3.4 Gonadal Steroid Hormones Influence the Development of the Song Control System

Treatment of newly hatched female zebra finches with estradiol masculinizes the structure of their song control system (reviewed in Arnold 1997). Such estrogenized females can produce male-typical song as adults. Hormones influence the incorporation of newly generated neurons into song control nuclei. Female zebra finches implanted with estradiol soon after hatching and injected with [³H]thymidine, a nucleotide incorporated into the DNA of mitotically dividing cells, show more newly generated neurons in HVC and area X than do females that do not receive hormone implants (reviewed in Alvarez-Buylla and Kirn 1997). We do not yet know whether hormones affect the birth and/or differentiation of new neurons, the migration of these neurons from the ventricular zone where they are born, or the survival of these cells in developing zebra finches. In adult female canaries, however, hormones do not alter the incorporation of [³H]thymidine into dividing cells in the ventricular zone, but do increase the migration and/or survival of new neurons (reviewed in Alvarez-Buylla and Kirn 1997).

Gonadal steroids also interact with neurotrophic factors to regulate the development of the song control regions. Lesions of lMAN in young zebra finches result in the death of efferent neurons in RA. If neurotrophins are infused into RA just after lMAN is lesioned, however, the death of RA neurons is completely suppressed (Johnson et al. 1997). Brain-derived neurotrophic factor (BDNF) is present in HVC, RA, area X, and lMAN during the development of the song system in juvenile male zebra finches, but is nearly absent from these regions in adult birds (Akutagawa and Konishi 1998). The gene for BDNF is normally expressed at high levels in HVC of male, but not female, zebra finches starting 30–35 days after hatching. If males are implanted with estradiol 15–25 days after hatching, the BDNF gene expression increases within 24 hours (Dittrich et al. 1999). Treatment of adult female canaries with testosterone increases the size of their song control nuclei, and also increases the level of BDNF in HVC. Infusion of an antibody that neutralizes BDNF blocks the testosterone-induced growth of HVC (Rasika et al. 1999).

These studies indicate that steroid hormones are important in the masculinization of song control regions. Furthermore, gonadal steroids and their metabolites are accumulated by receptors in the nuclei of target cells in most of the song nuclei. Therefore, hormones may act directly on cells in song regions through a classic steroid receptor mechanism. Masculinization of song system anatomy may also occur through indirect actions of steroid hormones. As an example, area X in the forebrain of female zebra finches is masculinized in size by early systemic treatment with testosterone or estradiol. But as described previously, area X itself has few or no receptors for these hormones. This growth of area X appears to result from afferent input from HVC (Fig. 6.2). Area X fails to develop in juvenile female zebra finches in which HVC is lesioned just before the birds receive implants of estradiol (Herrmann and Arnold 1991).

NMDA receptors are present in several song nuclei and play a role in the development of song circuits and song learning (reviewed in Nordeen and Nordeen 1997; White 2001). Binding of the NMDA receptor antagonist MK-801 in lMAN is greater 30 days after hatching (P30) in male zebra finches beginning to learn song than in adults producing stereotyped song. Juvenile male finches in which MK-801 was injected to block NMDA receptors on the days on which they were exposed to a song tutor showed little evidence of song learning, whereas males injected on the days following tutoring learned song normally (reviewed in Nordeen and Nordeen 1997). The mRNA for the NMDA receptor 2B subunit (NR2B) in lMAN was expressed at twice the level in P30 male finches as in adult males, and the binding of the NR2B-associated ligand [^{3}H]ifenprodil showed a similar developmental decrease (Basham et al. 1999). Treatment of P20 male finches with testosterone, which accelerates song development, decreased NR2B mRNA expression in lMAN at P35, relative to age-matched controls (Singh et al. 2000). Concomitant with the developmental change in NMDA receptor subtype is an increase with age in the decay rate of NMDA excitatory postsynaptic currents (EPSC) in lMAN (White et al. 1999). Treatment of fledgling (P21–P32) and juvenile (P38–P49), but not adult (>P90), male finches with testosterone accelerated the increase in decay rate of the NMDA–EPSC, and increased dendritic length and spine density, in lMAN (White et al. 1999). Taken together, these results suggest that circulating androgens may limit sensitive periods for song learning by altering synaptic transmission in the song nuclei.

4. Auditory Responses in the Song Control System

Hearing is essential for song learning and, in some species, maintenance of stable adult song. Information from the ascending auditory system influences the functioning of brain regions involved in song learning and production. Neurons within the song control nuclei, HVC, RA, lMAN, area X, DLM, and NIf, are responsive to acoustic stimuli. Some neurons within each of these nuclei respond maximally, although not exclusively, to presentation of the bird's own song (BOS). Neurons that respond more robustly (measured by spike rate) to the BOS than to the songs of other birds or acoustically manipulated versions of the BOS are called "song selective." Selective responses to the BOS are particularly interesting because birdsong is acoustically complex and is learned. Thus, song selective neurons demonstrate sensitivity to both the timing and frequency characteristics of highly complex acoustic stimuli and are tuned by experience. For these reasons, song selectivity has become a major focus in studies of the song system. Neurons showing song selectivity have now been examined in numerous brain regions, at the multiunit, single-unit, and intracellular levels (Table 6.1). The response properties of song selective neurons in the descending motor control pathway and the anterior forebrain pathway, as well as their development and potential functional significance, are discussed here.

4.1 The Descending Vocal Motor Pathway

HVc and RA form the primary efferent circuit for song production and are essential for both song learning and production (Fig. 6.5). Input from the auditory system first reaches the song control circuits at HVC and RA (Fig. 6.6; see Section 3). Neurons within both HVC and RA show song selectivity, which may develop through interactions between auditory feedback and motor commands during song learning.

4.1.1 HVC

The majority of neurons in HVC are responsive to sound in general, and some of these cells respond preferentially (measured by spike rate) to presentation of the BOS (see Margoliash 1997 for review). Margoliash (1983) found that a subset of single units in the HVC of anesthetized white-crowned sparrows respond maximally to the BOS and less well or not at all to tone pips, noise bursts, or other songs. Song selectivity is also evident when multiunit clusters are sampled; strong selectivity for the BOS over other song stimuli in HVC has been demonstrated in anesthetized white-crowned sparrows and zebra finches (Margoliash 1986; Volman 1993, 1996; Schmidt and Konishi 1998).

Song-selective neurons demonstrate sensitivity to both the spectral and temporal features of the song. Manipulated versions of the BOS such as the BOS played in reverse and the BOS with the syllable order reversed do not elicit strong responses (Margoliash 1983, 1986). Degrading the temporal and/or the spectral precision of the BOS also reduces the selectivity of neuronal responses to playback (Theunissen and Doupe 1998). Playing the entire song is not always required. Strong responses in some song-selective neurons can be elicited by playing only portions of the BOS. In white-crowned sparrows, presentation of multiple song phrases evoke selective responses and, in zebra finches, multiple syllables elicit similar responses (Margoliash 1983; Margoliash and Fortune

TABLE 6.1. Auditory responses in song control nuclei.

Nucleus	Species	Recording method	Stimuli	References
HVC	Zebra finch	intracellular/patch	BOS, REV, RO, SYL noise	Katz and Gurney (1981) Lewicki and Konishi (1995) Lewicki (1996) Mooney (2000)
		Single unit	CON, BOS, REV, SYL modified BOS	Margoliash and Fortune (1992) Lewicki and Arthur (1996) Theunissen and Doupe (1998) Cardin and Schmidt (2003)
		Single unit (unanesthetized)	BOS	Dave et al. (1998) Cardin and Schmidt (2003)
		Multiunit	CON, BOS, REV noise	Margoliash (1986) Volman (1996) Schmidt and Konishi (1998)
		Multiunit (unanesthetized)	HET, CON, BOS, REV	McCasland and Konishi (1981) Schmidt and Konishi (1998) Nick and Konishi (2001)
	White-crowned sparrow	Single unit	CON, BOS, REV synthesized songs song whistle pairs tones	Margoliash (1983) Whaling et al. (1997)
		Multiunit	CON, TUT, BOS, REV	Margoliash (1986)
	Juvenile WcS	Multiunit (unanesthetized)	CON, TUT, BOS, REV	McCasland and Konishi (1981)
	Canary	Multiunit	CON, TUT, BOS, REV	Volman (1993)
		Multiunit (unanesthetized)	CON, BOS, REV	McCasland and Konishi (1981)
	Zebra finch	Single unit	CON, BOS, REV, RO	Doupe and Konishi (1991)
Robust nucleus of the arcopallium (RA)		Single unit (unanesthetized)	BOS	Dave et al. (1998) Dave and Margoliash (2000)
		Multiunit	CON, BOS, REV	Vicario and Yohay (1993)
		Multiunit (unanesthetized)	CON, BOS, REV	Vicario and Yohay (1993)
Area X (X)	Zebra finch	Single unit	CON, TUT, BOS, REV RO, SYL, tones, noise abnormal BOS	Doupe and Konishi (1991) Doupe (1997) Solis and Doupe (2000)

TABLE 6.1. *Continued*

Nucleus	Species	Recording method	Stimuli	References
	Juvenile ZF	Single unit	CON, TUT, BOS, REV RO, SYL, tones, noise abnormal BOS	Doupe (1997) Solis and Doupe (1997, 1999)
Medial portion of the dorsolateral nucleus of the thalamus (DLM)	Zebra finch	Single unit	CON, BOS, REV, RO tones, noise	Doupe and Konishi (1991)
Lateral portion of the magnocellular nucleus of the anterior nidopallium (lMAN)	Zebra finch	Intracellular Single unit	CON, BOS, REV CON, TUT, BOS, REV RO, SYL, tones, noise abnormal BOS	Rosen and Mooney (2000) Doupe and Konishi (1991) Doupe (1997) Solis and Doupe (2000)
	Juvenile ZF	Multiunit (unanesthetized) Single unit	CON, BOS CON, BOS, REV, RO TUT, tones, noise abnormal BOS	Hessler and Doupe (1999a,b) Doupe (1997) Solis and Doupe (1997, 1999)
Nucleus interfacialis (Nif)	Zebra finch	Single unit	CON, BOS, REV	Janata and Margoliash (1999) Lewicki and Arthur (1996)
Field L	Zebra finch	Single unit	CON, BOS, REV, SYL	Janata and Margoliash (1999) Sen et al. (2001) Grace et al. (2002)
		Multiunit	CON, BOS tone ensembles	Schmidt and Konishi (1998) Theunissen and Doupe (2000) Sen et al. (2001) Grace et al. (2003)
		Multiunit (unanesthetized)	CON, BOS, tones	Schmidt and Konishi (1998)
	Juvenile ZF	Multiunit (unanesthetized)	CON, TUT, tones	Gehr et al. (1999)
	White-crowned sparrow	Multiunit	CON, TUT, BOS, REV	Gehr et al. (2000) Margoliash (1986)
	European starling	Single unit (unanesthetized)	CON, whistles, tones	Hausberger et al. (2000) Capsius and Leppelsack (1999)
		Multiunit	CON, tones, noise	George et al. (2002)
		Multiunit (unanesthetized)	Tones, noise	Nieder and Klump (1999) Nieder and Klump (2001)

HET, Heterospecific song; CON, conspecific song; TUT, tutor's song; BOS, bird's own song; REV, bird's own song played in reverse; RO, bird's own song reversed; SYL, song syllables; ZF, zebra finch; WcS, white-crowned sparrow.

1992). Strong responses require the presentation of multiple song segments in the "correct" order (i.e., that normally found in the BOS). If only one phrase or syllable is presented, responses are greatly diminished or absent (Margoliash 1986; Margoliash and Fortune 1992; Lewicki and Arthur 1996). The selectivity in HVC indicates that responses to these stimuli are nonlinear and selectivity is based on the sequential combination of sounds found in a bird's song. Similar sensitivity to the sequentially "correct" presentation of communication sounds has been demonstrated in the bat auditory cortex (Esser et al. 1997). Such temporal combination sensitive cells in the zebra finch HVC have now been described at the extracellular (Margoliash and Fortune 1992; Lewicki and Arthur 1996) and the intracellular levels (Lewicki and Konishi 1995; Lewicki 1996; Mooney 2000). Mooney (2000) demonstrated that both RA-projecting and X-projecting neurons within HVC are song selective. The subthreshold responses to BOS are quite different between these two cell types, however, suggesting that song selectivity can be generated by different mechanisms. Interestingly, single HVC neurons in swamp sparrows, which sing multiple song types, show subthreshold responses to multiple song types but often spike only to one song type. This suggests that selectivity for a song type could be generated within HVC, rather than by its presynaptic inputs (Mooney et al. 2001).

Given that selective responses require presentation of multiple and sequential segments of the BOS, it has been suggested that these neurons are able to integrate inputs for up to hundreds of milliseconds. Margoliash and Fortune (1992) estimated the lower limits of integration times for song selective units to be between 80 and 350 millisec. Integration times this long have also been estimated for mammalian cortical neurons. Facilitated responses to tones presented up to 600 millisec apart have been reported in the cat auditory cortex, for example (McKenna et al. 1989). HVC neurons do exhibit long-lasting hyperpolarization in response to song stimuli (Lewicki 1996; Mooney 2000). It is possible that the long integration times observed in song-selective HVC units could result from multiple, integrated inputs onto these cells. The cellular basis of input integration in these cells remains to be understood.

Neurons in field L, which have long been presumed to provide the primary auditory input to HVC, respond strongly to song stimuli but do not show response preferences for the BOS over conspecific song, heterospecific song, or songs played in reverse (Margoliash 1986; Lewicki and Arthur 1996; Janata and Margoliash 1999). This lack of selectivity for BOS in the auditory forebrain suggests that song-selective responses arise in HVC or other brain regions (other than field L) known to input to HVC (see Fortune and Margoliash 1995). Song-selective responses in NIf, for example, which projects to HVC, have been reported (Janata and Margoliash 1999).

Although field L neurons do not show selectivity for BOS, they do show selectivity for song over synthetic acoustic stimuli with similar characteristics. This also appears to be true for two additional auditory forebrain regions, CM and NCM (see Fig. 6.6). Grace et al. (2003) showed that, in anesthetized zebra finches, field L and CM neurons respond better to conspecific song than to synthetic sounds that match the acoustic parameters of zebra finch songs such as their power spectra and amplitude modulations. NCM neurons also appear to respond better to song over simple sounds such as tones (Stripling et al. 1997). The selectivity for song but not for BOS found in these forebrain regions that are presynaptic to the song system suggests that auditory response selectivity may be hierarchically constructed as sounds progress through the auditory processing stream.

4.1.2 Development of BOS Selectivity in HVC

What is the behavioral relevance of song selective neurons in HVC? One way this question has been addressed is by studying the development of song selectivity in HVC. Song-selective neurons could provide information about the memorized song model to which ongoing vocal output is matched during sensory–motor learning. If so, then song selectivity should arise during the sensitive period for song learning, as a bird memorizes a song model. Alternatively, the development of song selectivity in HVC neurons could parallel the motor development of the bird's own song, potentially acting as a representation of the current motor commands which could then be adjusted or reinforced in the song system premotor nucleus (RA). In addition, the neural representation of BOS could function in recognizing the songs of conspecifics. Volman (1993) addressed this issue by examining song selectivity in the HVC of anesthetized juvenile white-crowned sparrows. In this species, the memorization and sensory–motor (vocal practice) stages of song learning are separated by 8–10 months under laboratory tutoring conditions (Marler 1970). Volman showed that song selectivity is not found in the HVC of juvenile birds that have been exposed to a song tutor but have not yet begun to sing. Rather, these specialized responses emerge as juvenile birds develop song motor skills. In juveniles singing immature, plastic songs, HVC neurons respond selectively to both the BOS and the tutor's song (TUT). This finding suggests that the development of song selectivity in HVC may be related to vocal motor practice, such that neurons respond selectively only when juveniles have begun to practice their own vocalizations. HVC auditory selectivity, therefore, appears to parallel the bird's own vocal output in juveniles with changing songs and in adults with stable songs. Selectivity for both the BOS and the TUT may reflect the acquisition of a motor program that produces an accurate copy of the TUT, rather than memorization of the song model. Selectivity for the TUT, however, may be related to memory formation in that it could provide a neural mechanism for the comparison of vocal output and the memorized song model during vocal practice.

4.1.3 The Robust Nucleus of the Arcopallium (RA)

Multiunit activity in RA correlates well with presentation of the BOS, and many single units within RA respond selectively to the BOS, in anesthetized zebra finches (Doupe and Konishi 1991; Vicario and Yohay 1993; Dave et al. 1998). The auditory response characteristics of song-selective cells in RA do differ somewhat from those in HVC. For example, most song-selective units in RA will respond to broadband noise bursts (Doupe and Konishi 1991), whereas HVC units will not (Margoliash 1983). The second input to RA is from lMAN, which also exhibits song selectivity (see later). Interestingly, the patterned responses exhibited by RA neurons appear to be driven primarily by input from HVC rather than input from the AFP. Selective inactivation of lMAN has no apparent effect on RA responses to song stimuli, in anesthetized birds. Inactivation of HVC (which also deprives the AFP of patterned activity originating in HVC) abolishes auditory responsiveness in RA (Doupe and Konishi 1991; Vicario and Yohay 1993).

Song selectivity in HVC and RA appears to depend on the behavioral state of the bird. RA neurons, for example, are not song selective in awake, behaving zebra finches (Dave et al. 1998). This topic is discussed in Section 4.3.

4.2 The Anterior Forebrain Pathway (AFP)

The emerging understanding of cell types and the interplay among excitatory and inhibitory inputs in the AFP suggests that this circuit is similar to a mammalian cortical–basal ganglia–thalamic loop (see Perkel and Farries 2000 for review). This understanding is consistent with a role for the AFP in song learning. Each nucleus in the AFP contains neurons that are responsive to acoustic stimuli, and many neurons respond maximally to presentation of the BOS, in anesthetized birds (see Doupe and Solis 1997). Given the complexity of excitatory and inhibitory processing through the AFP, the characteristics of song selectivity are surprisingly similar in each AFP nucleus. There are some differences in selectivity, however, which offer clues as to the functions of the different AFP nuclei.

4.2.1 Area X and lMAN

Some single units and multiunit clusters within area X and lMAN of adult anesthetized zebra finches respond robustly to the BOS, and less well or not at all to that song played in reverse, conspecific song, noise, and tone bursts (Fig. 6.7; Doupe and Konishi 1991; Doupe 1997; Rosen and Mooney 2000; Solis and Doupe 2000). Neurons in area X and lMAN show similar degrees of selectivity for the BOS. One notable difference in responsivity between these two areas is that lMAN neurons are generally unresponsive or inhibited by noise, whereas most cells in area X (and DLM) do respond to noise (Doupe and Konishi 1991; Doupe 1997). This difference suggests a potential role for lMAN in refining responses in favor of highly structured or specific acoustic stimuli.

Song selective neurons in area X and lMAN are, like those in HVC, "order selective" and "combination sensitive." Most song-selective area X and lMAN neurons show significantly diminished responses to the BOS with syllable order reversed compared to the normal BOS (Doupe 1997). Also like in HVC, subsets of syllables from the BOS that are presented in the correct order can elicit strong

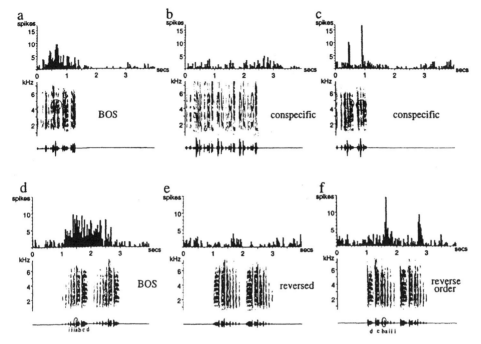

FIGURE 6.7. Response selectivity of auditory forebrain neurons for a bird's own song (BOS). Responses of two single units in the lMAN of two anesthetized adult male zebra finches to presentation of different songs are shown. Peristimulus time histograms (PSTHs) show a single neuron's response to song presentation. Spectrograms and amplitude waveforms of the song are shown below the PSTHs. (A) BOS evokes a strong response. (B) The same neuron shows little change in spike rate when presented with the song of another adult zebra finch. (C) the same neuron responds to a specific portion of another zebra finch song. In the spectrograms, *circles* mark the portions of the BOS and the conspecific song that elicit strong responses from the neuron. Each of these song portions contains more energy at 4.5 kHz than other portions of the song. (D) A second neuron in a different bird responds strongly to the BOS. (E) The same song played in reverse elicits a much weaker response. (F) The same song again but with syllable order reversed elicits a response that is unlike the response to the normal BOS, suggesting that the strong response to BOS requires the cumulative processing of correctly ordered syllables. [Reprinted with permission from Doupe and Solis (1997).]

responses, while other subsets or individual syllables elicit no (or greatly reduced) responses. Thus, in normal adult zebra finches, individual song-selective neurons within HVC, area X, and lMAN respond to song stimuli in similar ways.

4.2.2 Development of BOS Selectivity in Area X and lMAN

The development of song selectivity in the area X and lMAN of anesthetized birds suggests that song-selective units in the AFP could play a role in song learning; song selectivity in AFP neurons develops during song learning, and disrupting song acquisition alters the response properties of AFP neurons. As in the descending motor pathway, neurons within area X and lMAN of young zebra finches develop selectivity for the BOS while birds are singing plastic song. Neurons in area X and lMAN from birds aged 25–49 days (before or early in sensory–motor learning) respond to auditory stimuli including song but are not selective for the BOS over other songs (Doupe 1997). At 60 days of age, young zebra finches reach the end of the sensitive period for song memorization and have been singing plastic song for roughly 1 month (Immelmann 1969; Eales 1985). By this time, song selectivity is beginning to emerge in area X and lMAN (Solis and Doupe 1997). The degree of selectivity is less than that of adults, however. Unlike adult neurons, lMAN neurons in 60-day-old birds respond well to the BOS with syllable order reversed. This observation suggests that temporal sensitivity to syllable sequence may develop after sensitivity to the spectral and temporal modulations within a syllable.

4.2.3 Effects of Abnormal Song Development on Area X and lMAN Selectivity

Some juvenile area X and lMAN neurons are selective for both the plastic BOS and tutor's song (TUT). This observation raises the question as to whether selectivity for both stimuli is due to the acoustic similarities between the BOS and TUT or whether the bird's experience with each song results in tuning for each one separately. Juvenile birds can be experimentally induced to produce songs that are acoustically dissimilar to the TUT by cutting the tracheosyringeal nerves innervating the syrinx (avian vocal organ). This interferes with the control of the syringeal musculature and can cause highly abnormal songs to develop (Williams and McKibbin 1992; Solis and Doupe 1999). These nerve cuts can drastically alter the bird's experience during sensory–motor learning; not only are the resultant songs atypical but a mismatch between the BOS and the memorized TUT is maintained into adulthood. Under these circumstances, some area X and lMAN neurons in 60-day-old birds respond better to the abnormal BOS than to the TUT (Solis and Doupe 1999). This result demonstrates that the experience of self-generated song can shape the selectivity of neural responses in both area X and lMAN. Most area X and lMAN neurons in birds with nerve cuts respond equally well to the abnormal BOS and the TUT. Dual selectivity for acoustically dissimilar BOS and TUT in a single cell suggests that the cell is not simply tuned to a specific set of acoustic features. Instead, it appears to be tuned to often-heard songs that contain dissimilar acoustic features. Similar tuning to multiple acoustically dissimilar species-specific vocalizations has been reported for auditory neurons in squirrel monkey cortex (Newman and Wollberg 1973). This selectivity has not been shown to result from learning, however. In adults with songs that share acoustic properties with the TUT despite nerve cuts, the responses to song stimuli in area X and lMAN are like those in normal adults. In adults with songs that bear little resemblance to the TUT, responses in both area X and lMAN are unlike those in normal birds; most area X neurons exhibit less selectivity overall and respond equally well to the abnormal BOS and the TUT. Fewer than 5% of lMAN neurons in birds with nerve cuts respond to any song stimuli (Solis and Doupe 2000). Because the effects of nerve cuts on responsiveness and selectivity are significant only when the BOS differs from the TUT, a match between behavioral output and the TUT memory appears to be required for normal BOS selectivity to develop in these nuclei. The origin of the abnormal auditory responses in area X and lMAN remains unknown. It may be that the development and functioning of auditory regions, such as field L and HVC, could be affected by the abnormal auditory feedback caused by nerve cuts (see Gehr et al. 2000).

Given that the auditory response properties of many area X and lMAN neurons parallel the changing acoustic composition of song during development, it would be interesting to assess the song selectivity of HVC, area X, and lMAN neurons in adult birds that have changed their songs because of temporary manipulations in auditory feedback. For example, if an adult zebra finch's song has degraded because it has been exposed to delayed auditory feedback, do the song-selective neurons become tuned to the degraded song? Or, do they remain tuned to the original song, potentially providing an internal reference that can be used for recovering the original song after normal feedback is reinstated?

4.3 Auditory Responses in Awake Behaving Birds

Some controversy over the functional significance of song-selective neurons has developed recently. The acoustic response characteristics of neurons in HVc and RA differ depending on the bird's behavioral state. Dave et al. (1998) reported that there are no responses to presented song stimuli in RA neurons of awake unrestrained zebra finches, but auditory responses to songs become evident in RA once birds fall asleep or are anesthetized (but see Vicario and Yohay 1993). Not only do RA neurons in sleeping birds respond to BOS, but also their firing patterns appear to match the motor activity of those same neurons during awake singing (Dave and Margoliash 2000; Hahnloser et al. 2002). Unlike in RA, some single HVC neurons in awake birds do appear to exhibit responses to songs (Dave et al. 1998; although see Cardin and Schmidt 2003). Strong auditory responses are not present at the multiunit level, but do appear once the bird falls asleep or is anesthetized (Schmidt and Konishi 1998; Nick and Konishi 2001). The finding that HVC is not responsive to acoustic stimuli in awake zebra finches is in contrast with an earlier report by McCasland and Konishi (1981) showing that multiunit clusters in HVC of awake canaries and a white-crowned sparrow were responsive to song playback. It is possible that fewer HVC cells respond selectively to song stimuli in awake birds. Therefore, the activity of those neurons in combination with the nonselective activity of other neurons could fail to generate song-selective responses at the multiunit level, even though some single units and small multiunit clusters show selectivity. In addition, different recording techniques could sample different populations of neurons as the nucleus is known to be made up of at least three cell types.

Auditory responses of lMAN neurons also appear to differ in awake and anesthetized male zebra finches (Hessler and Doupe 1999b). In multiunit recordings from awake birds, some but not all sites in lMAN are song selective. Auditory responses to BOS are more variable than in anesthetized birds and auditory evoked activity in these recordings showed only a small increase above spontaneous levels. Similar studies for the other AFP nuclei remain to be conducted.

Studying awake behaving birds will no doubt be a valuable approach to discovering the function of song selective responses during song learning, production, and perception. The studies published to date indicate that auditory responses in HVC and RA are at least somewhat gated by a bird's state of consciousness. There is also evidence that social context affects neural activity in lMAN during singing; the patterns of activity are different depending on whether a bird is singing directed (in the presence of a female) or undirected (alone) song (Hessler and Doupe 1999a,b). This topic deserves further investigation before conclusions about how awake birds use selective auditory responses to control song behavior can be made.

4.4 Song Perception

The selective auditory responses of neurons within the song nuclei suggest that these and related telencephalic regions may play a role in the perception of biologically relevant sounds. This hypothesis is supported by several studies. The lordosis-like copulation solicitation display performed by female birds in response to male song has been used as an assay of selectivity; females generally respond more strongly or exclusively to the songs of conspecific males (reviewed in Catchpole and Slater 1995). Lesions of HVC in female canaries, however, resulted in their performing copulation solicitations to heterospecific songs as strongly as to conspecific songs (Brenowitz 1991; Del Negro et al. 1998). This result suggests that HVC plays an important role in the recognition of conspecific song. Female zebra finches that received lesions of the caudomedial mesopallium (CMM) (see Fig. 6.6) similarly performed copulation solicitations to both conspecific and heterospecific songs (MacDougall-Shackleton et al. 1998). Lesions of HVC in that study, most of which spared part of the nucleus, did not induce responses to heterospecific song. The authors suggested that the apparent difference in the effect of HVC lesions between zebra finches and canaries may be due to the fact that HVC is quite small in female zebra finches and does not establish synaptic contact with RA, unlike in female canaries. A different study, however, suggests that HVC may play a role in song perception in male zebra finches. Sakaguchi et al. (1999) found that phosphorylation of the cAMP response element binding protein (CREB) occurred in area X–projecting neurons in HVC when birds were exposed to playback of conspecific song, but not in response to heterospecific song or white noise.

Area X and lMAN also appear to function in the perception of auditory stimuli. Lesions directed at area X (that may have also invaded lMAN) caused male zebra finches to require more trials than control birds to learn a discrimination between their own song and that of another conspecific in an operant conditioning paradigm (Scharff et al. 1998). In another study (Burt et al. 2000), female canaries were trained using operant methods to discriminate among different exemplars of synthetic sound stimuli, conspecific songs, and heterospecific songs. Lesions of lMAN caused deficits in the discrimination of all classes of auditory stimuli, not just conspecific song. The lesions had no effect on a visual discrimination task used as a control for generalized perceptual disruption. These results suggest that nuclei in the canary AFP play a role in the perception of auditory stimuli in general. These results seem inconsistent with the selective auditory responses of lMAN neurons in anesthetized male zebra finches that were described in the preceding section. The difference may be accounted for by a difference in responsivity between neurons in awake vs. anesthetized birds (see earlier). The generalized pattern of lesion effects is consistent with the observation that neurons in lMAN of *female* zebra finches have more general response properties than those in males (Maekawa and Uno 1996). Thus, the general discrimination deficits reported by Burt et al. might also reflect the use of female canaries. It has been suggested that the AFP may affect song perception (and learning) by mediating auditory attention (M. Konishi, personal communication). If lMAN's role in auditory processing is more general in females than in males, then lesions might be expected to cause general auditory attentional deficits. Lesions of lMAN in male canaries might result in a more specific pattern of discrimination deficits.

There is also *indirect* evidence that females may use lMAN for song perception. The size of lMAN in females of different species of European warblers correlates with conspecific male repertoire size (DeVoogd et al. 1996). In female cowbirds (*Molothorus ater*), the volume of lMAN is correlated with selectivity in performing copulation solicitations to male song (Hamilton et al. 1997). These correlations may reflect a constraint imposed on perceptual ability by the number of neurons in female lMAN.

5. Adult Song Plasticity

5.1 Song Stability and Auditory Feedback

The degree to which adult birds depend on auditory experience to maintain normal song behavior varies widely among species (Table 6.2). Adult open-ended learners depend on auditory experience to learn new song seasonally and therefore require intact hearing to maintain normal song patterns throughout life. The songs of adult canaries, which are open-ended learners, deteriorate within 1 week of deafening, indicating that these birds require auditory feedback to maintain stable song output (Nottebohm et al. 1976). In contrast to open-ended

TABLE 6.2. Effects of auditory feedback manipulations on adult song.

Species	Type of learner	Manipulation	Effects	Time course	References
White-crowned sparrow	Age-limited	Surgical deafening	None	No effect after 18 months	Konishi (1965)
Chaffinch	Age-limited	Surgical deafening	None	No effect after 7 months	Nottebohm (1968)
Canary	Open-ended	Surgical deafening	Degradation	1 week	Nottebohm et al. (1976)
Zebra finch	Age-limited	Surgical deafening	Degradation	4–16 weeks	Price (1979); Botjer et al. (1984)
					Nordeen and Nordeen (1992, 1993)
					Wang et al. (1999)
					Lombardino and Nottebohm (2000)
					Scott et al. (2000)
		Delayed auditory feedback (DAF)	Degradation	6 weeks	Brainard and Doupe (2000a,b)
					Leonardo and Konishi (1999)
Bengalese finch	Age-limited	Surgical deafening	Degradation	1 week	Woolley and Rubel (1997)
					Okanoya and Yamaguchi (1997)
					Scott et al. (2000)
		Auditory hair cell destruction	Degradation	1 week	Woolley and Rubel (1999, 2002)

learners, age-limited song learners do not normally acquire new song as adults (Immelmann 1969; Price 1979; Dietrich 1980; Eales 1985; Clayton 1987; Woolley and Rubel 1997; Lombardino and Nottebohm 2000; Nordby et al. 2002). In seasonally breeding age-limited learners, such as the white-crowned sparrow and song sparrow, songs are less acoustically stereotyped in the nonbreeding season than they are in the breeding season (Smith et al. 1995, 1997; Hough et al. 2000). Stable song produced during the breeding season, however, does not change from year to year.

Continually stable adult song in age-limited learners could be the product of motor circuits that maintain fixed output independent of sensory feedback. In several species of age-limited learners, removal of auditory feedback does not appear to affect adult song production. Surgical deafening, for example, has no significant effects on adult song production in white-crowned sparrows and chaffinches (Konishi 1965; Nottebohm 1968; see Table 6.2). In contrast, adult zebra finches and Bengalese finches require auditory feedback for the maintenance of stable song. Song degradation in surgically deafened adult zebra finches begins to occur 1 month to 1 year following the surgery (Nordeen and Nordeen 1992; Lombardino and Nottebohm 2000). Bengalese finches, also age-limited learners, show adult song degradation within only 1 week of surgical deafening (Fig. 6.8; Okanoya and Yamaguchi 1997; Woolley and Rubel 1997). Thus, the time course of song deterioration in the Bengalese finch appears to be more similar to that of the canary, which is an open-ended learner, than to other species. These results suggest that the normal ability to learn song in adulthood and the degree to which stable adult singing depends on auditory feedback may not be correlated.

These deafening studies have yielded three somewhat unexpected conclusions. First, age-limited learners and open-ended learners can rely similarly on hearing for the maintenance of stable song. Second, the time course of adult song deterioration following deafening can differ significantly between taxonomically and behaviorally similar species. Bengalese finches and zebra finches are so similar that the two species can be cross-fostered and cross-tutored (Immelmann 1969; Clayton 1989). Yet, these two species show large differences in the extent to which they depend on auditory feedback to maintain stable adult song. Third (and not so unexpectedly), age appears to play a role in how rapidly removal of auditory feedback results in degradation of adult song. Lombardino and Nottebohm (2000) showed that the length of time an adult zebra finch continues to sing stable song after deafening is correlated with age; older birds retain stable vocal output longer than younger birds.

Surgical deafening and removing or disrupting auditory feedback by other less invasive methods have similar effects on song. In Bengalese finches, destruction of auditory hair cells causes song degradation comparable to that seen after surgical deafening (Woolley and Rubel 1997, 1999). Using a different technique, Leonardo and Konishi (1999) created a computer-controlled delayed auditory feedback (DAF) environment in which slightly delayed overlays of a bird's song are externally presented while the bird sings. Under DAF conditions,

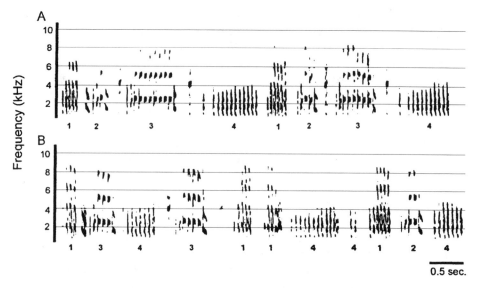

FIGURE 6.8. Some adult songbirds require auditory feedback to maintain their normal song behavior. Two spectrograms of adult song from the same male Bengalese finch are shown. **(A)** The repetitive sequence of 4 syllables (marked with *numbers* below the time axis) is characteristic of this bird's normal stereotyped song. **(B)** song recorded from the same bird 1 week after surgical deafening. The *bottom* spectrogram shows that the deafened bird no longer sings a stereotyped song but, instead, sings a randomly ordered sequence of syllables. Slight changes in the acoustic structure of individual syllables are detectable by 1 week after deafening and worsen dramatically over time. [Reprinted with permission from Woolley and Rubel (1997).]

the bird's auditory feedback becomes a jumbled sequence of temporally overlapping syllables. After several weeks in the DAF environment, the structure of adult zebra finch song deteriorates significantly; syllables are "stuttered" and deleted, and syllable structure becomes more variable than in normal song. Interestingly, presentation of DAF to humans during speech production also results in disfluency, including stuttering (see Timmons 1982 for review).

One question of interest concerns the nature of the acoustic information that is necessary for song maintenance. Do adult songbirds require all of the spectral and temporal information in the song to retain stable song patterns? Or, is just a portion of the information within song feedback crucial for stable singing? Selective destruction of auditory hair cells along the basilar papilla provides a way of manipulating the frequency range of feedback available to birds while they sing. In Bengalese finches, preservation of only the apical hair cells that encode low-frequency sound prevents the song degradation seen when hair cells across the entire papilla are destroyed (Woolley and Rubel 1999). Under these conditions, birds have access to only a small portion of the spectral feedback they normally receive. They continue to produce normal song, however. This observation suggests that low-frequency information and/or the temporal amplitude envelope of the song are the critical feedback for song maintenance, and that the full spectrum of the song is less important.

5.2 Induced Song Plasticity—Degradation and Recovery

In species where song degrades when birds can no longer hear themselves sing, what happens to the songs if auditory feedback is restored? Both DAF and hair cell loss are methods of reversibly disrupting normal auditory feedback to examine the plasticity of adult song patterns. Normal auditory feedback can be restored to birds in DAF environments by turning off the DAF. Deafness caused by auditory hair cell loss due to noise exposure or ototoxic drugs is reversible because birds regenerate new hair cells (Corwin and Cotanche 1988; Ryals and Rubel 1988), and the regenerated cells restore hearing (Marean et al. 1993; Woolley et al. 2001). With these techniques, questions about song memory and learning can be addressed in adults. For example, do adult birds store memories of their own songs? If birds that sing degraded songs can recover their original songs after normal feedback is returned, then they must store neural representations of their normal songs which can be used to guide vocal recovery. Zebra finches with degraded songs due to DAF recover their original songs once the DAF is turned off. Syllable variability decreases, syllable order returns to normal, and songs stabilize over 2–4 months (Leonardo and Konishi 1999). Bengalese finches with degraded songs due to hair cell loss also reconstruct their original songs as hair cells regenerate and hearing is restored (Woolley and Rubel 2002). Syllable order recovers to match the original song over 1 month following hair cell loss. Syllable structure improves toward that of the original song and, in most cases, matches well with normal syllables after 2 months of hair cell regeneration. The findings that zebra and Bengalese finches recover their original songs after the restoration of normal auditory feedback indicates that memories of normal adult songs are stored in the adult brain, regardless of ongoing vocal output, and that those models can be used to reshape normal vocal behavior.

In some Bengalese finches, hair cell loss and regeneration result in a renewed ability to learn song from other birds (Woolley and Rubel 2002). The songs of some birds undergoing hair cell regeneration stop redeveloping toward a match with the original songs and are gradually modified to match the songs of other birds to which they are exposed. These changes in song suggest that either hair cell loss and regeneration or hearing loss and recovery can induce a period of unprecedented vocal plasticity; birds that normally learn only as juveniles can learn song as adults. Adult song plasticity has also been demonstrated in zebra finches using auditory masking techniques (Zevin et al. 2000) and long-term perturbation of vocal function (Hough and Volman 2002). These changes in song, however, do not appear to be guided by learning. It is important to note that adult song learning in Bengalese finches is not the same as the learning that occurs in juvenile birds. The new song syllables that adults produce are the result of modifying and rearranging previously existing song material. In contrast, juveniles develop highly structured copies of their tutors' songs with a previous vocal repertoire of only begging calls. Therefore, the song learning exhibited by adults appears to be constrained compared to song acquisition in juvenile birds.

The studies described here indicate that adult song is more plastic than was previously recognized. First, in some species, maintenance of stable adult song appears to be an active process of vocal correction and reinforcement that relies on auditory feedback. Second, adult zebra and Bengalese finches appear to store memories of their own originally stable songs for a while but can make permanent modifications to song if auditory feedback is disrupted for a long period. These findings support the idea that stored representations of stable songs can be used as templates to which ongoing vocal output can be matched over time. Third, Bengalese finches, which are age-limited learners, can be experimentally induced to modify their adult songs based on experience. This last conclusion suggests that at least some neural circuits required for song learning are retained in the adult brain.

5.3 Neural Basis of Adult Song Plasticity

The neural mechanisms whereby auditory feedback is used to regulate adult song production have only begun to be investigated. Questions that have been addressed so far include: (1) What brain regions are necessary for the active maintenance of stable song? (2) How might neurons that code song motor commands be reinforced or altered by auditory feedback information?

5.3.1 Influence of Anterior Forebrain Circuitry

The forebrain nucleus lMAN is a likely candidate for conveying auditory feedback information to the song motor output system, providing a mechanism for the adjustment of ongoing vocalizations (Nordeen and Nordeen 1993). lMAN is positioned within the anterior forebrain pathway (AFP) and projects to RA, a premotor nucleus in the descending vocal control pathway (Fig. 6.5). lMAN and HVc synapse onto the same RA neurons (Mooney and Konishi 1991; Mooney 1992). This connectivity suggests that information from lMAN could influence the motor commands coded in RA.

As described in Section 3, lMAN lesions in juvenile zebra finches result in the early stabilization of immature songs (Scharff and Nottebohm 1991). The same lesions have been reported to have no effect on stable adult song (Nottebohm et al. 1976; Morrison and Nottebohm 1993; Nordeen and Nordeen 1993). Previous studies, however, may not have analyzed song structure on a large enough scale to detect relatively subtle effects of these lesions on adult song. In addition, it is possible that lMAN lesions may have no apparent effect because the lesions fix song in its normal (stable) state. lMAN may participate in adult song maintenance by providing a feedback signal to be matched with the expected vocal output coded as motor commands in RA. Lesions of lMAN in deafened adult zebra finches prevent the long-term changes in song expected to occur after tracheosyringeal nerve injury and deafening (Williams and Mehta 1999; Brainard and Doupe 2000a,b). Thus, the effects of lMAN lesions on adult zebra finch song may be obvious only when manipulations that ordinarily induce large changes in song structure (e.g., nerve injury, deafening) fail to do so.

Converging evidence from these studies suggests that lMAN may provide input to the vocal motor system that guides vocal output in both juveniles and adults. In the zebra finch, lMAN appears to function in the flow of information that supports developmental and experimentally induced changes in song. Although, if this is the case, then the changes in the output of lMAN neurons after removal of auditory feedback must take time because the firing patterns of lMAN neurons are not different between normal singing birds and deaf birds singing soon after deafening (Hessler and Doupe 1999b). Species differences in the effects of lMAN lesions on adult song have also been reported. For example, lMAN lesions in laboratory-raised white-crowned sparrows result in addition and modification of song elements, rather than the lack of vocal change seen in the zebra finch (Benton et al. 1998).

Lesions of area X in adult male zebra finches were reported to not disrupt the production of stable song (Sohrabji et al. 1990; Scharff and Nottebohm 1991). More recent studies, however, have found that lesions of area X in adults do affect song production. In adult male zebra finches, song becomes more stereotyped in structure and accelerated in tempo several weeks to months after area X is lesioned (Scharff and Jarvis, personal communication). Kobayashi et al. (2001) found that area X lesions in adult Bengalese finches caused a transient but substantial change in song production. Birds with even small, incomplete lesions of this nucleus repeated some of their song notes as many as 50–60 times, rather than the normal 8–10 times. This effect was reminiscent of stuttering in humans. The "stuttering" continued for up to 20 days postlesion but then disappeared. The lesions did not apparently disrupt other aspects of song syntax or stereotypy. These results suggest that area X does influence aspects of the motor production of song in adult birds, though its normal role in this context is still not well understood.

5.3.2 Influence of Vocal Motor Circuitry

The active maintenance of stable song using auditory feedback requires a vocal control system that is flexible enough to alter its output in response to sensory feedback. In zebra finches, projection neurons between HVC and RA are added in adulthood (Alvarez-Buylla et al. 1990; Kirn et al. 1991; Nottebohm et al. 1994; Alvarez-Buylla and Kirn 1997), and are sensitive to manipulations in auditory experience (Burek et al. 1991; Wang et al. 1999). Thus, the continually changing population of HVc-RA projection neurons is one mechanism whereby vocal motor commands could be altered.

If the turnover of HVC-RA neurons provides a mechanism for vocal change, then a correlation might exist between the rate of vocal change after deafening and the rate of RA-projecting neuron addition to HVC. One study suggests such a correlation exists. Scott et al. (2000) compared the addition of new neurons to HVC in Bengalese finches and zebra finches. These species were compared because Bengalese finch song degrades more rapidly following deafening than does zebra finch song. In the month following deafening, the addition of new neurons to the Bengalese finch HVC is double that in the zebra finch HVC, and the proportion of new HVC-RA projection neurons is three times higher in Bengalese finches. One interpretation is that the new neurons may not be successfully "tutored" by the normal song in hearing impaired birds and thus may not participate in the motor production of the predeafening song. A consequence may be that, as more "untutored" neurons are incorporated into HVC over time, song structure progressively deteriorates. One way to test this hypothesis further would be to examine the addition of new HVC-RA neurons after deafening in a species that does not appear to alter song output based on changes in auditory feedback, such as the white-crowned sparrow (Konishi 1965).

The adaptive significance of a system that is plastic yet actively maintains stable output under normal circumstances is unclear. In most species, the adult song control system does not appear to exploit its capacity to alter song. Could adult song plasticity simply be a consequence of a system that depends on auditory feedback for song development? If so, why do some species appear to maintain stable adult song in the absence of auditory feedback? Auditory feedback could function in "tuning" new adult neurons to the bird's own song. Another possibility is that subtle changes in song do normally occur in the adult songs of birds such as zebra finches, as has been shown to occur seasonally in white-crowned sparrows and song sparrows. In this case, the use of auditory feedback and the presence of a stored internal model of a bird's stable song could guide the restabilization of songs with subtle seasonal decreases in stereotypy. Finally, plasticity of the song nuclei may be related to their role in song perception. Neural plasticity within the song nuclei may be required for adult birds to learn to recognize the songs of their territory neighbors or mates. In this context the ability to modify perceptual memories of song would continue throughout adulthood.

6. Plasticity of Adult Song Circuits

6.1 Neurogenesis

One of the most striking forms of plasticity in the adult song control system is the ongoing addition of new neurons to HVC. The recruitment of new neurons occurs widely throughout the telencephalon of songbirds, but HVC is the only song nucleus that receives new neurons at a high rate. As much as 1.5% of the HVC neurons in an adult female canary are generated each day (reviewed in Alvarez-Buylla and Kirn 1997; Scharff 2000). Paton and Nottebohm (1984) demonstrated that HVC cells labeled with [³H]thymidine show synaptic and action potentials and respond to auditory stimuli. This was the first demonstration that newly generated cells in HVC are neurons and that they are incorporated into functional circuits. These new neurons in HVC include many interneurons. But at least half of the new HVC neurons project to nucleus RA. Adult neurogenesis is also observed in the brains of other groups of birds, including the chicken (*Gallus domesticus*), Japanese quail (*Coturnix japonica*), parakeet (*Melopsittacus undulatus*), and ring dove (*Streptopelia risola*) (reviewed in Alvarez-Buylla and Kirn 1997). Other than the parakeet, none of these species learn their vocalizations, which indicates that adult neurogenesis is not limited to bird species that have specialized neural circuits for vocal learning. In all of these bird groups, the addition of new neurons is largely limited to the telencephalon.

Neuronal progenitor cells undergo their last division in the walls of the lateral ventricle (reviewed in Alvarez-Buylla and Kirn 1997). The new neurons migrate through the parenchyma, often for several millimeters, before their final differentiation. These young neurons are tightly apposed to radial glia cells during the initial portion of their migration. Most new neurons die relatively soon; the number of labeled neurons seen 40 days after [³H]thymidine injection is only one third of the total number of migrating cells that are labeled at day 20. The cues used by young neurons to find a path through the adult parenchyma, and how the site of final differentiation is determined, are not yet known.

HVC experiences much of its growth after hatching, and adds more than one half of its neurons during the first 4 months (reviewed in Alvarez-Buylla and Kirn 1997; Bottjer and Johnson 1997; Nordeen and Nordeen 1997). The neurons added to HVC after hatching are interneurons and RA-projecting neurons. HVC neurons that project to area X are born before hatching. New neurons are also added to area X after hatching (reviewed in Alvarez-Buylla and Kirn 1997; Nordeen and Nordeen 1997). Song nuclei other than HVC and area X do not incorporate new neurons in adulthood.

6.2 Neuronal Turnover

Neuronal addition in adult telencephalon appears to be related to cell loss. Even though neurons continue to be added to HVC in adult canaries, total neuron number does not increase with age beyond the first 4 months (reviewed in Alvarez-Buylla and Kirn 1997). Support for this suggestion comes from a study in which RA-projecting neurons in adult HVC were labeled in April with a vital retrograde marker; 30–50% of labeled cells were lost and replaced by October (Kirn and Nottebohm, reviewed in Alvarez-Buylla and Kirn 1997). The survival of new neurons varies with season. Most neurons born in the spring die within 4 months. Most neurons born in the fall, however, survive at least 8 months (Alvarez-Buylla and Kirn 1997).

The hypothesis that neuronal recruitment is causally related to cell loss received strong support from a recent study by Scharff and colleagues (reviewed in Scharff 2000). They used targeted photolysis to eliminate selectively either

RA- or area X–projecting neurons in adult male zebra finches and injected [³H]thymidine over the next 10 days. Three months after RA-projecting neurons were eliminated, there was a threefold increase in the number of new RA-projecting neurons in experimental birds, compared with controls. The total number of HVC neurons retrogradely labeled from RA did not differ between the lesioned and control birds, however. Lesions of area X–projecting neurons did not increase the recruitment of newly generated neurons in HVC; the lesioned and control birds did not differ in the number of new neurons in HVC, and the lesioned birds had 60% fewer HVC neurons retrogradely labeled from area X. In some of the birds with lesions of RA-projecting neurons, song structure deteriorated and became more variable within 4 days of the lesion. The quality of song structure in these birds improved, however, by 3 months after the lesion. Upon recovery of song, two birds produced motifs that had not been observed before the lesion. The increased neuronal recruitment to HVC may have allowed adults to learn new song motifs, or to "recover" motifs produced during the plastic song phase but not retained on crystallization.

The incorporation and survival of new neurons to HVC are influenced by auditory experience. In adult zebra finches injected with [³H]thymidine 2–3 weeks after deafening, the number of labeled neurons in HVC 1 month after the injection was 70% lower in deafened birds compared with controls (Wang et al., reviewed in Scharff 2000). Those neurons that were incorporated into HVC of deafened birds, however, survived over the next 3 months, whereas two thirds of the new neurons recruited to HVC in control birds died over this same interval. These results suggest that auditory stimulation increases the short-term recruitment of new neurons. It is not yet known what stage of neuronal incorporation is affected by auditory experience.

The survival of new neurons in adult HVC may also be influenced by song production. In adult male canaries there is a correlation between the rate of song production, expression of the mRNA for BDNF, and the survival of new neurons in HVC over approximately 30 days (Li et al. 2000). As will be discussed below, Rasika et al. (1999) found that BDNF influences neuronal survival in HVC. Li et al. proposed an intriguing scenario in which increased singing up-regulates the synthesis of BDNF, which in turn fosters the survival of HVC neurons. They did not, however, measure circulating testosterone in their birds. As discussed elsewhere in this chapter, T is known to affect song rate, BDNF mRNA expression, and neuronal survival. It is therefore possible that individual differences in plasma testosterone regulated all three of these traits in HVC. Consistent with this suggestion, Alvarez-Borda and Nottebohm (2002) showed that the contribution of T to neuronal recruitment in canary HVC is much more pronounced than that of singing.

The seasonal patterns of neuronal survival discussed in the preceding section are correlated with seasonal changes in the levels of gonadal steroids. The highest rates of HVC neuronal death are preceded by decreases in testosterone levels (Fig. 6.9). Each peak of neuronal death is followed by a peak of new neuron addition when testosterone levels begin to rise again (reviewed in

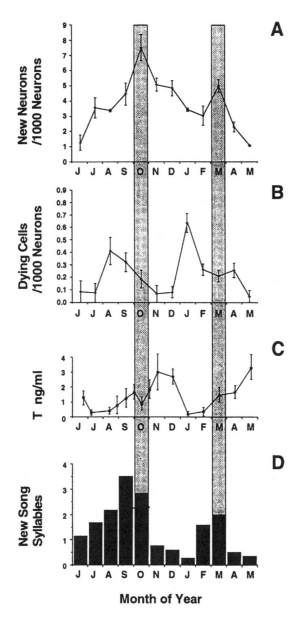

FIGURE 6.9. Seasonal changes in new neuron recruitment to HVC (**A**), cell death in HVC (**B**), testosterone levels (**C**), and the addition of new song syllables to the song repertoire of adult male canaries (**D**). *Shaded vertical bars* highlight the timing of peaks in the ratio of new neurons relative to changes in cell death, hormone levels, and song plasticity. These results suggest a relationship between seasonal HVC cell turnover, testosterone, and the production of new song syllables. [Reprinted with permission from Alvarez-Buylla and Kirn (1997).]

Alvarez-Buylla and Kirn 1997). Decreases in plasma testosterone may therefore lead to cell death, which creates "vacancies" for the subsequent incorporation of new neurons. The subsequent increases of testosterone may provide trophic support that maintains these new neurons (Alvarez-Buylla and Kirn 1997). Support for a trophic role of testosterone comes from the observation that treating adult female canaries with testosterone *after* [³H]thymidine injection triples the number of labeled neurons incorporated into HVC (Rasika et al., reviewed in Alvarez-Buylla and Kirn 1997; Scharff 2000). Hormone treatment does not alter the rate of cell birth in the ventricular zone (reviewed in Alvarez-Buylla and Kirn 1997). It is therefore likely that hormones influence postmitotic events (Nordeen and Nordeen 1997). Hormones could affect, for example, the migration, differentiation, establishment of synaptic connections, and/or survival of new neurons.

6.3 Potential Functions for Neuronal Turnover

The functional significance of neuron recruitment to adult HVC is not clear. It may be related to the ability to learn new songs in adulthood (Nottebohm 1987). The seasonal changes in the HVC neuron turnover are correlated with seasonal changes in song behavior. Canaries produce new song elements at the highest rate during the nonbreeding season when song syllables are produced with less temporal and spectral stereotypy (Fig. 6.9) (reviewed in Alvarez-Buylla and Kirn 1997). These peak periods of song learning coincide with peaks in the incorporation of new neurons into HVC. This correlation between song plasticity and neuronal recruitment suggests that seasonal patterns of neuronal replacement in HVC may provide the neural substrate for seasonal song learning in adult birds; the incorporation of new "naïve" neurons into functional circuits may be a source of plasticity for this adult learning (Nottebohm 1987). Seasonal changes in the recruitment of new neurons to HVC are also observed, however, in the song sparrow, a species that does not learn new songs in adulthood (reviewed in Tramontin and Brenowitz 2000). Like in the canary, neuronal recruitment is greater during the nonbreeding season in the sparrow HVC. Also, sparrows' song structure becomes less stereotyped. This increased song variability in the nonbreeding season, however, does not lead to the development of new song patterns in song sparrows (Nordby et al. 2002). These observations suggest that seasonal changes in adult neuronal recruitment may be functionally related to song stereotypy rather than song learning per se. The decreased stereotypy of song outside the breeding season may have served as a preadaptation that enabled the evolution of adult song learning in some species of birds.

A second functional hypothesis is that the incorporation of new neurons in adult HVC is related to song perception (Nottebohm et al. 1990). As discussed in the preceding sections, HVC neurons receive auditory input and show selective responses to conspecific song. Lesions of HVC disrupt the perception of song. Neuronal addition may provide plasticity for acquiring new perceptual memories of songs each year, which is important for males in the contexts of learning to recognize the songs of their territorial neighbors and for females in learning to recognize the songs of their mates and adjacent territorial males.

A third functional hypothesis is that neuronal turnover in HVC may be a compromise adaptation between two conflicting selective pressures (Nottebohm 1989). On the one hand, birds are relatively long-lived and this favors a brain large enough for the formation and storage of new memories throughout life. On the other hand, flight imposes severe energetic constraints that favor minimizing body (and brain) weight, as shown by the evolution of hollow bones that contain air sacs. Incurring the metabolic costs of neuronal replacement may represent a strategy for balancing these factors. In the breeding season, when song is produced at high rates for mate attraction and territorial defense, birds increase neuron number in HVC. Outside the breeding season, birds conserve energy by decreasing neuron number in HVC (Wennstrom et al. 2001). Song production is greatly reduced or absent in the fall and winter. Territorial defense is relaxed or absent at this time, so there is likely to be little cost to decreased song stereotypy due to regression of HVC (and other song nuclei).

6.4 Seasonal and Social Plasticity

There are pronounced seasonal changes in the morphology of song nuclei in adults of every seasonally breeding species examined (Nottebohm 1981; reviewed in Ball 1999; Tramontin and Brenowitz 2000). The entire volumes of several song nuclei including HVC, area X, RA, and nXIIts are larger during the spring breeding season than during the autumn and winter. In the most extreme case investigated, the volume of HVC in spotted towhees (*Pipilo maculatus*) nearly triples during the breeding season (Smith 1996). lMAN, however, does not change in volume between seasons (reviewed in Tramontin and Brenowitz 2000).

The seasonal change in HVC volume is primarily due to a large increase in neuron number (Tramontin and Brenowitz 2000). In one study of wild song sparrows, for example, neuron number in HVC increased from about 150,000 in late autumn to 250,000 in early spring (Smith et al., reviewed in Tramontin and Brenowitz 2000). The breeding season increase in neuron number results from the interaction between ongoing neurogenesis and seasonal patterns of neuronal turnover in HVC, as discussed in the preceding section. Neuronal turnover is greatest during the nonbreeding season (Alvarez-Buylla and Kirn 1997; Tramontin and Brenowitz 2000). Elevated levels of testosterone and/or its metabolites seem to decrease the turnover and increase the survival of HVC neurons, thus increasing their numbers during the breeding season (reviewed in Alvarez-Buylla and Kirn 1997; Tramontin and Brenowitz 2000).

The cellular basis of volumetric growth of RA differs from that seen in HVC. Neuron number does not change seasonally in RA, but neuron size, spacing, dendritic arborizations, and the sizes of pre- and postsynaptic profiles are greater in the breeding season (reviewed in Tramontin and Brenowitz 2000). These seasonal patterns of dendritic change suggest that synaptic efficacy in RA is enhanced during the breeding season (DeVoogd 1991).

Several lines of evidence suggest that testosterone (or its active metabolites) is the primary physiological cue that mediates seasonal changes in the song nuclei. As discussed previously, gonadal steroid receptors are present in HVC, RA, and nXIIts. Seasonal patterns of circulating T correlate positively with the seasonal growth pattern of the song nuclei (reviewed in Tramontin and Brenowitz 2000; Schlinger and Brenowitz 2002). Castration severely attenuates the seasonal growth of the song regions (e.g., Smith et al. 1997). Exogenous T induces growth of the song nuclei in adult female birds, in castrated males, and in nonbreeding males in the fall and winter (reviewed in Nottebohm 1987; Wennstrom et al. 2001; Schlinger and Brenowitz 2002).

T appears to induce growth of the adult song circuits by acting directly on HVC, which then stimulates growth of HVC's efferent nuclei via transsynaptic effects. HVC grows rapidly in response to exposure to breeding levels of T, whereas RA and area X grow more slowly (reviewed in Tramontin and Brenowitz 2000). Unilateral lesion of HVC selectively blocks the growth of the ipsilateral, but not contralateral, RA and area X in response to exposure to breeding T levels and photoperiod (Brenowitz and Lent 2001). Implanting T adjacent to HVC unilaterally stimulates the growth of the ipsilateral, but not contralateral, HVC, RA, and area X (Brenowitz and Lent 2002). Implanting T adjacent to RA, however, does not stimulate growth of any song nuclei. These results suggest that direct T stimulation of HVC is both necessary and sufficient for growth of the song control circuits. It is notable that RA does not grow in response to high plasma T in the absence of afferent input from HVC, even though RA neurons have abundant ARs. Also noteworthy is the observation that nXIIts grows in response to systemic T implants even when the ipsilateral HVC is lesioned, and does not grow when T is implanted adjacent to the ipsilateral HVC. These results may indicate that nXIIts, which has high levels of AR, does not require an intact HVC to grow in response to high plasma T. T may act directly on the motor neurons of nXIIts and/or on the AR-containing syringeal muscles innervated by these neurons, which might then have a retrograde trophic effect on the motor neurons.

The trophic effect of T on HVC may be mediated, at least partially, through BDNF. Treatment of adult female canaries with T increases protein synthesis and BDNF-like immunoreactivity in HVC (Konishi 1985; Rasika et al. 1999). Infusion of BDNF in the parenchyma adjacent to HVC mimics the effects of T, increasing neuronal survival in HVC and increasing its volume. Also, infusing neutralizing antibodies to BDNF blocks the effects of T on neuronal survival within and volumetric growth of HVC (Rasika et al. 1999). It will be interesting to determine whether BDNF or other growth factors similarly influence the effects of gonadal steroids on seasonal growth of the song circuits.

The seasonal growth of the song nuclei can be modulated by factors other than photoperiod. In the laboratory, social cues from sexually receptive female white-crowned sparrows enhanced the photo-induced growth of two song nuclei in their male cagemates (reviewed in Tramontin and Brenowitz 2000). HVC and RA were 20% and 15% larger, respectively, in males housed with females on long springlike days than in males housed similarly without females.

7. Summary and Conclusions

Over the 50 years that birdsong has been studied in detail, it has become clear that song is characterized by extensive plasticity that spans from early juvenile life into adulthood. The plasticity observed in the song system has stimulated the search for comparable patterns of plasticity in mammalian brains. For example, the first observation of extensive ongoing replacement of projection neurons in the forebrain of adult endothermic vertebrates was in the song system (Goldman and Nottebohm 1983). Spurred by this observation, recent research has suggested ongoing neuronal turnover in the cortex of nonhuman primates and humans (reviewed in Gross 2000). Another example involves the occurrence of seasonal changes in the morphology of song nuclei. Following the initial report of such seasonal plasticity in the song system (Nottebohm 1981), seasonal changes in morphology, pharmacology, or physiology have also been observed in the brains of every vertebrate class, including humans (Tramontin and Brenowitz 2000). The song system has thus emerged as a leading model of plasticity in the vertebrate brain.

A number of open questions should be addressed by future studies of plasticity in the song system: (1) What is the neural basis of "innate" auditory preferences for conspecific song? (2) What physiological and experiential factors determine the opening and closing of the sensitive period for memorization of song models? (3) How do age-limited and open-ended song learning species differ in the mechanisms underlying plasticity of song development? (4) How do steroid hormones, neurotrophins, neurotransmitters, and their receptors interact during developmental and adult plasticity? (5) What is the functional role of selective responses to a bird's own song observed in the song nuclei of anesthetized birds? (6) What is the role of nuclei in the anterior forebrain pathway in mediating adult song plasticity? (7) What is the role of ongoing auditory feedback in the maintenance of crystallized adult song? (8) What is the functional significance of ongoing neuronal replacement in the adult song system? (9) How and why do seasonal and social cues mediate plasticity of the song circuits? (10) What are the molecular determinants of song system plasticity at different stages of life? These questions suggest that the song system will continue to thrive as a model for understanding changes in neural structure and function in years to come.

Acknowledgments. We thank Pablo Monsivais and Michelle Solis for helpful discussion of the manuscript. EAB is supported by NIH MH53032 and the

Bloedel Hearing Research Center. SMNW is supported by NIH DC00287 and DC05087, and the University of Washington Royalty Research Fund.

References

Adret P (1993) Operant conditioning, song learning and imprinting to taped song in the zebra finch. Anim Behav 46:149–159.

Airey DC, Kroodsma DE, DeVoogd TJ (2000) Differences in the complexity of song tutoring cause differences in the amount learned and in dendritic spine density in a songbird telencephalic song control nucleus. Neurobiol Learn Mem 73:274–281.

Akutagawa E, Konishi M (1998) Transient expression and transport of brain-derived neurotrophic factor in male zebra finch's song system during vocal development. Proc Natl Acad Sci USA 95:11429–11434.

Alvarez-Borda B, Nottebohm F (2002) Gonads and singing play separate, additive roles in new neuron recruitment in adult canary brain. J Neurosci 22:8684–8690.

Alvarez-Buylla A, Kirn JR (1997) Birth, migration, incorporation, and death of vocal control neurons in adult songbirds. J Neurobiol 33:585–601.

Alvarez-Buylla A, Kirn JR, Nottebohm F (1990) Birth of projection neurons in adult avian brain may be related to perceptual or motor learning [published erratum appears in Science 1990 Oct 19;250(4979):360]. Science 249:1444–1446.

Arnold AP (1997) Sexual differentiation of the zebra finch song system: positive evidence, negative evidence, null hypotheses, and a paradigm shift. J Neurobiol 33:572–584.

Ball GF (1999) Neuroendocrine basis of seasonal changes in vocal behavior among songbirds. In: Konishi M, Hauser M (eds), The Design of Animal Communication. Cambridge, MA: The MIT Press, pp. 213–253.

Baptista LF, Morton ML (1981) Interspecific song acquisition by a white-crowned sparrow. Auk 98:383–385.

Baptista LF, Morton ML (1988) Song learning in montane white-crowned sparrows: from whom and when. Anim Behav 36:1753–1764.

Baptista LF, Petrinovich L (1984) Social interaction, sensitive phases and the song template hypothesis in the white-crowned sparrow. Anim Behav 32:172–181.

Baptista LF, Petrinovich L (1986) Song development in the white-crowned sparrow: social factors and sex differences. Anim Behav 34:1359–1371.

Basham ME, Sohrabji F, Singh TD, Nordeen EJ, Nordeen KW (1999) Developmental regulation of NMDA receptor 2B subunit mRNA and ifenprodil binding in the zebra finch anterior forebrain. J Neurobiol 39:155–167.

Beecher MD, Campbell SE, Nordby JC (2000a) Territory tenure in song sparrows is related to song sharing with neighbours, but not to repertoire size. Anim Behav 59:29–37.

Beecher, MD, Campbell ES, Burt J, Hill C, Nordby JC (2000b) Song-type matching between neighboring sparrows. Anim Behav 59:21–27.

Benton S, Nelson DA, Marler P, DeVoogd TJ (1998) Anterior forebrain pathway is needed for stable song expression in adult male white-crowned sparrows (*Zonotrichia leucophrys*). Behav Brain Res 96:135–150.

Bottjer, SW (2002) Neural strategies for learning during sensitive periods of development. J Comp Physiol A 188:917–928.

Bottjer SW, Johnson F (1997) Circuits, hormones, and learning: vocal behavior in song-birds. J Neurobiol 33:602–618.

Bottjer SW, Miesner EA, Arnold AP (1984) Forebrain lesions disrupt development but not maintenance of song in passerine birds. Science 224:901–903.

Braaten RF, Reynolds K (1999) Auditory preference for conspecific song in isolation-reared zebra finches. Anim Behav 58:105–111.

Brainard MS, Doupe AJ (2000a) Interruption of a basal ganglia-forebrain circuit prevents plasticity of learned vocalizations. Nature 404:762–766.

Brainard MS, Doupe AJ (2000b) Auditory feedback in learning and maintenance of vocal behaviour. Nat Rev Neurosci 1:31–40.

Brenowitz EA (1991) Altered perception of species-specific song by female birds after lesions of a forebrain nucleus. Science 251:303–305.

Brenowitz EA (1997) Comparative approaches to the avian song system. J Neurobiol 33:517–531.

Brenowitz EA, Kroodsma DE (1996) The neuroethology of birdsong. In: Kroodsma DE, Miller EH (eds), Ecology and Evolution of Acoustic Communication in Birds. Ithaca: Cornell University Press, pp. 269–281.

Brenowitz EA, Lent K (2001) Afferent input is necessary for seasonal growth and main-tenance of adult avian song control circuits. J Neurosci 21:2320–2329.

Brenowitz EA, Lent K (2002) Act locally and think globally: intracerebral testosterone implants induce seasonal-like growth of adult avian song control circuits. Proc Natl Acad Sci USA 99:12421–12426.

Brenowitz EA, Margoliash D, Nordeen KW (1997) An introduction to birdsong and the avian song system. J Neurobiol 33:495–500.

Burek MJ, Nordeen KW, Nordeen EJ (1991) Neuron loss and addition in developing zebra finch song nuclei are independent of auditory experience during song learning. J Neurobiol 22:215–223.

Burt JM, Lent KL, Beecher MD, Brenowitz EA (2000) Lesions of the anterior forebrain song control pathway in female canaries affect song perception in an operant task. J Neurobiol 42:1–13.

Capsius B, Leppelsack H (1999) Response patterns and their relationship to frequency analysis in auditory forebrain centers of a songbird. Hear Res 136:91–99.

Cardin JA, Schmidt MF (2003) Song system auditory responses are stable and highly tuned during sedation, rapidly modulated and unselective during wakefulness, and suppressed by arousal. J Neurophysiol 90:2884–2899.

Catchpole CK, Slater PJB (1995) Bird song: biological themes and variations. Cam-bridge, UK: Cambridge University Press.

Clayton NS (1987) Song learning in Bengalese finches: a comparison with zebra finches. Ethology 76:247–255.

Clayton NS (1989) The effects of cross-fostering on selective song learning in estrildid finches. Behaviour 109:163–175.

Corwin JT, Cotanche DA (1988) Regeneration of sensory hair cells after acoustic trauma. Science 240:1772–1774.

Dave AS, Margoliash D (2000) Song replay during sleep and computational rules for sensorimotor vocal learning. Science 290:812–816.

Dave AS, Yu AC, Margoliash D (1998) Behavioral state modulation of auditory activity in a vocal motor system. Science 282:2250–2254.

Del Negro C, Gahr M, Leboucher G, Kreutzer M (1998) The selectivity of sexual responses to song displays: effects of partial chemical lesion of the HVC in female canaries. Behav Brain Res 96:151–159.

DeVoogd TJ (1991) Endocrine modulation of the development and adult function of the avian song system. Psychoneuroendocrinology 16:41–66.

DeVoogd TJ, Cardin JA, Szekely T, Buki J, Newman SW (1996) Relative volume of lMAN in female warbler species varies with the number of songs produced by conspecific males. Soc Neurosci Abstr 22:755.

DeWolfe BB, Baptista LF, Petrinovich L (1989) Song development and territory establishment in Nuttall's white-crowned sparrows. Condor 91:397–407.

Dietrich K (1980) Model choice in the song development of young male Bengalese finches. Z Tierpsychol 52:57–76.

Dittrich F, Feng Y, Metzdorf R, Gahr M (1999) Estrogen-inducible, sex-specific expression of brain-derived neurotrophic factor mRNA in a forebrain song control nucleus of the juvenile zebra finch. Proc Natl Acad Sci USA 96:8241–8246.

Dooling R, Searcy M (1980) Early perceptual selectivity in the swamp sparrow. Dev Psychobiol 13:499–506.

Dooling RJ, Searcy MH (1981) A comparison of auditory evoked potentials in two species of sparrow. Physiol Psychol 9:293–298.

Doupe AJ (1997) Song- and order-selective neurons in the songbird anterior forebrain and their emergence during vocal development. J Neurosci 17:1147–1167.

Doupe AJ, Konishi M (1991) Song-selective auditory circuits in the vocal control system of the zebra finch. Proc Natl Acad Sci USA 88:11339–11343.

Doupe AJ, Kuhl PK (1999) Birdsong and human speech: common themes and mechanisms. Annu Rev Neurosci 22:567–631.

Doupe AJ, Solis MM (1997) Song- and order-selective neurons develop in the songbird anterior forebrain during vocal learning. J Neurobiol 33:694–709.

Eales LA (1985) Song learning in zebra finches: some effects of song model availability on what is learnt and when. Anim Behav 33:1293–1300.

Eales LA (1989) The influences of visual and vocal interaction on song learning in zebra finches. Anim Behav 37:507–508.

Eens M (1992) Song learning in captive European starlings, Sturnus vulgaris. Anim Behav 44:1131–1143.

Esser KH, Condon CJ, Suga N, Kanwal JS (1997) Syntax processing by auditory cortical neurons in the FM-FM area of the mustached bat Pteronotus parnellii. Proc Natl Acad Sci USA 94:14019–14024.

Fortune ES, Margoliash D (1995) Parallel pathways and convergence onto HVc and adjacent neostriatum of adult zebra finches (Taeniopygia guttata). J Comp Neurol 360: 413–441.

Foster EF, Bottjer SW (1998) Axonal connections of the high vocal center and surrounding cortical regions in juvenile and adult male zebra finches. J Comp Neurol 397: 118–138.

Gahr M, Wild JM (1997) Localization of androgen receptor mRNA-containing cells in avian respiratory-vocal nuclei: an in situ hybridization study. J Neurobiol 33:865–876.

Gahr M, Flugge G, Guttinger HR (1987) Immunocytochemical localization of estrogen-binding neurons in the songbird brain. Brain Res 402:173–177.

Gahr M, Guttinger HR, Kroodsma DE (1993) Estrogen receptors in the avian brain: survey reveals general distribution and forebrain areas unique to songbirds. J Comp Neurol 327:112–122.

Gehr DD, Capsius B, Grabner P, Gahr M, Leppelsack HJ (1999) Functional organisation of the field-L-complex of adult male zebra finches. NeuroReport 10:375–380.

Gehr DD, Hofer SB, Marquardt D, Leppelsack H (2000) Functional changes in field L complex during song development of juvenile male zebra finches. Brain Res Dev Brain Res 125:153–165.

George I, Cousillas H, Richard JP, Hausberger M (2002) Song perception in the European starling: hemispheric specialisation and individual variations. C R Biol 325:197–204.

Goldman SA, Nottebohm F (1983) Neuronal production, migration, and differentiation in a vocal control nucleus of the adult female canary brain. Proc Natl Acad Sci USA 80:2390–2394.

Grace JA, Amin N, Singh NC, Theunissen FE (2003) Selectivity for conspecific song in the zebra finch auditory forebrain. J Neurophysiol 89:472–487.

Gross CG (2000) Neurogenesis in the adult brain: death of a dogma. Nat Rev Neurosci 1:67–73.

Hahnloser RH, Kozhevnikov AA, Fee MS (2002) An ultra-sparse code underlies the generation of neural sequences in a songbird. Nature 419:65–70.

Hamilton KS, King AP, Sengelaub DR, West MJ (1997) A brain of her own: a neural correlate of song assessment in a female songbird. Neurobiol Learn Mem 68:325–332.

Hausberger M, Leppelsack E, Richard J, Leppelsack HJ (2000) Neuronal bases of categorization in starling song. Behav Brain Res 114:89–95.

Herrmann K, Arnold AP (1991) Lesions of HVc block the developmental masculinizing effects of estradiol in the female zebra finch song system. J Neurobiol 22:29–39.

Hessler NA, Doupe AJ (1999a) Social context modulates singing-related neural activity in the songbird forebrain. Nat Neurosci 2:209–211.

Hessler NA, Doupe AJ (1999b) Singing-related neural activity in a dorsal forebrain-basal ganglia circuit of adult zebra finches. J Neurosci 19:10461–10481.

Hough GE, II, Volman SF (2002) Short-term and long-term effects of vocal distortion on song maintenance in zebra finches. J Neurosci 22:1177–1186.

Hough GE, II, Nelson DA, Volman SF (2000) Re-expression of songs deleted during vocal development in white-crowned sparrows, Zonotrichia leucophrys. Anim Behav 60:279–287.

Immelmann K (1969) Song development in the zebra finch and other estrildid finches. In: Hinde RA (ed), Bird Vocalizations. Cambridge, UK: Cambridge University Press, pp. 61–77.

Jacobs EC, Arnold AP, Campagnoni AT (1999) Developmental regulation of the distribution of aromatase- and estrogen-receptor- mRNA-expressing cells in the zebra finch brain. Dev Neurosci 21:453–472.

Janata P, Margoliash D (1999) Gradual emergence of song selectivity in sensorimotor structures of the male zebra finch song system. J Neurosci 19:5108–5118.

Jenkins PF (1978) Cultural transmission of song patterns and dialect development in a free-living bird population. Anim Behav 26:50–78.

Johnson F, Hohmann SE, DiStefano PS, Bottjer SW (1997) Neurotrophins suppress apoptosis induced by deafferentation of an avian motor-cortical region. J Neurosci 17:2101–2111.

Johnson F, Soderstrom K, Whitney O (2002) Quantifying song bout production during zebra finch sensory-motor learning suggests a sensitive period for vocal practice. Behav Brain Res 131:57–65.

Katz LC, Gurney ME (1981) Auditory responses in the zebra finch's motor system for song. Brain Res 221:192–197.

Kirn JR, Alvarez-Buylla A, Nottebohm F (1991) Production and survival of projection neurons in a forebrain vocal center of adult male canaries. J Neurosci 11:1756–1762.

Kobayashi K, Uno H, Okanoya K (2001) Partial lesions in the anterior forebrain pathway affect song production in adult Bengalese finches. NeuroReport 12:353–358.

Konishi M (1964) Effects of deafening on song development in two species of Juncos. Condor 66:85–102.

Konishi M (1965) The role of auditory feedback in the control of vocalization in the white-crowned sparrow. Z Tierpsychol 22:770–783.

Konishi M (1985) Birdsong: from behavior to neuron. Annu Rev Neurosci 8:125–170.

Konishi M, Nottebohm F (1969) Experimental studies in the ontogeny of avian vocalizations. In: Hinde RA (ed), Bird Vocalizations: Their Relations to Current Problems in Biology and Psychology; Essays Presented to W.H. Thorpe. Cambridge, UK: Cambridge University Press.

Kroodsma DE (1978) Aspects of learning in the ontogeny of birdsong. In: Burghart GM, Bekoff, MC (eds), The Development of Behavior: Comparative and Evolutionary Aspects. New York: Garland Press, pp. 215–230.

Kroodsma DE, Pickert R (1980) Environmentally dependent sensitive periods for avian vocal learning. Nature 288:477–479.

Leonardo A, Konishi M (1999) Decrystallization of adult birdsong by perturbation of auditory feedback. Nature 399:466–470.

Lewicki MS (1996) Intracellular characterization of song-specific neurons in the zebra finch auditory forebrain. J Neurosci 16:5855–5863.

Lewicki MS, Arthur BJ (1996) Hierarchical organization of auditory temporal context sensitivity. J Neurosci 16:6987–6998.

Lewicki MS, Konishi M (1995) Mechanisms underlying the sensitivity of songbird forebrain neurons to temporal order. Proc Natl Acad Sci USA 92:5582–5586.

Li XC, Jarvis ED, Alvarez-Borda B, Lim DA, Nottebohm F (2000) A relationship between behavior, neurotrophin expression, and new neuron survival. Proc Natl Acad Sci USA 97:8584–8589.

Lombardino AJ, Nottebohm F (2000) Age at deafening affects the stability of learned song in adult male zebra finches. J Neurosci 20:5054–5064.

MacDougall-Shackleton SA, Hulse SH, Ball GF (1998) Neural bases of song preferences in female zebra finches (*Taeniopygia guttata*). NeuroReport 9:3047–3052.

Maekawa M, Uno H (1996) Difference in selectivity to song note properties between the vocal nuclei of the zebra finch. Neurosci Res 18(Suppl):709.

Marean GC, Burt JM, Beecher MD, Rubel EW (1993) Hair cell regeneration in the European starling (*Sturnus vulgaris*): recovery of pure-tone detection thresholds. Hear Res 71:125–136.

Margoliash D (1983) Acoustic parameters underlying the responses of song-specific neurons in the white-crowned sparrow. J Neurosci 3:1039–1057.

Margoliash D (1986) Preference for autogenous song by auditory neurons in a song system nucleus of the white-crowned sparrow. J Neurosci 6:1643–1661.

Margoliash D (1997) Functional organization of forebrain pathways for song production and perception. J Neurobiol 33:671–693.

Margoliash D, Fortune ES (1992) Temporal and harmonic combination-sensitive neurons in the zebra finch's HVc. J Neurosci 12:4309–4326.

Marler P (1952) Variations in the song of the chaffinch *Fringilla ceolebs*. Ibis 94:458–472.

Marler P (1970) A comparative approach to vocal learning: song development in white-crowned sparrows. J Comp Physiol Psychol 71:1–25.

Marler P (1997) Three models of song learning: evidence from behavior. J Neurobiol 33:501–516.

Marler P, Peters S (1977) Selective vocal learning in a sparrow. Science 198:519–521.

Marler P, Peters S (1981) Sparrows learn adult song and more from memory. Science 213:780–782.

Marler P, Peters, S (1982a) Developmental overproduction and selective attrition: new processes in the epigenesis of birdsong. Dev Psychobiol 15:369–378.

Marler P, Peters, S (1982b) Subsong and plastic song: their role in the vocal learning process. In: Acoustic Communication in Birds, Volume 2 (Kroodsma, DE and Miller, EH, eds), pp. 25–50. New York: Academic Press.

Marler P, Peters S (1987) A sensitive period for song acquisition in the song sparrow, *Melospiza melodia*: a case of age-limited learning. Ethol (formerly Z Tierpsychol) 76: 89–100.

Marler P, Peters S (1988a) Sensitive periods for song acquisition from tape recordings and live tutors in the swamp sparrow, *Melospizia georgiana*. Ethol (formerly Z Tierpsychol) 77:76–84.

Marler P, Peters S (1988b) The role of song phonology and syntax in vocal learning preferences in the song sparrow, *Melospiza melodia*. Ethol (formerly Z Tierpsychol) 77:125–149.

Marler P, Nelson, D (1992) Neuroselection and song learning in birds: species universals in culturally transmitted behavior. Semin Neurosci 4:415–423.

Marler P, Sherman V (1983) Song structure without auditory feedback: emendations of the auditory template hypothesis. J Neurosci 3:517–531.

Marler P, Tamura M (1964) Culturally transmitted patterns of vocal behavior in a sparrow. Science 146:1483–1486.

Marler P, Waser MS (1977) Role of auditory feedback in canary song development. J Comp Physiol Psychol 91:8–16.

Marler P, Peters S, Ball GF, Dufty AM Jr, Wingfield JC (1988) The role of sex steroids in the acquisition and production of birdsong. Nature 336:770–772.

McCasland JS, Konishi M (1981) Interaction between auditory and motor activities in an avian song control nucleus. Proc Natl Acad Sci USA 78:7815–7819.

McGregor PK, Krebs JR (1989) Song learning in adult great tits (*Parus major*): effects of neighbors. Behav 108:139–159.

McKenna TM, Weinberger, NM, Diamond DM (1989) Responses of single auditory cortical neurons to tone sequences. Brain Res 27:142–153.

Metzdorf R, Gahr M, Fusani L (1999) Distribution of aromatase, estrogen receptor, and androgen receptor mRNA in the forebrain of songbirds and nonsongbirds. J Comp Neurol 407:115–129.

Mooney R (1992) Synaptic basis for developmental plasticity in a birdsong nucleus. J Neurosci 12:2464–2477.

Mooney R (2000) Different subthreshold mechanisms underlie song selectivity in identified HVc neurons of the zebra finch [In Process Citation]. J Neurosci 20:5420–5436.

Mooney R, Konishi M (1991) Two distinct inputs to an avian song control nucleus activate different glutamate receptor subtypes on individual neurons. PNAS 88:4075–4079.

Mooney R, Hoese W, Nowicki S (2001) Auditory representation of the vocal repertoire in a songbird with multiple song types. Proc Natl Acad Sci USA 98:12778–12783.

Morrison RG, Nottebohm F (1993) Role of a telencephalic nucleus in the delayed song learning of socially isolated zebra finches. J Neurobiol 24:1045–1064.

Mountjoy DJ, Lemon RE (1995) Extended song learning in wild European starlings. Anim Behav 49:357–366.

Nelson DA (1992) Song overproduction and selective attrition lead to song sharing in the field sparrow (Spizella pusilla). Behav Ecol Sociobiol 30:415–424.

Nelson DA (1998) External validity and experimental design: the sensitive phase for song learning. Anim Behav 56:487–491.

Nelson DA, Marler P (1993) Innate recognition of song in white-crowned sparrows: a role in selective vocal learning. Anim Behav 46:806–808.

Newman, JD, Wollberg Z (1973) Responses of single neurons in the auditory cortex of squirrel monkey to variants of a single cell type. Exp Neurol 40:821–824.

Nick TA, Konishi M (2001) Dynamic control of auditory activity during sleep: correlation between song response and EEG. Proc Natl Acad Sci USA 98:14012–14016.

Nieder A, Klump GM (1999) Adjustable frequency selectivity of auditory forebrain neurons recorded in a freely moving songbird via radiotelemetry. Hear Res 127:41–54.

Nieder A, Klump GM (2001) Signal detection in amplitude-modulated maskers. II. Processing in the songbird's auditory forebrain. Eur J Neurosci 13:1033–1044.

Nordby JC, Campbell ES, Burt JM, Beecher MD (2000) Social influences during song development in the song sparrow: a laboratory experiment simulating field conditions. Anim Behav 59:1187–1197.

Nordby JC, Campbell ES, Beecher MD (2001) Late song learning in song sparrows. Anim Behav 61:835–846.

Nordby JC, Campbell ES, Beecher MD (2002) Adult song sparrows do not alter their song repertoires. Ethology 108:39–50.

Nordeen KW, Nordeen EJ (1992) Auditory feedback is necessary for the maintenance of stereotyped song in adult zebra finches. Behav Neural Biol 57:58–66.

Nordeen KW, Nordeen EJ (1993) Long-term maintenance of song in adult zebra finches is not affected by lesions of a forebrain region involved in song learning. Behav Neural Biol 59:79–82.

Nordeen KW, Nordeen EJ (1997) Anatomical and synaptic substrates for avian song learning. J Neurobiol 33:532–548.

Nottebohm F (1968) Auditory experience and song development in the chaffinch Fringilla coelebs. Ibis 110:549–568.

Nottebohm F (1981) A brain for all seasons: cyclical anatomical changes in song control nuclei of the canary brain. Science 214:1368–1370.

Nottebohm F (1987) Plasticity in adult avian central nervous system: possible relation between hormones, learning, and brain repair. In: Plum F (ed), Handbook of Physiology, Section 1 Baltimore: Williams & Wilkins, pp. 85–108.

Nottebohm F (1989) From bird song to neurogenesis. Sci Am 260:74–79.

Nottebohm F, Stokes TM, Leonard CM (1976) Central control of song in the canary, Serinus canarius. J Comp Neurol 165:457–486.

Nottebohm F, Kasparian S, Pandazis C (1981) Brain space for a learned task. Brain Res 213:99–109.

Nottebohm F, Nottebohm, M, Crane L (1986) Developmental and seasonal changes in canary song and their relation to changes in the anatomy of song-control nuclei. Behav Neural Biol 46:445–471.

Nottebohm F, Alvarez-Buylla A, Cynx J, Kirn J, Ling CY, Nottebohm M, Suter R, Tolles A, Williams H (1990) Song learning in birds: the relation between perception and production. Philos Trans R Soc Lond B Biol Sci 329:115–124.

Nottebohm F, O'Loughlin B, Gould K, Yohay K, Alvarez-Buylla A (1994) The life span of new neurons in a song control nucleus of the adult canary brain depends on time of year when these cells are born. Proc Natl Acad Sci USA 91:7849–7853.

Okanoya K, Dooling RJ (1987) Hearing in passerine and psittacine birds: a comparative study of absolute and masked auditory thresholds. J Comp Psychol 101:7–15.

Okanoya K, Yamaguchi A (1997) Adult Bengalese finches (*Lonchura striata* var. *domestica*) require real-time auditory feedback to produce normal song syntax. J Neurobiol 33:343–356.

Paton JA, Nottebohm FN (1984) Neurons generated in the adult brain are recruited into functional circuits. Science 225:1046–1048.

Perkel DJ, Farries MA (2000) Complementary 'bottom-up' and 'top-down' approaches to basal ganglia function. Curr Opin Neurobiol 10:725–731.

Perlman WR, Ramachandran B, Arnold AP (2003) Expression of androgen receptor mRNA in the late embryonic and early posthatch zebra finch brain. J Comp Neurol 455:513–530.

Petrinovich L, Baptista LF (1987) Song development in the white-crowned sparrow: modification of learned song. Anim Behav 35:961–974.

Price PH (1979) Developmental determinants of structure in zebra finch song. J Comp Physiol Psychol 93:268–277.

Raikow RJ (1982) Monophyly of the passeriformes: test of a phylogenetic hypothesis. Auk 99:431–445.

Rasika S, Alvarez-Buylla A, Nottebohm F (1999) BDNF mediates the effects of testosterone on the survival of new neurons in an adult brain. Neuron 22:53–62.

Ribeiro S, Mello CV (2000) Gene expression and synaptic plasticity in the auditory forebrain of songbirds. Learn Mem 7:235–243.

Rosen MJ, Mooney R (2000) Intrinsic and extrinsic contributions to auditory selectivity in a song nucleus critical for vocal plasticity. J Neurosci 20:5437–5448.

Ryals BM, Rubel EW (1988) Hair cell regeneration after acoustic trauma in adult Coturnix quail. Science 240:1774–1776.

Sakaguchi H, Wada K, Maekawa M, Watsuji T, Hagiwara M (1999) Song-induced phosphorylation of cAMP response element-binding protein in the songbird brain. J Neurosci 19:3973–3981.

Scharff C (2000) Chasing fate and function of new neurons in adult brains. Curr Opin Neurobiol 10:774–783.

Scharff C, Nottebohm F (1991) A comparative study of the behavioral deficits following lesions of various parts of the zebra finch song system: implications for vocal learning. J Neurosci 11:2896–2913.

Scharff C, Nottebohm F, Cynx J (1998) Conspecific and heterospecific song discrimination in male zebra finches with lesions in the anterior forebrain pathway. J Neurobiol 36:81–90.

Scharff C, Kirn JR, Grossman M, Macklis JD, Nottebohm F (2000) Targeted neuronal death affects neuronal replacement and vocal behavior in adult songbirds. Neuron 25:481–492.

Schlinger BA, Brenowitz EA (2002) Neural and hormonal control of birdsong. In: Phaff DW (ed), Hormones, Brain and Behavior. New York: Academic Press.

Schmidt MF, Konishi M (1998) Gating of auditory responses in the vocal control system of awake songbirds. Nat Neurosci 1:513–518.

Scott LL, Nordeen EJ, Nordeen KW (2000) The relationship between rates of HVc neuron addition and vocal plasticity in adult songbirds. J Neurobiol 43:79–88.

Sen K, Theunissen FE, Doupe AJ (2001) Feature analysis of natural sounds in the songbird auditory forebrain. J Neurophysiol 86:1445–1458.

Sibley C, Ahlquist J, Monroe BJ (1988) A classification of the living birds of the world based on DNA–DNA hybridization studies. Auk 105:409–423.

Singh TD, Basham ME, Nordeen EJ, Nordeen KW (2000) Early sensory and hormonal experience modulate age-related changes in NR2B mRNA within a forebrain region controlling avian vocal learning. J Neurobiol 44:82–94.

Slater PB, Eales LA, Clayton NS (1988) Song learning in zebra finches: progress and prospects. Adv Study Behav 18:1–34.

Slater PJ, Richards C, Mann NI (1991) Song learning in zebra finches exposed to a series of tutors during the sensitive phase. Ethology (formerly Z Tierpsychol) 88:163–171.

Smith GT (1996) Seasonal plasticity in the song nuclei of wild rufous-sided towhees. Brain Res 734:79–85.

Smith GT, Brenowitz EA, Wingfield JC, Baptista LF (1995) Seasonal changes in song nuclei and song behavior in Gambel's white-crowned sparrows. J Neurobiol 28:114–125.

Smith GT, Brenowitz EA, Wingfield JC (1997) Roles of photoperiod and testosterone in seasonal plasticity of the avian song control system. J Neurobiol 32:426–442.

Soha JA, Marler P (2000) A species-specific acoustic cue for selective song learning in the white-crowned sparrow. Anim Behav 60:297–306.

Sohrabji F, Nordeen EJ, Nordeen KW (1990) Selective impairment of song learning following lesions of a forebrain nucleus in the juvenile zebra finch. Behav Neural Biol 53:51–63.

Solis MM, Doupe AJ (1997) Anterior forebrain neurons develop selectivity by an intermediate stage of birdsong learning [published erratum appears in J Neurosci 1999 Jan 1;19(1):preceding I]. J Neurosci 17:6447–6462.

Solis MM, Doupe AJ (1999) Contributions of tutor and bird's own song experience to neural selectivity in the songbird anterior forebrain. J Neurosci 19:4559–4584.

Solis MM, Doupe AJ (2000) Compromised neural selectivity for song in birds with impaired sensorimotor learning. Neuron 25:109–121.

Stripling R, Volman SF, Clayton DF (1997) Response modulation in the zebra finch neostriatum: relationship to nuclear gene regulation. J Neurosci 17:3883–3893.

Suthers RA, Goller F, Pytte C (1999) The neuromuscular control of birdsong. Philos Trans Roy Soc Lond B Biol Sci 354:927–939.

Tchernichovski O, Nottebohm F (1998) Social inhibition of song imitation among sibling male zebra finches. Proc Natl Acad Sci USA 95:8951–8956.

Tchernichovski O, Lints T, Mitra PP, Nottebohm F (1999) Vocal imitation in zebra finches is inversely related to model abundance. Proc Natl Acad Sci USA 96:12901–12904.

ten Cate C (1991) Behaviour-contingent exposure to taped song and zebra finch song learning. Anim Behav 42:857–859.

Theunissen FE, Doupe AJ (1998) Temporal and spectral sensitivity of complex auditory neurons in the nucleus HVc of male zebra finches. J Neurosci 18:3786–3802.

Theunissen FE, Doupe AJ (2000) Spectral-temporal receptive fields of nonlinear auditory neurons obtained using natural sounds. J Neurosci 20:2315–2331.

Thorpe WH (1958) The learning of song patterns by birds, with especial reference to the song of the chaffinch *Fringilla coelebs*. Ibis 100:535–570.

Thorpe WH (1961) Bird Song. Cambridge, UK: Cambridge University Press.

Timmons BA (1982) Physiological factors related to delayed auditory feedback and stuttering: a review. Percept Mot Skills 55:1179–1189.

Tramontin AD, Brenowitz EA (2000) Seasonal plasticity in the adult brain. Trends Neurosci 23:251–258.

Vates GE, Broome BM, Mello CV, Nottebohm F (1996) Auditory pathways of caudal telencephalon and their relation to the song system of adult male zebra finches. J Comp Neurol 366:613–642.

Vicario DS (1991) Organization of the zebra finch song control system: II. Functional organization of outputs from nucleus *Robustus archistriatalis*. J Comp Neurol 309: 486–494.

Vicario DS, Yohay KH (1993) Song-selective auditory input to a forebrain vocal control nucleus in the zebra finch. J Neurobiol 24:488–505.

Volman SF (1993) Development of neural selectivity for birdsong during vocal learning. J Neurosci 13:4737–4747.

Volman SF (1996) Quantitative assessment of song-selectivity in the zebra finch "high vocal center." Comp Physiol A 178:849–862.

Wang N, Aviram R, Kirn JR (1999) Deafening alters neuron turnover within the telencephalic motor pathway for song control in adult zebra finches. J Neurosci 19:10554–10561.

Waser MS, Marler P (1977) Song learning in canaries. J Comp Physiol Psychol 91:1–7.

Wennstrom KL, Reeves BJ, Brenowitz EA (2001) Testosterone treatment increases the metabolic capacity of adult avian song control nuclei. J Neurobiol 48:256–264.

Whaling CS, Solis MM, Doupe AJ, Soha JA, Marler P (1997) Acoustic and neural bases for innate recognition of song. Proc Natl Acad Sci USA 94:12694–12698.

White SA (2001) Learning to communicate. Curr Opin Neurobiol 11:510–520.

White SA, Livingston FS, Mooney R (1999) Androgens modulate NMDA receptor-mediated EPSCs in the zebra finch song system. J Neurophysiol 82:2221–2234.

Wild JM (1997) Neural pathways for the control of birdsong production. J Neurobiol 33:653–670.

Williams H, McKibben JR (1992) Changes in stereotyped central motor patterns controlling vocalization are induced by peripheral nerve injury. Behav Neural Biol 57: 67–78.

Williams H, Mehta N (1999) Changes in adult zebra finch song require a forebrain nucleus that is not necessary for song production. J Neurobiol 39:14–28.

Woolley SM, Rubel EW (1997) Bengalese finches Lonchura *Striata domestica* depend upon auditory feedback for the maintenance of adult song. J Neurosci 17:6380–6390.

Woolley SM, Rubel EW (1999) High-frequency auditory feedback is not required for adult song maintenance in Bengalese finches. J Neurosci 19:358–371.

Woolley SM, Rubel EW (2002) Vocal memory and learning in adult Bengalese finches with regenerated hair cells. J Neurosci 22:7774–7787.

Woolley SM, Wissman AM, Rubel EW (2001) Hair cell regeneration and recovery of auditory thresholds following aminoglycoside ototoxicity in Bengalese finches. Hear Res 153:181–195.

Yu AC, Margoliash D (1996) Temporal hierarchical control of singing in birds. Science 273:1871–1875.

Zevin JD, Seidenberg MS, Bottjer SW (2000) Song plasticity in adult zebra finches exposed to white noise. Soc Neurosci Abstr 26:723.

7

Plasticity in the Auditory System of Insects

Reinhard Lakes-Harlan

1. Introduction

The term "plasticity" covers different aspects and mechanisms in the nervous system. It is, however, still not readily associated with insects for a number of reasons. First, insects often seem to have a stereotyped behavior. Second, scientists established the identified neuron concept, which implies that the very same nerve cells can be identified repeatedly by their function and morphology from individual to individual (Hoyle 1983). Both views led to the belief that information processing in the nervous systems of insects is hardwired rather than plastic. Despite this belief, early studies pointed out that at least some insects, such as bees, are capable of remarkable learning tasks (von Frisch 1914; Thorpe 1939; Horridge 1962; Alloway 1972). Learning, however, requires plastic changes in the nervous system. At present learning and memory are not only well established for different insects (for recent reviews see Hammer and Menzel 1995; Menzel and Müller 1996; Menzel 2001) but also genetically accessible in *Drosophila* (Dubnau and Tully 1998), making insects favorable model organisms, at least for this aspect of plasticity.

Given the many similarities of nervous system development between vertebrates and invertebrates (Goodman 1996), it might not be a surprise that insects also exhibit different aspects of plasticity on various levels within the nervous system (Palka 1984; Murphey 1986b; Rössler and Lakes-Harlan 1999; Meinertzhagen 2001). In this chapter "plasticity" is used to describe functional and structural adjustments made within the neural system in response to internal and external requirements for adaptive auditory information processing. Four aspects of plasticity in the auditory system of insects that are related to each other and that may or may not comprise similar mechanisms are selected:

- Plasticity in the developing auditory system. During embryonic and postembryonic development the nervous system adjusts itself to the individual internal and external environment.
- Activity-dependent modifications of the auditory system. These plastic events

comprise any form of activity related to functional and structural modifications from development to adult life, including learning and memory.

- Modulatory effects in auditory networks. Neuronal information processing can be modified in relation to the behavioral or hormonal status of the individual.
- Plasticity to restore functions. Lesions and losses can be compensated for in the auditory system, not only during ontogenetic development, but also during adult life. These compensations require plastic changes in the wiring of neurons as well as regenerative effects.

The separation into these four sections is artificial, and some aspects, such as activity-dependent changes, might be dealt with in one section or another. After a short introduction into insect auditory systems [for further details see Hoy (1998) in this series], which is important for a discussion of plasticity, the chapter closes with some general thoughts on plasticity in insects.

2. Auditory System of Insects

Insects have independently evolved auditory systems many times (Fullard and Yack 1993; Hoy and Robert 1996; Yager 1999; Stumpner and von Helversen 2001). Tympanal auditory organs are found on various body parts (except for the head) of different insect taxa. Most of these auditory organs contain similar features such as a thin tympanal membrane backed by an air-filled chamber (acting as a peripheral transduction apparatus) and primary sensory cells within a scolopidial sense organ.

The number of sensory cells in an insect ear is generally between 30 and 80, although some species have only two sensory cells and others have several thousand (Fonseca et al. 2000). The sensory cells project their axons into distinct neuropil areas of the thoracic central nervous system (Fig. 7.1). At least in the species studied, the sensory cells have a precise somatotopic and tonotopic projection order within the central nervous system (CNS).

Information processing, such as extraction of temporal and frequency parameters, can already be performed at the first synaptic station within the thoracic nervous system where afferents synapse onto local and multiganglionic interneurons (Fig. 7.1). Because processing of auditory information, such as determining sound directions, starts at this level, modulatory changes might also be expected in the thoracic ganglia. Interneurons relay the information to other ganglia and to the brain. The number of identified interneurons ranges up to perhaps 30, but in most species many fewer neurons have been described so far.

The brain is an important station for recognizing species-specific sound signals (Schildberger 1984; von Helversen and von Helversen 1997), but not much is known about auditory centers in the brain. Furthermore, neural processing of auditory stimuli might be different among the insects, even within the Orthoptera (Pollack 1998): analysis of temporal parameters in crickets is done

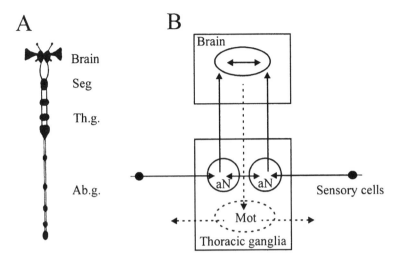

FIGURE 7.1. (**A**) Organization of the CNS of a grasshopper. The brain and subesophageal ganglion (Seg) is located within the head capsule; each thoracic segment contains a thoracic ganglion (Th.g.) and the abdomen is supplied by a chain of abdominal ganglia (Ab.g.). (**B**) Schematic view of many insect auditory systems. Peripherally located sensory cells project into auditory neuropils (aN) within the thoracic ganglia. Ascending interneurons transmit the information to the brain, whereas local interneurons process information at the level of the thoracic ganglia. Efferent information flows from the brain to motor centers in the thoracic ganglia (Mot) which supply the peripheral musculature.

mainly in the brain, whereas in grasshoppers the metathoracic ganglion plays an important role in such an analysis. As a consequence of the multiple evolutionary origin of ears among the insects the networks at which plasticity occurs are not identical, either anatomically or in terms of evolutionary history. Furthermore, evidence is accumulating that the auditory systems are derived from chordotonal proprioceptive systems (Meier and Reichert 1990; Boyan 1993; Fullard and Yack 1993; van Staaden and Römer 1998; Lakes-Harlan et al. 1999). Thus, similar plasticity might also occur in proprioceptive systems.

The variety of auditory systems among the insects will allow for a comparison of plasticity at different evolutionary levels and for different functional tasks (intraspecific communication, predator detection, and host finding) of the system (Pollack 1998). Generally, the acoustic signal must be recognized against a wealth of surrounding noises and localized in space. The intraspecific communication signals seem to be invariant and the overall auditory behavior is genetically fixed. Plasticity, however, might be needed for adjustment of auditory networks including behavior during development, for interaction with other neuronal networks, for repair, and perhaps for evolution of the diverse acoustic communication signals.

3. Plasticity in the Developing Auditory System

Developmental plasticity involves the adjustment of the nervous system to external and/or internal conditions during embryonic and postembryonic stages. Typical examples that might occur during postembryonic maturation are the adjustment of the neuronal cell number (due to proliferation or cell death) or adjustments of neuronal morphology and synaptic connections (pre- and postsynaptic modifications). In insects, adjustment of neuronal cell number seems to occur only rarely in the thoracic nervous system. The cell lineages seem to be rather fixed (Goodman 1982; Doe and Technau 1993) and even cell death is found in a fixed pattern (Truman et al. 1992). Presumably, these neurons do not compete for limited factors, such as neurotrophic factors, which have not yet been found in the insect nervous system [despite some growth-promoting effects in vitro of neurotrophic factors on insect neurons, including auditory sensory cells (Pfahlert and Lakes-Harlan 1997)].

Adjustment of cell number has been found during postembryonic development, especially in the brains of insects (see Section 3.1). In many insects, sensory system receptor cells are constantly added postembryonically and the underlying neural circuits are modified (Chiba et al. 1988; Meinertzhagen 1993). For example, the synaptic strength of "old" sensory neurons is weakened in favor of newly born sensory cells (Chiba et al. 1988). Modifications of neuronal networks are even more evident during metamorphosis in holometabolous (insects with larvae quite unlike the adult and a pupal stage) insects (Weeks and Levine, 1990). It could be shown that during metamorphosis some motorneurons of the larvae survive, whereas some die and other motorneurons undergo remodeling and form new synaptic connections. These neurons are direct targets of the steroid hormone ecdysone (Weeks et al. 1997).

The development of auditory systems has been reviewed by Boyan (1998) in this series and is not repeated here, but some aspects are discussed here from the standpoint of plasticity. Generally, these developmental studies (except for some activity-dependent modifications) do not show a direct plastic internal or external influence on the generation of the auditory system. However, variability in auditory systems can be observed between adult individuals, which might reflect developmental plasticity. Even parthenogenetic clones of grasshoppers exhibit variability of nonauditory neurons in their cell number, axon pathways, and other structural parameters, with some of them clearly epigenetically regulated (Goodman 1978).

3.1 Adjustment of Neuronal Cell Number

Neurons arise during embryonic development from neuroblasts (neurons of the CNS) or from sensory mother cells (peripheral sensory cells). The genetically controlled cell lineage is fixed with a distinct number of progenitors. The auditory receptor cells probably arise from a group of sensory mother cells, similar to recent findings of chordotonal organs in *Drosophila* (zur Lage and Jarman 1999). In most species sensory cells are likely to be born during embryogenesis (Klose 1990; Meier and Reichert 1990). Consequently, their number is usually not altered during postembryonic development (Rössler 1992) despite the gradual maturation of the sense organ (see Section 3.3). Thus, observed variations in the number of species-specific auditory cells arise from variations in the embryonic cell lineages. Katydids (see also Hoy 1998) have an almost linear array of auditory sensory cells within their foreleg tibia. In this array individual cells can be precisely counted and a variation of about 3–8% has been described (Young and Ball 1974; Lakes and Schikorski 1990; Stölting and Stumpner 1998). In auditory organs of other species the number is more difficult to determine, but an electron microscopic study suggests similar variations in grasshoppers (Jacobs and Lakes-Harlan 1999).

Studies on the cell lineage of central auditory interneurons exist only for very few neurons (Boyan 1998). There is still no way to label all of the distributed auditory cells in one individual, resulting in a lack of information about their total number and detailed development. Concerning plasticity, one would expect some variation in cell number, similar to that encountered in the sense organs. Probably most of the proliferation (and apoptosis) takes place during embryogenesis and is therefore terminated long before the auditory system starts functioning (see Section 3.3). Thus, no hints for an adjustment of neuronal cell number in the central auditory system have been found with respect to environmental requirements.

3.2 Embryonic Development of Neuronal Fibers

The early embryonic development of the morphology of sensory afferents as well as interneuronal fibers also seems to follow a fixed pattern. A first study shows that axons of auditory receptor cells of locusts extend along distinct pathways in the periphery as well as within the CNS (Schäffer and Lakes-Harlan 2001). Within the CNS, the axons arborize in their target area and, interestingly, form specific projections according to the different receptor types. Thus, the tonotopic order corresponding to the different types of receptor cells already develops during embryogenesis and without any auditory function. Furthermore, synaptic connections between afferents and target interneurons are almost certainly formed before electrical activity and definitely before *auditory* activity, as has been studied by an expression of synapse-specific protein synapsin in the auditory neuropil (Schäffer and Lakes-Harlan 2001). Thus, the initial outgrowth of afferents and interneurons is activity independent—a feature seen in many other systems as well (Goodman and Shatz 1993). Activity is often required for synaptic wiring (Goodman and Shatz 1993). In insects, however, synaptogenesis also occurs independently of activity (Chiba et al. 1988; Chiba and Murphey 1991). Corresponding to the afferents, the formation of the neurite, axon, and dendrites of auditory interneurons follows a rigid pattern of outgrowth resulting in a rather stereotyped morphology (Boyan 1998).

The ontogenetic development results in the neuronal structure seen in the adult. By comparing different individuals some variation in the neuronal morphology has been described despite their seemingly rigid early development (Fig. 7.2). Receptor fibers of sensory cells at identical peripheral positions in bushcrickets have a slightly variable field of projection in the neuropil (Stölting and Stumpner 1998). Morphological variation has also been seen in individually identifiable auditory interneurons (Lakes et al. 1990). Such variations comprise the position of the cell body, the density of dendritic and axonal processes, and the formation of collateral sprouts (Fig. 7.2). On the other hand, some morphologically indistinguishable auditory interneurons are referred to as "twins." However, such twin neurons can exhibit clear physiological differences (Stumpner 1989) that are a consequence of different synaptic connections despite over-

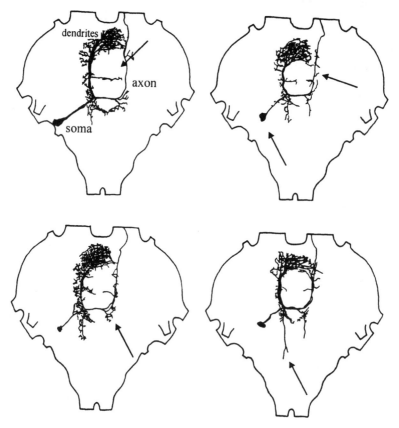

FIGURE 7.2. Morphological variability of an ascending auditory interneuron (AN3) in four individuals of the grasshopper *Chorthippus biguttulus*. The variations (*arrows*) comprise slight differences in soma position (despite probably identical cell lineage), density of collaterals in the dendritic region, and formation of axonal sprouts. [Modified after Stumpner (1989), with permission.]

all similarity. All these variations, including the minor ones, are likely to have some effect on the synaptic connections in the dense neuropil, but so far no study has been performed on the correlation of pre- and postsynaptic neurons. Concerning the overall structure, one should keep in mind that some neurons such as the G neuron have a similar morphology in a variety of species, including those without an auditory organ (Thomas et al. 1984). Thus, the basic structure might be partly independent from the function. During evolution this might allow recruitment of new neurons into auditory networks, perhaps as in the case of an unusual auditory interneuron in bushcrickets (Stumpner 1999): the auditory interneuron AN5–AG7 in the katydid *Ancistrura nigrovitatta* has its dendrites in the prothoracic ganglion (corresponding to the terminals of auditory receptor fibers), but the cell body is located far away in the seventh abdominal ganglion. Except for the prothoracic dendrites, the neuron resembles cercal giant interneurons of insects, and it might be speculated whether and why a change in function from the abdominal cercal system to the prothoracic auditory system has taken place in evolution.

3.3 Postembryonic Maturation

The postembryonic development of the auditory system seems to be a perfect stage for plastic changes because of the gradual maturation of the peripheral sense organ and the acquisition of the hearing capability in hemimetabolous (insects with larva resembling the adult in form) insects, such as crickets and grasshoppers. The gradual development comprises the differentiation of the peripheral structures (trachea, cuticle, tympanal membrane) associated with the tympanal organ (Michel and Petersen 1982; Rössler 1992; Popov et al. 1994). In crickets some sensory neurons might be added during larval development (Ball and Young 1974), although this has not been related to functional changes in the auditory system. The postembryonic differentiation results in gradual changes in auditory sensitivity and tuning from molt to molt (Petersen et al. 1982; Boyan 1983; Rössler 1992), leading to a sensitive hearing capability in the adult insect when tympanal membranes are thin. As far as we know, these postembryonic functional changes are not reflected in synaptic rearrangements such as the projection of receptor cells (therefore their projection has to be called somatotopic and tonotopy is only a consequence of this process) or of changing morphology of interneurons. The G neuron of grasshoppers has the same basic structure in all developmental stages (Boyan 1983), but growth-related plastic changes of synaptic connection, as found in many other systems (Budnik 1996; Davies and Goodman 1998), have still to be analyzed in the auditory system. An age-dependent modification of neuronal structure has been described in the adult auditory system (after the final molt). The ON1 neuron in crickets has an ascending axon in 75% of young adults but only in 30% of older adults (Atkins and Pollack 1986); thus, although this cell is usually characterized as a local interneuron in adults, in some individuals this is not accurate. The function of this axon is unclear, and a regression of axons has otherwise been described previously only for some cells during embryogenesis (Goodman and Spitzer 1979) or during metamorphosis of holometabolous insects—but never during postembryonic life.

3.4 Functional Changes

A large degree of developmental plasticity is known from insect metamorphosis in holometabolous insects (Weeks and Levine 1990; Weeks et al. 1997). Motorneurons can change their synaptic connections and innervate different muscles, with different functions, in the larva and adult. The auditory system of holometabolous insects, such as moths, beetles, and flies, is present only in the adult. Considering the evolutionary descent of the auditory system from chordotonal organs we might expect distinct changes during development. The question arises of whether the adult auditory system has precursor structures (sensory cells, interneurons) in the larvae. In flies, the peripheral auditory system including the sensory neurons develops de novo during metamorphosis (Lakes-Harlan, unpublished results). This fits into the findings that most larval sensory neurons of other fly species, including *Drosophila*, degenerate during metamorphosis (Lakes and Pollack 1990; Williams and Shepherd 1999). However, it seems likely that in moths a larval chordotonal organ is an ontogenetic precursor organ of the adult auditory organ (Blänsdorf and Lakes-Harlan 1993; Hintze-Podufal and Hermanni 1996; Lewis and Fullard 1996). The sensory cells of this organ are not altered in their number but they are altered in function. In the larva the chordotonal organ is a segmental proprioreceptor insensitive to acoustic stimuli, whereas the sensory cells of the adult moth react to airborne sound. Further analysis of the neuronal network behind the sense organs is of major interest for the plasticity in the neuronal network owing to the change in function and for the understanding of evolution of the auditory systems.

4. Activity-Dependent Modifications

Activity- and experience-related structural and behavioral plasticity is an important phenomenon during maturation of vertebrate sensory systems. In the auditory system, for example, adjustments must be made to permit establishment of an auditory space map and phonetic representation (Rauschecker 1999). These processes, which are largely developmental, do not seem to apply to insect auditory systems as auditory function occurs only in adult insects. Thus, for example, stereopsis does not change because the auditory organs are present only in one stage. Furthermore, auditory space might not represent a problem to insects because they use simpler rules for locating a sound source than do vertebrates. It has been suggested, and often experimentally confirmed, that crickets localize a calling male following the rule "turn to the side of the ear most strongly stimulated" (Horseman and Huber 1994), whereby the localization depends on previous recognition (Stabel et al. 1989). For aversive auditory behavior similar rules apply with a sign reversal. Nevertheless, activity also influences the auditory system in insects. The following distinctions partly overlap, but for practical reasons they are separated into three points.

4.1 Habituation

Habituation is one aspect of nonassociative learning and short-term memory that occurs in auditory systems. It is commonly understood as a decrease in the response magnitude during repetitive stimuli; this applies for neurons as well as for behavior.

Many, but not all, auditory neurons show a decrease in response magnitude to repetitive auditory stimuli (Kalmring et al. 1978; Popov and Markovich 1982; Boyan and Altman 1985; Atkins and Pollack 1987). Habituation can already be found at the level of sensory neurons (Givois and Pollack 2000). In locusts habituation of the G neuron depends on the stimulus intensity, duration, and repetition rate (Wolf 1986). This habituation is at least partly a central process, although the mechanism by which this happens is not understood in detail (Ocker and Hedwig 1993). Some neurons with bilateral input, such as the SA3 neurons in the subesophageal ganglion of locusts and the TN2 neuron in crickets, independently habituate for each side of input (Boyan and Altman 1985; Atkins and Pollack 1987).

In auditory behavior dishabituation is seen in addition to habituation. During flight, crickets turn toward low-frequency calling song and away from ultrasonic tones that resemble bat hunting cries (Nolen and Hoy 1984). Using this behavior, it was shown that the response of crickets habituate (May and Hoy 1991), fulfilling most of the proposed criteria for habituation (Thompson and Spencer 1966). The response declines exponentially, recovers spontaneously, and depends on the stimulus repetition rate (Fig. 7.3; May and Hoy 1991). Detailed investigations of the temporal parameters show that different mechanisms might contribute to habituation. Initial recovery from habituation can be in the range of a few seconds; however, further responses suggest that the behavior status is still not completely reset (Fig. 7.3). Using this reaction it could be shown that the response is the outcome of categorical perception (Wyttenbach et al. 1996), whereby the response could be dishabituated with a low-frequency tone. The neuronal correlate of this habituation is still somewhat unclear. The ultrasound sensitive Int-1 neuron (mediating avoidance reaction) is a neuron that does not show habituation (Moiseff and Hoy 1983). Thus, habituation must occur at other levels or on different neurons.

4.2 Activity- and Experience-Related Plasticity

Auditory experience can influence insect behavior. *Drosophila melanogaster* males produce species-specific courtship songs by wing vibration. Although the pattern itself is genetically fixed, males that experience this behavior in early adult life have different courtship indices than nonexperienced males (Hirsch

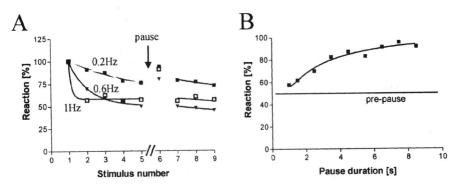

FIGURE 7.3. Habituation of the auditory avoidance behavior of tethered flying crickets. Female *Gryllus bimaculatus* have been stimulated with a train of nine sound pulses of 10 msec duration each, 16 kHz carrier frequency and 80 dB SPL. After the fifth stimulus the train was disrupted with a pause. The avoidance reaction was registered with an optoelectronic camera and the response to the first stimulus was set to 100%. **(A)** Habituation to three different stimulus repetition rates (0.2 Hz, 0.6 Hz, and 1 Hz), showing that the velocity of habituation (to first five stimuli) depends on the repetition rate. After a pause of 5 sec the reaction (to the sixth stimulus) recovers and the amplitude reaches 80–100% of the first reaction. However, the reaction to the following stimuli shows that this dishabituation is not complete: despite the pause and the reaction to stimulus 6, the amplitudes to stimuli 7–9 fit almost perfectly into the nonlinear regression curves of the first five reactions. **(B)** Influence of a pause on the habituation. The reaction to stimulus 6 depends on the pause duration (shown for a repetition rate of 2 Hz). The mean prepause reaction was 51%. Extrapolation of the nonlinear regression curve indicates that a complete reset could be expected after 21 sec. [Modified after Fölsch and Lakes-Harlan (2003).]

and Tompkins 1994). The effect is even more pronounced in animals with a combination of auditory and visual deprivation. This result is in concordance with other findings on sensory deprivation or dense population rearing of *Drosophila* in which various structural (especially the mushroom bodies in the brain) and behavioral parameters are affected (Meinertzhagen 2001; Technau 1984; Heisenberg et al. 1995; Barth et al. 1997). Isolated males exhibit higher locomotor activity after playback of simulated courtship songs than males reared in groups (von Schilcher 1976). Early experience in adult life also affects the selectivity of female crickets to different song models (Shuvalov 1990). Adult female crickets start to perform phonotaxis a few days after the final molt. During that time, they often have the chance to listen to the species-specific calling song in nature. If experimentally deprived of acoustic stimuli or stimulated with non-species-specific acoustic signals, females are rather inselective and show auditory behavior toward different acoustic signals. Females that experience the male song (or a close model of it) neglect other models and prefer the male song.

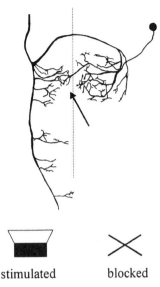

FIGURE 7.4. Activity dependence of neuronal morphology. Dendrites (*arrow*) of a BSN1 neuron of *Locusta migratoria* prune across the midline after chronic acoustic stimulation and blocking of the ear ipsilateral to the dendrites with high-viscosity silicone. The animal was chronically stimulated for 5 days (two sound pulses per second, 100 msec duration each, white noise, approximately 80 dB SPL, depending on the position of the freely moving animal within the cage). (Lakes-Harlan and Pfahlert, unpublished data.)

stimulated   blocked

The influences of neuronal activity on growth, morphology, and synaptic connections of insect neurons seems to be controversial. Enhancing or decreasing activity by mutational changes in axonal conductances in *Drosophila* does not change the overall structure of neurons very much (Burg and Wu 1986, 1989). Also blocking of activity with tetrodotoxin (TTX) does not change the regenerative growth and stage specific synaptogenesis of mechanosensory sensilla in crickets (Chiba and Murphey 1991). On the other hand, the projection of filiform hairs of locusts changes if their activity is blocked (Pflüger et al. 1994). In the visual system of flies, rearing affects neuronal structure, down to the level of the synapse (Meinertzhagen 2001). Thus, the question of whether activity influences neuronal networks seems to depend on the sensory system. In the auditory system, experiments with increasing auditory activity due to chronic stimulation did not change the morphology and physiology of receptor fibers (Psilopolous and Lakes-Harlan 1999). By contrast, the morphology of auditory interneurons can change owing to left–right differences in the auditory activity. Dendritic growth across the midline into the contralateral neuropil has been observed if one ear is blocked with wax and the other chronically stimulated with sound over several days (Fig. 7.4). Thus, it seems that activity can have an effect that extends onto the auditory neuronal network, but this has to be studied in more detail.

4.3 Learning and Memory

Insects (and invertebrates in general, e.g., *Aplysia*) have abilities to learn environmental patterns. Higher forms of learning have been described for the olfactory and visual sense in bees and flies (Hammer and Menzel 1995; Menzel and Giurfa 1999; Menzel 2001). By contrast, the auditory repertoire of neither the sender or receiver has to be learned (except for higher selectivity; see Section 4.2). Larvae of grasshoppers, cicadas, and many other species grow up on their own and produce the species-specific song without hearing it before. Similarly the receiver "knows" how to decode the species-specific signal even if it has never been heard before. Both properties appear rather suddenly in the adult, although some larval stridulatory behavior has been reported (Halfmann and Elsner 1978), suggesting that the motor pattern matures during postembryonic development.

Can adult insects learn sounds? So far, no clear answer can be provided. Learning might be a predisposition that enables learning of some experiences but not others (Menzel and Müller 1996). Generally, insects possess the molecular machinery involved in learning. For example, the long-term-potentiation (LTP) studied in vertebrates has also been reported in the locust nervous system (Parker 1995); a Ca^{2+}-dependent LTP changes the output of motor circuits and is modulated by neurohormones such as octopamine and serotonin. Other recent evidence suggests that protein kinase C, which is essential to LTP in mammals, is important in *Drosophila* learning (Drier et al. 2002).

Auditory information is transmitted to the higher centers of the brain. It can be shown that, although they are not a primary visual center, the multimodal mushroom bodies of the brain are involved in context generalization during visual learning (Liu et al. 1999). Perhaps the involvement of the mushroom bodies and other CNS structures in auditory learning could be tested with mutants of *Drosophila* where a flight paradigm might be used for discrimination of different sound patterns.

5. Modulation of Auditory Networks

The neural response to stimuli often is context dependent. A large number of variables can influence the response, which might then be regarded as plastic. Neuromodulation changes the neuronal output and plays an important role in olfactory learning and task allocation in bees. Neuronal responses, however, can already be modified at the level of the sensory cells. Modulation can also change neuronal networks in a behavior and in a motivational (e.g., hormonal) state dependent manner.

5.1 Modulation of Sensory Responses

Insect auditory receptor cells usually respond tonically and show an intensity–response curve with a dynamic range of about 30 dB SPL and subsequent saturation. This response can be modulated in various systems by efferent control. In cicadas, the tympanal organ is shut off during sound production, probably to avoid damage due to the very high sound pressure (Hennig et al. 1994). In the ears of katydids and locusts peripheral synapses or a nerve plexus have been described ultrastructurally but nothing is known about their function (Popov and Svetlogorskaya 1971; Zhantiev and Korsunovskaja 1978). However, the response of auditory receptor cells of locusts changes after they have been separated from the CNS. Recordings of a receptor cell axon show an increased discharge after separation from the CNS during the recording (Fig. 7.5). Thus, this experiment confirms that central influences modify the peripheral sensory response.

The afferent responses are also modified by presynaptic influences within the CNS. A presynaptic inhibition might be important to protect sensory synapses from habituation and modulates the effectiveness of sensory spikes (Boyan 1988). Presynaptic inhibition probably acts via depolarizing inhibitory currents, with a reversal potential close to the resting potential, mediated mostly by γ-aminobutyric acid-ergic (GABAergic) synapses (Watson 1992). Such input

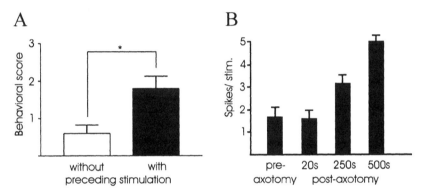

FIGURE 7.5. Modulatory influence on the auditory system. (**A**) Increase of the auditory response of the parasitic fly *Emblemasoma auditrix* after preceding acoustic stimulation. The phonotactic threshold toward the calling song of the host was determined for 10 individually marked animals (usually 60–65 dB SPL). Thereafter animals were tested 5 dB lower than individual threshold and their behavior was scored (e.g., initial reaction to the stimulus, movement into direction of the sound signal, and complete phonotaxis toward the loudspeaker; mean: *open bar*). The same animals were then stimulated with the calling song (80 dB SPL, 30 sec duration) immediately prior to testing again at 5 dB below threshold. The score increased significantly (*t*-test, $n = 10$) and some animals even completed phonotaxis. Thus, their auditory responsiveness was increased with the stimulation, a behavior also observed in the field (deVries and Lakes-Harlan, unpublished results). (**B**) Example of modulatory central influences on auditory afferents in grasshoppers. Recording of a type 1 receptor fiber of *Schistocerca gregaria* stimulated with pulses of 50 msec duration, 80 dB SPL, 4 kHz carrier frequency, and 2 Hz repetition rate. After adaptation the receptor cell responds with 1.7 action potential per stimulus (pre-axotomy). Thereafter the tympanal nerve was cut close to the CNS and direct central influences to the auditory organ were eliminated: the response of the receptor cell increased with time post-axotomy. [Modified after Jacobs (1997), with permission.]

synapses onto auditory receptor terminals have been described ultrastructurally in crickets (Hardt and Watson 1999) and locusts (Jacobs and Lakes-Harlan 2000). Thus, synapses that form the substrate for presynaptic inhibition and facilitation are universally present in vertebrates and invertebrates and modulate sensory information processing.

5.2 *Modulatory Effects on Auditory Networks*

Most electrophysiological studies of the central auditory network have been performed on animals that were dissected to various degrees. However, in locust no descending influence of the head on spike propagation in ascending auditory interneurons could be found (Kalmring et al. 1978). Wolf (1986) could show in almost intact, freely moving grasshoppers that large ascending auditory interneurons show the same characteristics as in dissected animals. Thus no motivational influence could be detected at this thoracic level of auditory information processing. Auditory neurons, however, can be influenced in their response properties by motor programs. During stridulation they can be switched off, owing partly to proprioceptive feedback and partly to central influences (Wolf and von Helversen 1986; Hedwig and Meyer 1994). On the other hand, descending neuronal activity in response to ultrasound is enhanced in flying animals (Brodfuehrer and Hoy 1989). The phonotaxis selectivity is partly different in walking and flying animals (Schul and Schulze 2001). A neuronal substrate of such central modulation could be the dorsal unpaired median neurons (DUM neurons). DUM neurons participate in motor movements and are activated during behavior (Burrows and Pflüger 1995; Duch et al. 1999). These neurons are either GABAergic or octopaminergic. GABAergic DUM neurons, which are perhaps involved in presynaptic inhibition, also process auditory information (Thompson and Siegler 1991). Octopaminergic DUM neurons not only might modulate electrical properties of auditory neurons, but also change neuronal assemblies as has been proposed for learning in bees (Hammer and Menzel 1995).

Auditory information processing is modulated by the history of sensory stimuli (see also Section 4.1). For example, an ongoing auditory signal results not only in trains of action potentials but also in a long-lasting inhibition, as seen in the omega cell of crickets and bushcrickets (Römer 1993; Sobel and Tank 1994). This inhibition improves the signal-to-noise ratio by subtraction of background noise and represents a modifiable mechanism for selective attention (Pollack 1988). The long-lasting inhibition probably depends on calcium influx into the cell (Sobel and Tank 1994) and is context dependent, thereby modifying the auditory response by forward masking. This mechanism might be restricted to some species or parts of the auditory network, because in locusts such a mechanism could not yet be found (Lang 1996).

Acoustic signals also might facilitate auditory behavior. Such "arousal" has been observed in grasshoppers and has been studied in the parasitic fly *Emblemasoma auditrix*. The fly performs phonotaxis specifically toward the calling song of the host (Lakes-Harlan et al. 2000). However, flies that have been attracted by this signal react to a broad range of otherwise ineffective acoustic signals and their behavioral threshold decreases (Fig. 7.5).

Hormone-mediated plasticity has been proposed for the phonotactic responsiveness of crickets (reviewed by Stout et al. 2002). The threshold for female phonotaxis decreases within the first days after the final molt. This decrease relates to an increase of juvenile hormone III production. It could be shown that topical application of juvenile hormone III decreases the threshold also in the first days. Furthermore, electrophysiological experiments suggest that the hormone acts directly on some auditory interneurons, changing their responsiveness probably by regulation of neurotransmitter receptors (Stout et al. 2002).

6. Lesions—Functional Compensations

Major aspects of plasticity in the nervous system are functional and structural compensations after injury. Experiments and treatments on nervous system injuries receive widespread attention, especially in vertebrates (Schwab 2002). Many different mechanisms contribute to recovery of function, for example, compensatory sprouting as well as regenerative fiber growth. They might be related to developmental mechanisms, although in many cases they cannot be simple recapitulations of development. Experimental lesioning of insect nervous systems has led to the changing view on the stability and dynamics of the insect nervous system (Murphey 1986b). A lesion might result in one or more of the following processes: replacement of lost structures; degeneration of damaged neurons; and reorganization of synaptic connections in unlesioned parts of the neuronal network, including collateral sprouting or regeneration of axonal processes (Fig. 7.6). In insects some processes take place in adults (compensatory growth, regeneration of fibers), whereas others, such as a replacement of cells, occur only during molts in postembryonic development.

6.1 Regeneration of Complete Cells

Most insects are capable of regenerating a lost structure. However, regeneration depends on the species and on the stage, with younger animals more capable of replacing lost structures than older ones (reviews Edwards 1969; Edwards and Palka 1976; Edwards 1988). Different species of insects have different capabilities for regeneration (Bullière and Bullière 1985). In crickets and bushcrickets, auditory sense organs can be replaced if the loss occurs early in postembryonic life (Ball 1979; Biggin 1981; Huber 1987; Lakes and Mücke 1989). However, an auditory organ is never completely replaced, but only some structural elements as well as a largely reduced number of sensory units can be found in regenerated legs. In locusts, experimentally removed auditory organs or their anlagen are not replaced (Lakes 1988), but transplanted auditory anlagen are able to develop into ectopic tympanal organs (Otte and Lakes-Harlan 1997).

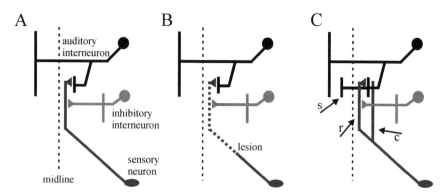

FIGURE 7.6. Schema of the first auditory synapse in the auditory neuropil of grasshoppers and their changes after a lesion. The receptor fiber synapses ipsilateral on a first-order interneuron, which transmits the information to the contralateral side. The receptor fiber itself is postsynaptic to an inhibitory interneuron, whose input connections are unknown. After lesion and regeneration of the sensory fiber the neuronal connections are changed: dendrites of the temporarily deafferented auditory interneuron sprout into the intact neuropil (s), the regenerated fiber form new synaptic contacts (r), and during regeneration additional collaterals enter the auditory neuropil (c).

These organs are functional and their sensory axons extend into the primary target areas of the CNS. It has not been tested yet whether single receptor cells can be replaced after damage as has been shown in vertebrates (Staecker and Van De Water 1998). But as for this a sensory mother cell would have to be formed first, which has never been found during postembryonic development, a replacement seems unlikely.

6.2 Regrowth of Fibers

Parts of auditory sensory cells can easily be replaced in insects. After axotomy of sensory axons, the distal parts of the auditory fibers degenerate within a few days (Jacobs and Lakes-Harlan 1999). This Wallerian degeneration is not self-evident, as other sensory fibers show a much slower time course of degeneration, and lesioned processes of interneurons and motorneurons often take weeks to degenerate (Boulton 1969; Jacobs and Lakes-Harlan 1999). Starting at the site of the lesion, auditory axons regrow and reenter the CNS. This can be seen in larvae as well as in adults (Pallas and Hoy 1988; Lakes and Kalmring 1991; Lakes-Harlan and Pfahlert 1995; Jacobs and Lakes-Harlan 2000). A detailed single cell analysis revealed that the specificity of these sensory fibers is high (Fig. 7.6); they arborize in their somatotopically organized target area in the metathoracic ganglion (Jacobs and Lakes-Harlan 2000). This specificity is independent of the pathway that the fibers take. Even if the tympanal nerve is completely cut, the regrown fibers can find their target area by using different nerves to access the CNS (Lakes and Kalmring 1991). In the target area the regenerated afferents synapse onto target interneurons but the response properties of the interneurons are changed (Lakes and Kalmring 1991). Single-cell recordings show that the tuning of the neurons with regenerated synapses is variable in respect to best frequency, intensity response, and latency. This might be due to incomplete regeneration of functional synapses, but might also be caused by complex plastic changes of interneurons and rearranged neuronal networks (see Section 6.2; Fig. 7.6).

Fortunately, it was possible to test in grasshoppers if these regenerated synapses permit normal behavior. Males of *Chorthippus biguttulus* exhibit a turning reaction toward the side from which they hear the female calling song loudest. This highly stereotyped and quantifiable reaction was used as an experimental paradigm in adult animals that received a unilateral axotomy of tympanal receptor fibers (Lakes-Harlan and Pfahlert 1995). After axotomy the animals turned only to the side with the intact tympanal nerve, regardless of the side of the stimulus. However, some days after the lesion the animals also turned to the lesioned side. This regeneration of the behavior improves with time after the lesion and is correlated with the amount of regenerated fibers in the main auditory neuropil in the metathoracic ganglion (Fig. 7.7). A bilateral axotomy,

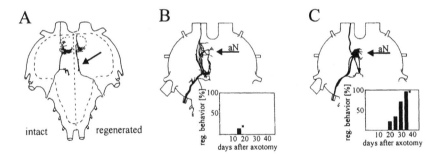

FIGURE 7.7. Regeneration of sensory fibers after axotomy and restoration of behavior. (**A**) Single-cell recordings show that the regenerated fibers might follow different pathways (*arrow*, pathway close to the midline); however, their synaptic arborizations are in the correct area of the tonotopic ordered neuropil (compare to intact cell on the left). [Sixteen days after axotomy, type 2 high-frequency receptor of *Schistocerca gregaria*; Jacobs and Lakes-Harlan 2000, with permission.] (**B,C**) Correlation of collateral regeneration within the auditory neuropil and auditory behavior in *Chorthippus biguttulus*. Animals were tested for their lateralization behavior after axotomy. The *bars* in the *insets* represent the percentage of restored behavioral reactions (reg. behavior). Animals were sacrificed at different stages of regeneration and the sensory projection was labeled. At early stages only a few axonal collaterals are seen within the auditory neuropil (aN, *arrow*, B). Animals with fully regenerated behavior have dense arborizations in the neuropil (*arrow*, C). [Modified after Pfahlert and Lakes-Harlan (1997), with permission.]

however, did not lead to a regeneration of normal behavior. Furthermore, recovered behavior was lost if the intact side was axotomized after regeneration following the first lesion.

This paradoxical result can be explained by the finding that song recognition and localization (lateralization) are two different processes computed in parallel neuronal networks in the CNS of the grasshopper (von Helversen and von Helversen 1997). Obviously song localization could be reestablished by regenerated fibers, but not song recognition for which an intact tympanal nerve is required. It seems likely that the song recognition network requires more precise information processing (e.g., synaptic connections) than the localization network. For localization it would be sufficient if one side is more excited than the other, independent of precise synapses. It might further be deduced that the neuronal circuits are changed after the injury and are not restored to the original state after regeneration. In crickets regeneration of afferent fibers did not eliminate collateral sprouts of an interneuron, which pruned in a first response to the lesion (Pallas and Hoy 1988).

Central auditory interneurons have not been tested yet for their ability to regrow lesioned fibers. However, studies on interneurons in general suggest that the regrowth capability is rather low (Jacobs and Lakes-Harlan 1999). After lesioning the distal part of the axon seems not to degenerate, but to be compartmentalized by invaginating glial cell processes. Thereby the distal parts are maintained for some weeks (keeping in mind that the life span of adults of the species that have been investigated for their auditory sense might range to roughly 2 months). The proximal parts of interneurons often fail to grow out again, although this might depend on the distance of the lesion to the cell body.

6.3 Plasticity of Unlesioned Neurons

Plastic modifications of unlesioned neurons have been a surprising finding of recent years. A unilateral peripheral lesion in the auditory system leads to changes in the central auditory network. First, in contrast to many vertebrate systems, the survival of postsynaptic neurons seems not to depend on afferent innervation. After deprivation no degeneration of interneurons has been confirmed, although some cells are more difficult to label (Pallas and Hoy 1988). Second, contralateral afferents can sprout collaterals into the lesioned side of the neuropil, even though only to a small extent (Schmitz 1989; Lakes et al. 1990). In other insect sensory systems, such as the cercal system of crickets, contralateral afferent sprouting is a major response to injury (Murphey 1986a). Third, in the auditory system interneurons are triggered for a compensatory growth.

In crickets dendritic sprouting has been observed in various neurons, such as Int-1, AN1, AN2, and ON1 neurons (Fig. 7.8; Hoy et al. 1985; Schildberger et al. 1986; Huber 1987; Schmitz 1989). In grasshoppers, growth has been seen in the BSN1 neuron (Fig. 7.8) and some other local interneurons (Römer and Büngers 1988; Lakes et al. 1990), but not in other multisegmental and multi-

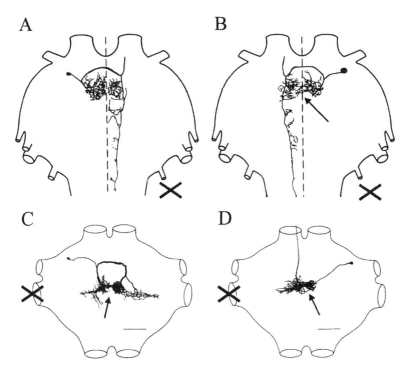

FIGURE 7.8. Changes in central auditory interneurons after deafferentation. (**A,B**) Local BSN1 interneuron of *Locusta migratoria*. The neuron contralateral (in respect to position of soma and dendrites) to the deafferentation (x) remains unchanged in morphology (A). Dendritic sprouting across the midline (*arrow*) is found at the ipsilateral, deafferented interneuron (B). [Modified after Lakes-Harlan et al. (1991), with permission.] (**C,D**) Dendritic sprouting (*arrows*) occurs at first order interneurons, independent of the function in the auditory network of *Gryllus bimaculatus*. Local interneuron: Omega neuron (C); ascending interneuron: AN2 (D). [Modified after Schildberger et al. (1986), with permission.]

modal auditory neurons. Common to all these changed neurons are monosynaptic connections to the afferents which have been lost after removal of the sensory cell bodies. This compensatory pruning results in regaining synaptic input from the intact contralateral afferents. Sprouting, however, might be suppressed if the contralateral afferents are also missing (Pallas and Hoy 1988). There is little, if any, competition because the response of intact neurons connected to the same afferents is not changed; one would expect a compensatory reduction after reorganization (Brodfuehrer and Hoy 1988). The response of the deafferented neurons is largely normal (response curve, threshold). However, some parameters, such as two-tone suppression, are changed (Brodfuehrer and Hoy 1988; Lakes et al. 1990). This is probably due to changes in the inhibitory connections (Schmitz 1989).

The growth is dependent on the time after the operation (Brodfuehrer and Hoy 1988; Schmitz 1989), but also of the developmental stage of the experimental animal. Sprouting is relatively slow and in the cricket complete by 28 days after lesion (Brodfuehrer and Hoy 1988). In the BSN1 neuron, first excitatory postsynaptic potentials could be recorded after 6 days and this is associated with sprouting of dendrites, while action potentials are seen later (Pfahlert and Lakes-Harlan, unpublished observations). Adult insects show sprouting but not to the extent that larval crickets do (Brodfuehrer and Hoy 1988). Also the behavior—the orientation performance—improves with the interval between lesion and test (Schmitz 1989). The amount of midline crossing fibers and the amount of contralateral formed dendrites are usually much lower than the dendritic area of intact neurons. Interestingly, one-eared crickets can also localize a sound source (Schildberger and Kleindienst 1989).

7. Summary

Plastic phenomena in the auditory system of insects occur as variations of neuronal morphology during development, activity-dependent modifications in the adult, or modulatory influences on auditory networks, and they are pronounced in functional compensations after lesions. Therefore, plasticity is likely to have an adaptive value not only in vertebrates, but also in the insect nervous system. Insects can play an important role in increasing our understanding of plasticity, especially with genetically accessible model organisms. A first mutational study on the auditory system in *Drosophila* has been published (Eberl et al. 1997) indicating that molecular dissection of auditory system including plasticity should be possible in the near future. Furthermore, the diverse but well known auditory systems of different species will allow us to extract general mechanisms for plasticity.

Developmental plasticity adjusts the auditory system to internal and external requirements. This includes homeostatic plasticity during larval growth when neurons increase in size and consequently change electrical properties. Plasticity could then regulate synaptic efficacy to keep connections in a functional range. This process is still independent of auditory function which largely develops in the adult, but spontaneous activity might be a decisive factor. For the auditory system, one should keep in mind that most of the plasticity described occurs at the level of the thoracic nervous system, which is an important center of auditory information processing. Nevertheless, plastic changes might be even more pronounced in the brain, in which little is known about the auditory processing. In other insects the brain is a major site for structural plasticity (Meinertzhagen 2001) and in some species even cell proliferation takes place in adult brains (Cayre et al. 1994, 1996).

Plasticity after lesions can have a biological background, as in crickets where a loss of a foreleg might actually happen in the field. Regeneration after lesions within the CNS is rather unlikely to have any specific selective value, as such a lesion will be deadly in the field. However, it might be that the ability for regeneration and for plastic changes is linked to other processes such as development and does not need a specific selective advantage to be present in the insect. Future studies on lesioned animals might also allow identification of the mechanisms that stabilize the auditory system in "intact" animals.

Plasticity is also important for the evolution of auditory systems. These systems are almost certainly derived from proprioceptive predecessor systems, and one can expect that the neuronal networks are modified for the auditory function. These processes, such as recruitment of new neurons, seem to be possible if neurons exhibit pre- and postsynaptic plasticity. In addition, the evolution of the acoustic communication systems might depend on plasticity of either the sender or the receiver. A mutational change in the signal properties of the sender can be expected to be still recognizable by the receiver, if it either reacts to a wide range of signals or if plasticity allows one to widen the range. Of course, at least basic parts of the acoustic communication system develop without auditory function and the overall auditory behavior is genetically fixed. Intraspecific acoustic signals are not learned, either by the sender or the receiver. However, specificity can increase with experience and it cannot be completely ruled out that sound recognition is adjusted to the environment. These adjustments might correspond to observed, although low, variabilities in songs of orthopterans (Ritchie 1992; Souroukis et al. 1992; von Helversen and von Helversen 1997). Therefore, comparative studies from behavior to molecular biology are needed for an understanding of plasticity.

Acknowledgments. I thank my colleagues A. Stumpner and K. Jacobs for many discussions and comments on parts of the manuscript.

References

Alloway TM (1972) Learning and memory in insects. Annu Rev Entomol 17:43–56.

Atkins G, Pollack GS (1986) Age dependent occurrence of an ascending axon on the omega neuron of the cricket, *Teleogryllus oceanicus*. J Comp Neurol 243:527–534.

Atkins G, Pollack GS (1987) Response properties of prothoracic, interganglionic, sound-activated interneurons in the cricket *Teleogryllus oceanicus*. J Comp Physiol A 161: 681–693.

Ball E (1979) Development of the auditory tympanum in the cricket *Teleogryllus commodus*: experiments on regeneration and transplantation. Experienta 35:324–325.

Ball E, Young D (1974) Structure and development of the auditory system in the prothoracic leg of the cricket *Teleogryllus commodus* (Walker). II. Postembryonic development. Z Zellforsch 147:313–324.

Barth M, Hirsch HVB, Meinertzhagen IA, Heisenberg M (1997) Experience-dependent developmental plasticity in the optic lobe of *Drosophila melanogaster*. J Neurosci 17: 1493–1504.

Biggin RJ (1981) Pattern re-establishment: transplantation and regeneration of the leg on the cricket *Teleogryllus commodus*. J Embryol Exp Morphol 61:87–101.

Blänsdorf I, Lakes-Harlan R (1993) Development of the scolophorous cells of the tympanal organ of *Galleria mellonella* L. In: Elsner N, Heisenberg M (eds), Gene–Brain–Behaviour. Stuttgart and New York: Thieme, p. 216.

Boulton PS (1969) Degeneration and regeneration in the insect central nervous system. Z Zellforsch Mikros Anat 101:98–118.

Boyan GS (1983) Postembryonic development in the auditory system of the locust. J Comp Physiol 151:499–513.

Boyan GS (1988) Presynaptic inhibition of identified wind-sensitive afferents in the cercal system of the locust. J Neurosci 8:2748–2757.

Boyan GS (1993) Another look at insect audition: the tympanic receptors as an evolutionary specialization of the chordotonal system. J Insect Physiol 39:187–200.

Boyan GS (1998) Development of the Insect Auditory System. In: Hoy RR, Popper AN, Fay RR (eds), Comparative Hearing in Insects. New York: Springer-Verlag, pp. 97–138.

Boyan GS, Altman JS (1985) The suboesophageal ganglion: a "missing link" in the auditory pathway of the locust. J Comp Physiol A 156:413–428.

Brodfuehrer PD, Hoy RR (1988) Effect of auditory deafferentation on the synaptic connectivity of a pair of identified interneurons in adult field crickets. J Neurobiol 19: 17–38.

Brodfuehrer PD, Hoy RR (1989) Integration of ultrasound and flight inputs on descending neurons in the cricket brain. J Exp Biol 145:157–171.

Budnik V (1996) Synapse maturation and structural plasticity at *Drosophila* neuromuscular junctions. Curr Opin Neurobiol 6:858–868.

Bullière D, Bullière F (1985) Regeneration. In: Kerkut GA, Gilbert LI (eds), Comprehensive Insect Physiology, Biochemistry and Pharmacology, Vol. 5 Oxford: Pergamon Press, pp. 372–424.

Burg MG, Wu C-F (1986) Differentiation and central projections of peripheral sensory cells with action-potential block in *Drosophila* mosaics. J Neurosci 6:2968–2976.

Burg MG, Wu C-F (1989) Central projections of peripheral mechanosensory cells with increased excitability in *Drosophila* mosaics. Dev Biol 131:505–514.

Burrows M, Pflüger HJ (1995) Action of locust neuromodulatory neurons is coupled to specific motor patterns. J Neurophysiol 74:347–357.

Cayre M, Strambi C, Strambi A (1994) Neurogenesis in an adult insect brain and its hormonal control. Nature 386:57–59.

Cayre M, Strambi C, Charpin P, Augier R, Meyer MR, Edwards JS, Strambi A (1996) Neurogenesis in adult insect mushroom bodies. J Comp Neurol 371:300–310.

Chiba A, Murphey RK (1991) Connectivity of identified central synapses in the cricket is normal following regeneration and blockade of presynaptic activity. J Neurobiol 22: 130–142.

Chiba A, Shepherd D, Murphey RK (1988) Synaptic rearrangement during postembryonic development in the cricket. Science 240:901–905.

Davies GW, Goodman CS (1998) Genetic analysis of synaptic development and plasticity: homeostatic regulation of synaptic efficacy. Curr Opin Neurobiol 8:149–156.

Doe CQ, Technau GM (1993) Identification and cell lineage of individual neural precursors in the *Drosophila* CNS. Trends Neurosci 16:510–514.

Drier EA, Tello MK, Cowan M, Wu P, Blace N, Sachtor TC, Yin JCP (2002) Memory enhancement and formation by atypical PKM activity in *Drosophila melanogaster*. Nat Neurosci 5:316–324.

Dubnau J, Tully T (1998) Gene discovery in *Drosophila*: new insights for learning and memory. Annu Rev Neurosci 21:401–444.

Duch C, Mentel T, Pflüger HJ (1999) Distribution and activation of different types of octopaminergic DUM neurons in the locust. J Comp Neurol 403:119–134.

Eberl DF, Duyk GM, Perrimon N (1997) A genetic screen for mutations that disrupt an auditory response in *Drosophila melanogaster*. Proc Natl Acad Sci USA 94:14837–14842.

Edwards JS (1969) Postembryonic development and regeneration in the insect nervous system. Adv Insect Physiol 6:97–137.

Edwards JS (1988) Sensory regeneration in arthropods: implications of homeosis and of ectopic sensilla. Am Zool 28:1155–1164.

Edwards JS, Palka J (1976) Neural generation and regeneration. In: Fentress J (ed), Simpler Networks and Behaviour. Sunderland, MA: Sinauer, pp. 167–185.

Fölsch A, Lakes-Harlan R (2003) Habituation of the startle response of *Gryllus bimaculatus* (Orthoptera). In: Elsner N, Zimmermann H (eds), Göttingen Neurobiology Report. Stuttgart and New York: Thieme (in press).

Fonseca PJ, Münch D, Hennig RM (2000) How cicadas interpret acoustic signals. Nature 405:297–298.

Fullard JH, Yack JE (1993) The evolutionary biology of insect hearing. Trends Ecol Evol 8:248–252.

Givois V, Pollack GS (2000) Sensory habituation of auditory receptor neurons: implications for sound localization. J Exp Biol 203:2529–2537.

Goodman CS (1978) Isogenic grasshoppers: genetic variability in the morphology of identified neurons. J Comp Neurol 182:681–706.

Goodman CS (1982) Embryonic development of identified neurons in the grasshopper. In: Spitzer NC (ed), Neuronal Development. New York: Plenum Press, pp. 171–212.

Goodman CS (1996) Mechanisms and molecules that control growth cone guidance. Annu Rev Neurosci 19:341–377.

Goodman CS, Shatz CJ (1993) Developmental mechanisms that generate precise patterns of neuronal connectivity. Cell 72 (Suppl.):77–98.

Goodman CS, Spitzer NC (1979) Embryonic development of identified neurones: differentiation from neuroblast to neurone. Nature 280:208–214.

Halfmann K, Elsner N (1978) Larval stridulation in acridid grasshoppers. Naturwissenschaften 65:265.

Hammer M, Menzel R (1995) Learning and memory in the honeybee. J Neurosci 15:1617–1630.

Hardt M, Watson AH (1999) Distribution of input and output synapses of the central branches of bushcricket and cricket auditory afferent neurones: immuncytochemical evidence for GABA and glutamate in different populations of presynaptic boutons. J Comp Neurol 403:281–294.

Hedwig B, Meyer J (1994) Auditory information processing in stridulating grasshoppers: tympanic membrane vibrations and neurophysiology. J Comp Physiol A 174:121–131.

Heisenberg M, Heusipp M, Wanke C (1995) Structural plasticity in the *Drosophila* brain. J Neurosci 15:1951–1960.

Hennig RM, Weber T, Huber F, Kleindienst H-U, Moore TE, Popov AV (1994) Auditory threshold change in singing cicadas. J Exp Biol 187:45–55.

Hintze-Podufal C, von Hermanni G (1996) The development of the tympanic organs of wax moth species and their inverted scolopidia (Lepidoptera: Pyralidae: Galleriinae). Entomol Gen 20:195–201.

Hirsch HVB, Tompkins L (1994) The flexible fly: experience-dependent development of complex behaviors in *Drosophila melanogaster*. J Exp Biol 195:1–18.

Horridge GA (1962) Learning of leg position by headless insects. Nature 193:697–698.

Horseman G, Huber F (1994) Sound localisation in crickets. II. Modeling the role of a simple neural network in the prothoracic ganglion. J Comp Physiol A 175:399–413.

Hoy RR (1998) Acute as a bug's ear: An informal discussion of hearing in insects. In: Hoy RR, Popper AN, Fay RR (eds), Comparative Hearing in Insects. New York: Springer-Verlag, pp. 1–18.

Hoy RR, Robert D (1996) Tympanal hearing in insects. Annu Rev Entomol 41:433–450.

Hoy RR, Nolen TG, Casaday GC (1985) Dendritic sprouting and compensatory synaptogenesis in an identified interneuron follow auditory deprivation in a cricket. Proc Natl Acad Sci USA 82:7772–7776.

Hoyle G (1983) On the way to neuroethology: the identified neuron approach. In: Huber F, Markl H (eds), Neuroethology and Behavioral Physiology. Berlin: Springer-Verlag, pp. 9–25.

Huber F (1987) Plasticity in the auditory system of crickets: phonotaxis with one ear and neuronal reorganization within the auditory pathway. J Comp Physiol A 161:583–604.

Jacobs K (1997) Axonale Degeneration und Regeneration tympanaler Sinneszellen von *Schistocerca gregaria*. Dissertation, Universität Göttingen.

Jacobs K, Lakes-Harlan R (1999) Axonal degeneration within the tympanal nerve of *Schistocerca gregaria*. Cell Tissue Res 298:167–178.

Jacobs K, Lakes-Harlan R (2000) Pathfinding, target recognition and synapse formation of single regenerating fibres in the adult grasshopper *Schistocerca gregaria*. J Neurobiol 42:394–409.

Kalmring K, Kühne R, Moysich F (1978) The auditory pathway in the ventral cord of the migratory locust (*Locusta migratoria*): response transmission in the axons. J Comp Physiol 126:25–33.

Klose M (1990) Development of nerve pathways and receptor organs in cricket legs. In: Elsner N, Roth G (eds), Brain–Perception–Cognition; Proceedings of the 18th Göttingen Neurobiology Conference. Stuttgart and New York: Thieme, p. 104.

Lakes R (1988) Postembryonic determination and plasticity in the auditory system of *Locusta migratoria*. Monogr Dev Biol 21:214–221.

Lakes R, Kalmring K (1991) Regeneration of the projection and synaptic connections of tympanic receptor fibers of *Locusta migratoria* (Orthoptera) after axotomy. J Neurobiol 22:169–181.

Lakes R, Mücke A (1989) Regeneration of the foreleg tibia and tarsi of *Ephippiger ephippiger* (Orthoptera, Tettigoniidae). J Exp Zool 250:176–187.

Lakes R, Pollack GS (1990) The development of the sensory organs of the legs in the blowfly, *Phormia regina*. Cell Tissue Res 259:93–104.

Lakes R, Schikorski T (1990) Neuroanatomy of the Tettigoniids. In: Bailey WJ, Rentz, DCF (eds), The Tettigoniidae: Biology, Systematics and Evolution. Bathurst: Crawford House Press, pp. 166–190.

Lakes R, Kalmring K, Engelhardt K-H (1990) Changes in the auditory system of locusts (*Locusta migratoria and Schistocerca gregaria*) after deafferentation. J Comp Physiol A 166:553–563.

Lakes-Harlan R, Pfahlert C (1995) Regeneration of axotomized tympanal nerve fibres in the adult grasshopper *Chorthippus biguttulus* (L.) (Orthoptera: Acrididae) correlates with regaining the localization ability. J Comp Physiol A 176:797–807.

Lakes-Harlan R, Stölting H, Stumpner A (1999) Convergent evolution of insect hearing organs from a preadaptive structure. Proc Roy Soc B 266:1161–1167.

Lakes-Harlan R, Stölting H, Moore T (2000) Phonotactic behavior of a parasitoid fly (*Emblemasoma auditrix*, Diptera, Sarcophagidae) in response to the calling song of the host cicada (*Okanagana rimosa*, Homoptera, Cicadidae). Zoology 103:31–39.

Lang F (1996) Noise filtering in the auditory system of *Locusta migratoria* L. J Comp Physiol A 179:575–585.

Lewis FP, Fullard JH (1996) Neurometamorphosis of the ear in the gypsy moth, *Lymantria dispar*, and its homologue in the earless forest tent caterpillar moth, *Malacosoma disstria*. J Neurobiol 31:245–262.

Liu L, Wolf R, Ernst R, Heisenberg M (1999) Context generalization on *Drosophila* visual learning requires the mushroom bodies. Nature 400:753–756.

May ML, Hoy RR (1991) Habituation of the ultrasound-induced acoustic startle response in flying crickets. J Exp Biol 159:489–499.

Meier T, Reichert H (1990) Embryonic development and evolutionary origin of the orthopteran auditory organs. J Neurobiol 21:592–610.

Meinertzhagen IA (1993) The synaptic populations of the fly's optic neuropil and their dynamic regulation: parallels with the vertebrate retina. Prog Retin Res 12:13–39.

Meinertzhagen IA (2001) Plasticity in the insect nervous system. Adv Insect Physiol 28: 84–167.

Menzel R (2001) Searching for the memory trace in a mini-brain, the honeybee. Learn Mem 8:53–62.

Menzel R, Giurfa M (1999) Cognition by a mini brain. Nature 400:718–719.

Menzel R, Müller U (1996) Learning and memory in honeybees: from behavior to neural substrates. Annu Rev Neurosci 19:379–404.

Michel K, Petersen M (1982) Development of the tympanal organ in larvae of the migratory locust. Cell Tissue Res 222:667–676.

Moiseff A, Hoy RR (1983) Sensitivity to ultrasound in an identified auditory interneuron in the cricket: a possible neural link to phonotactic behavior. J Comp Physiol 152: 155–167.

Murphey RK (1986a) Competition and the dynamics of axon arbor growth in the cricket. J Comp Neurol 251:100–110.

Murphey RK (1986b) The myth of the inflexible invertebrate: competition and synaptic remodelling in the development of invertebrate nervous systems. J Neurobiol 17:585–591.

Nolen TG, Hoy RR (1984) Initiation of behavior by single neurons: the role of behavioral context. Science 226:992–994.

Ocker W-G, Hedwig B (1993) Serial response decrement in the auditory pathway of the locust. Zool Jb Physiol 97:312–326.

Otte B, Lakes-Harlan R (1997) Changes in the auditory system of *Schistocerca gregaria* induced by an implanted, additional ear. In: Elsner N, Wässle H (eds), Neurobiology: from Membrane to Mind; Proceedings of the 25th Göttingen Neurobiology Conference. Stuttgart and New York: Thieme, p. 98.

Palka J (1984) Precision and plasticity in the insect nervous system. Trends Neurosci 7: 455–456.

Pallas SL, Hoy RR (1988) Regeneration of normal afferent input does not eliminate aberrant synaptic connections of an identified interneuron in the cricket *Teleogryllus oceanicus*. J Comp Neurol 248:348–359.

Parker D (1995) Long-lasting potentiation of a direct central connection between identified motor neurons in the locust. Eur J Neurosci 7:1097–1106.

Petersen M, Kalmring K, Cokl A (1982) The auditory system in larvae of the migratory locust. Physiol Entomol 7:43–54.

Pfahlert C, Lakes-Harlan R (1997) Responses of insect neurons to neurotrophic factors *in vitro*. Naturwissenschaften 84:163–165.

Pflüger H-J, Hurdelbrink S, Czjzek A, Burrows M (1994) Activity-dependent structural dynamics of insect sensory fibers. J Neurosci 14:6946–6955.

Pollack GS (1988) Selective attention on an insect auditory neuron. J Neurosci 8:2635–2639.

Pollack GS (1998) Neural processing of acoustic signals. In: Hoy RR, Popper AN, Fay RR (eds), Comparative Hearing in Insects. New York: Springer-Verlag, pp. 139–196.

Popov AV, Markovich AM (1982) Auditory interneurons in the prothoracic ganglion of the cricket, *Gryllus bimaculatus*. II. A high-frequency ascending neuron (HF1AN). J Comp Physiol A 146:351–359.

Popov AV, Svetlogorskaya ID (1971) Ultrastructural organization of the auditory nerve in *Locusta migratoria*. J Evol Biochem Physiol 7:439–443.

Popov AV, Michelsen A, Lewis B (1994) Changes in the mechanics of the cricket ear during early days of adult life. J Comp Physiol A 175:165–170.

Psilopolous K, Lakes-Harlan R (1999) On the influence of chronic activity on the auditory system of *Schistocerca gregaria*. In: Elsner N, Eysel U (eds), From Molecular Neurobiology to Clinical Neuroscience. Stuttgart and New York: Thieme, p. 143.

Rauschecker JP (1999) Auditory cortical plasticity: a comparison with other sensory systems. Trends Neurosci 22:74–80.

Ritchie MG (1992) Variation in male song and female preference within a population of *Ephippiger ephippiger* (Orthoptera: Tettigonidae). Anim Behav 43:845–855.

Römer H (1993) Environmental and biological constraints for the evolution of long-range signalling and hearing in acoustic insects. Trans Roy Soc Lond B 226:179–185.

Römer H, Büngers D (1988) Plasticity of the locust auditory pathway following unilateral deafferentation in early larval development. In: Elsner N, Barth FG (eds), Sense Organs; Proceedings of the 16th Göttingen Neurobiology Conference. Stuttgart and New York: Georg Thieme, p. 152.

Rössler W (1992) Postembryonic development of the complex tibial organ in the foreleg of the bushcricket *Ephippiger ephippiger* (Orthoptera, Tettigoniidae). Cell Tissue Res 269:505–514.

Rössler W, Lakes-Harlan R (1999) Plasticity of the insect nervous system. In: Elsner N, Eysel U (eds), From Molecular Neurobiology to Clinical Neuroscience. Stuttgart and New York: Thieme, pp. 426–434.

Schäffer S, Lakes-Harlan R (2001) Embryonic development of the central projection of auditory afferents (*Schistocerca gregaria*, Orthoptera, Insecta). J Neurobiol 46:97–112.

Schildberger K (1984) Temporal selectivity of identified auditory neurons in the cricket brain. J Comp Physiol A 155:171–185.

Schildberger K, Kleindienst H-U (1989) Sound localization in intact and one-eared crickets. J Comp Physiol A 165:615–626.

Schildberger K, Wohlers DW, Schmitz B, Kleindienst H-U (1986) Morphological and physiological changes in central auditory neurons following unilateral amputation on larval crickets. J Comp Physiol A 158:291–300.

Schmitz B (1989) Neuroplasticity and phonotaxis in monaural adult female crickets (*Gryllus bimaculatus* de Geer). J Comp Physiol A 164:343–358.

Schul J, Schulze W (2001) Phonotaxis during walking and flight: are differences in selectivity due to predation pressure? Naturwissenschaften 88:438–442.

Schwab ME (2002) Repairing the injured spinal cord. Science 295:1029–1031.

Shuvalov AF (1990) Plasticity of phonotaxis specificity in crickets. In: Gribakin FG, Wiese K, Popov AV (eds), Sensory Systems and Communication in Arthropods. Boston: Birkhäuser, pp. 341–344.

Sobel EC, Tank DW (1994) In vivo Ca^{2+} dynamics in a cricket auditory neuron: an example of chemical computation. Science 263:823–826.

Souroukis K, Cade WH, Rowell G (1992) Factors that possibly influence variation in the calling song of field crickets: temperature, time. and male size, age and wing morphology. Can J Zool 70:950–955.

Stabel J, Wendler G, Scharstein H (1989) Cricket phonotaxis: localization depends on recognition of the calling song pattern. J Comp Physiol A 165:165–177.

Staecker H, Van De Water TR (1998) Factors controlling hair-cell regeneration/repair in the inner ear. Curr Opin Neurobiol 8:480–487.

Stölting H, Stumpner A (1998) Tonotopic organization of auditory receptors of the bushcricket *Pholidoptera griseoptera* (Tettigoniidae, Decticinae). Cell Tissue Res 294:377–386.

Stout J, Atkins G, Walikonis R, Hao J, Bronsert M (2002) Influence of juvenile hormone III on the development and plasticity of responsiveness of female crickets to calling males through control of the response properties of identified auditory neurons. In: Pfaff D, Arnold A, Etgen A, Fahrbach S, Rubin R (eds), Hormones, Brain and Behavior. San Diego: Academic Press, pp. 167–193.

Stumpner A (1989) Physiological variability of auditory neurons in a grasshopper. Naturwissenschaften 76:427–429.

Stumpner A (1999) An interneurone of unusual morphology is tuned to the female song frequency in the bushcricket *Ancistrura nigrovittata* (Orthoptera, Phaneropteridae). J Exp Biol 202:2071–2081.

Stumpner A, von Helversen D (2001) Evolution and function of auditory systems in insects. Naturwissenschaften 88:159–170.

Technau GM (1984) Fiber number in the mushroom bodies of adult *Drosophila melanogaster* depends on age, sex and experience. J Neurogenet 1:113–126.

Thomas JB, Bastiani MJ, Bate CM, Goodman CS (1984) From grasshopper to *Drosophila*: a common plan for neuronal development. Nature 310:203–207.

Thompson KJ, Siegler MV (1991) Anatomy and physiology of spiking local and intersegmental interneurons in the median neuroblast lineage of the grasshopper. J Comp Neurol 305:659–675.

Thompson RF, Spencer WA (1966) Habituation: a model phenomenon for the study of neuronal substrates of behaviour. Psychol Rev 173:16–43.

Thorpe WH (1939) Further studies on olfactory conditioning in a parasitic insect: the nature of the conditioning process. Proc R Soc Lond B 126:379–397.

Truman JW, Thorn RS, Robinow S (1992) Programmed neuronal cell death in insect development. J Neurobiol 23:1295–1311.

van Staaden MJ, Römer H (1998) Evolutionary transition from stretch to hearing organs in ancient grasshoppers. Nature 394:773–776.

von Frisch K (1914) Der Farbensinn und Formensinn der Biene. Zool Jb Physiol 37:1–238.

von Helversen D, von Helversen O (1997) Recognition of sex in the acoustic communication of the grasshopper *Chorthippus biguttulus* (Orthoptera, Acrididae). J Comp Physiol A 180:373–386.

von Schilcher F (1976) The role of auditory stimuli in the courtship of *Drosophila melanogaster*. Anim Behav 24:18–26.

Watson AHD (1992) Presynaptic modulation of sensory afferents in the invertebrate and vertebrate nervous system. Comp Biochem Physiol 103A:227–239.

Weeks JC, Levine RB (1990) Postembryonic neuronal plasticity and its control during insect metamorphosis. Annu Rev Neurosci 13:183–194.

Weeks JC, Jacobs GA, Pierce JT, Sandstrom DJ, Streichert LC, Trimmer BA, Wiel DE, Wood ER (1997) Neural mechanisms of behavioral plasticity: metamorphosis and learning in *Manduca sexta*. Brain Behav Evol 50 (Suppl.):69–80.

Williams DW, Shepherd D (1999) Persistent larval sensory neurons in adult *Drosophila*. J Neurobiol 39:275–286.

Wolf H (1986) Response patterns of two auditory interneurons in a freely moving grasshopper (*Chorthippus biguttulus* L.). J Comp Physiol A 158:689–696.

Wolf H, von Helversen O (1986) "Switching-off" of an auditory interneuron during stridulation in the acridid grasshopper *Chorthippus biguttulus* L. J Comp Physiol A 158:861–871.

Wyttenbach RA, May ML, Hoy RR (1996) Categorical perception of sound frequency by crickets. Science 273:1542–1544.

Yager DD (1999) Structure, development, and evolution of insect auditory systems. Microsc Res Tech 47:380–400.

Young D, Ball E (1974) Structure and development of the auditory system in the prothoracic leg of the cricket *Teleogryllus commodus* (Walker). I. Adult structure. Z Zellforsch 147:293–312.

Zhantiev RD, Korsunovskaja OS (1978) Morphofunctional organization of tympanal organs in *Tettigonia cantans* F. (Orthoptera, Tettigoniidae). Zool J 57:1012–1016.

zur Lage P, Jarman AP (1999) Antagonism of EGFR and Notch signalling in the reiterative recruitment of *Drosophila* adult chordotonal sense organ precursors. Development 126:3149–3157.

Index

SPRINGER HANDBOOK OF AUDITORY RESEARCH *(continued from page ii)*

Volume 22: Evolution of the Vertebrate Auditory System
Edited by Geoffrey A. Manley, Arthur N. Popper and Richard R. Fay

Volume 23: Plasticity of the Auditory System
Edited by Thomas N. Parks, Edwin W Rubel, Richard N. Popper, and Richard R. Fay

For more information about the series, please visit www.springer-ny.com/shar.